# Minkowski Space

## " A Four-Dimensional Manifold "

Edited by Paul F. Kisak

# Contents

# Chapter 1

# Minkowski space

In mathematical physics, **Minkowski space** or **Minkowski spacetime** is a combination of Euclidean space and time into a four-dimensional manifold where the spacetime interval between any two events is independent of the inertial frame of reference in which they are recorded. Although initially developed by mathematician Hermann Minkowski for Maxwell's equations of electromagnetism, the mathematical structure of Minkowski spacetime was shown to be an immediate consequence of the postulates of special relativity.[1]

Minkowski space is closely associated with Einstein's theory of special relativity, and is the most common mathematical structure on which special relativity is formulated. While the individual components in Euclidean space and time will often differ due to length contraction and time dilation, in Minkowski spacetime, all frames of reference will agree on the total distance in spacetime between events.[nb 1] Because it treats time differently than the three spatial dimensions, Minkowski space differs from four-dimensional Euclidean space.[nb 2]

In Euclidean space, the isometry group (the maps preserving the regular inner product) is the Euclidean group. The analogous isometry group for Minkowski space, preserving intervals of spacetime equipped with the associated nonpositive definite bilinear form (here called the **Minkowski inner product**,[nb 3]) is the Poincaré group. The Minkowski inner product is defined as to yield the spacetime interval between two events when given their coordinate difference vector as argument.

*Hermann Minkowski (1864 – 1909) found that the theory of special relativity, introduced by his former student Albert Einstein, could best be understood in a four-dimensional space, since known as the Minkowski spacetime.*

## 1.1 History

### 1.1.1 Four-dimensional Euclidean space-time

See also: Four-dimensional space

In 1905, and later published in 1906, Henri Poincaré showed that by taking time to be an imaginary fourth spacetime coordinate ($\sqrt{-1}\, c\, t$), a Lorentz transformation

1

can be regarded as a rotation of coordinates in a four-dimensional Euclidean space with three real coordinates representing space, and one imaginary coordinate, representing time, as the fourth dimension. Since the space is then a pseudo-Euclidean space, the rotation is a representation of a hyperbolic rotation, although Poincaré did not give this interpretation, his purpose being only to explain the Lorentz transformation in terms of the familiar Euclidean rotation.[2]

This idea was elaborated by Hermann Minkowski,[3] who used it to restate the Maxwell equations in four dimensions, showing directly their invariance under the Lorentz transformation. He further reformulated in four dimensions the then-recent theory of special relativity of Einstein. From this he concluded that time and space should be treated equally, and so arose his concept of events taking place in a unified four-dimensional spacetime continuum.

### 1.1.2   Minkowski space

In a further development,[4] he gave an alternative formulation of this idea that used a real time coordinate instead of an imaginary one, representing the four variables $(x, y, z, t)$ of space and time in coordinate form in a four dimensional affine space. Points in this space correspond to events in spacetime. In this space, there is a defined light-cone associated with each point, and events not on the light-cone are classified by their relation to the apex as *spacelike* or *timelike*. It is principally this view of spacetime that is current nowadays, although the older view involving imaginary time has also influenced special relativity. Minkowski, aware of the fundamental restatement of the theory which he had made, said

> The views of space and time which I wish to lay before you have sprung from the soil of experimental physics, and therein lies their strength. They are radical. Henceforth space by itself, and time by itself, are doomed to fade away into mere shadows, and only a kind of union of the two will preserve an independent reality.
> — Hermann Minkowski, 1907[4]

For further historical information see references Galison (1979), Corry (1997) and Walter (1999).

## 1.2   Mathematical structure

For an overview, Minkowski space is a 4-dimensional real vector space equipped with a nondegenerate, symmetric bi-

linear form on the tangent space at each point in spacetime, here simply called the Minkowski inner product, with signature either $(+,-,-,-)$ or $(-,+,+,+)$. In practice, one need not be concerned with the tangent spaces. The vector space nature of Minkowski space allows for the canonical identification of vectors in tangent spaces at points (events) with vectors (points, events) in Minkowski space itself.[5] For some purposes it is desirable to identify tangent vectors at a point $p$ with *displacement vectors* at $p$, which is, of course, admissible by essentially the same canonical identification.[6]

The signature refers to which sign the Minkowski inner product yields when given space and time basis vectors as arguments. In general, mathematicians and general relativists prefer the former while particle physicists tend to use the latter. Arguments for the former (pure space vectors yield positive "norm-squared") include "continuity" from the Euclidean case corresponding to the non-relativistic limit $c \to \infty$. Arguments for the latter (pure space vectors yield negative "norm-squared") include that otherwise ubiquitous minus signs in particle physics go away.

Mathematically associated to this bilinear form is a tensor of type (0,2) at each point in spacetime, called the Minkowski metric. The Minkowski metric, the bilinear form, and the Minkowski inner product are actually all the very same object. In coordinates, this is the 4×4 matrix representing the bilinear form. Keeping this in mind may facilitate reading what follows.

For comparison, in general relativity, a Lorentzian manifold $L$ is likewise equipped with a metric tensor $g$, which is a nondegenerate symmetric bilinear form on the tangent space $TpL$ at each point p of $L$. In coordinates, it may be represented by a 4×4 matrix *depending on spacetime position*. Minkowski space is thus a comparatively simple special case of a Lorentzian manifold. Its metric tensor, called the Minkowski metric, is in coordinates the same symmetric matrix at every point of $M$, and its arguments can, per above, be taken as vectors in spacetime itself.

Introducing more terminology (but not more structure), Minkowski space is thus a pseudo-Euclidean space with total dimension $n = 4$ and signature (3, 1) or (1, 3). Elements of Minkowski space are called events. Minkowski space is often denoted $\mathbf{R}^{3,1}$ or $\mathbf{R}^{1,3}$ to emphasize the chosen signature, or just $M$. It is perhaps the simplest example of a pseudo-Riemannian manifold.

### 1.2.1   Pseudo-Euclidean metric generalities

Main article: Pseudo-Euclidean space

The Minkowski metric[nb 4] η is the metric tensor of

Minkowski space. It is a Pseudo-Euclidean metric. As such it is a nondegenerate symmetric bilinear form, a type $(0,2)$ tensor. It accepts two arguments $up$, $vp$, vectors in $TpM$, $p \in M$, the tangent space at $p$ in $M$. Due to the above-mentioned canonical identification of $TpM$ with $M$ itself, it accepts arguments $u$, $v$ with both $u$ and $v$ in $M$.

As a notational convention, vectors v in M, called 4-vectors, are denoted in sans-serif italics, and not, as is common in the Euclidean setting, with boldface v. The latter is generally reserved for the 3-vector part (to be introduced below) of a 4-vector.

The definition

$$u \cdot v = \eta(u, v)$$

yields an inner product-like structure on $M$, previously and also henceforth, called the Minkowski inner product, similar to the Euclidean inner product, but it describes a different geometry. It has the following properties.

- $\eta(au + v, w) = a\eta(u, w) + \eta(v, w), \quad \forall u, v \in M, \forall a \in \mathbb{R}$     slot) first in (linearity

- $\eta(u, v) = \eta(v, u)$     (symmetry)

- $\eta(u, v) = 0 \ \forall v \in M \Rightarrow u = 0$     (non-degeneracy)

The first two conditions imply bilinearity. The defining *difference* between a pseudo-inner product and an inner product proper is that the former is *not* required to be positive definite, that is, $\eta(u, u) < 0$ is allowed.

Two vectors $v$ and $w$ are said to be orthogonal if $\eta(v, w) = 0$.

A vector $e$ is called a unit vector if $\eta(e, e) = \pm 1$. A basis for $M$ consisting of mutually orthogonal unit vectors is called an orthonormal basis.

For a given inertial frame, an orthonormal basis in space, combined by the unit time vector, forms an orthonormal basis in Minkowski space. The number of positive and negative unit vectors in any such basis is a fixed pair of numbers, equal to the signature of the bilinear form associated with the inner product. This is Sylvester's law of inertia.

More terminology (but not more structure): The Minkowski metric is a pseudo-Riemannian metric, more specifically, a Lorentzian metric, even more specifically, *the* Lorentz metric, reserved for 4-dimensional flat spacetime with the remaining ambiguity only being the signature convention.

## 1.2.2  Minkowski metric

From the second postulate of special relativity, together with homogeneity of spacetime and isotropy of space, follows that the spacetime interval between two events 1, 2,

$$\pm \left[ c^2(t_1 - t_2)^2 - (x_1 - x_2)^2 - (y_1 - y_2)^2 - (z_1 - z_2)^2 \right],$$

is independent of the inertial frame chosen, as is shown here. The factor $\pm$ simply means that the choice of signature is left open. The numerical values of $\eta$, viewed as a matrix representing the Minkowski inner product, follow from the theory of bilinear forms.

Just as the signature of the metric is differently defined in the literature, this quantity is not consistently named. The interval (as defined here) is sometimes referred to as the interval squared.[7] Even the square root of the present interval occurs.[8] When signature and interval are fixed, ambiguity still remains as which coordinate is the time coordinate. It may be the fourth, or it may be the zeroth. This is not an exhaustive list of notational inconsistencies. It is a fact of life that one has to check out the definitions first thing when one consults the relativity literature.

The invariance of the interval under coordinate transformations between inertial frames follows from the invariance of

$$\pm \left[ c^2 t^2 - x^2 - y^2 - z^2 \right]$$

(with either sign $\pm$ preserved), provided the transformations are linear. This quadratic form can be used to define a bilinear form

$$u \cdot v = \pm \left[ c^2 t_1 t_2 - x_1 x_2 - y_1 y_2 - z_1 z_2 \right].$$

via the polarization identity. This bilinear form can in turn be written as

$$u \cdot v = u^{\mathsf{T}}[\eta] v,$$

where $[\eta]$ is a 4×4 matrix associated with $\eta$. Possibly confusingly, denote $[\eta]$ with just $\eta$ as is common practice. The matrix is read off from the explicit bilinear form as

$$\eta = \pm \begin{pmatrix} -1 & 0 & 0 & 0 \\ 0 & 1 & 0 & 0 \\ 0 & 0 & 1 & 0 \\ 0 & 0 & 0 & 1 \end{pmatrix},$$

and the bilinear form

$$u \cdot v = \eta(u, v),$$

with which this section started by assuming its existence, is now identified.

For definiteness and shorter presentation, the signature $(-,+,+,+)$ is adopted below. The choice has no (known) physical implications. The symmetry group preserving the bilinear form with one choice of signature is isomorphic (under the map given here) with the symmetry group preserving the other choice of signature. This means that both choices are in accord with the two postulates of relativity.

### 1.2.3    Standard basis

A standard basis for Minkowski space is a set of four mutually orthogonal vectors $\{ e_0, e_1, e_2, e_3 \}$ such that

$$-\eta(e_0, e_0) = \eta(e_1, e_1) = \eta(e_2, e_2) = \eta(e_3, e_3) = 1.$$

These conditions can be written compactly in the form

$$\eta(e_\mu, e_\nu) = \eta_{\mu\nu}.$$

Relative to a standard basis, the components of a vector $v$ are written $(v^0, v^1, v^2, v^3)$ where the Einstein notation is used to write $v = v^\mu e_\mu$. The component $v^0$ is called the **timelike component** of $v$ while the other three components are called the **spatial components**. The spatial components of a 4-vector $v$ may be identified with a 3-vector $\mathbf{v} = (v_1, v_2, v_3)$.

In terms of components, the Minkowski inner product between two vectors $v$ and $w$ is given by

$$\eta(v, w) = \eta_{\mu\nu} v^\mu w^\nu = v^0 w_0 + v^1 w_1 + v^2 w_2 + v^3 w_3 = v^\mu w_\mu \cdot$$

and

$$\eta(v, v) = \eta_{\mu\nu} v^\mu v^\nu = v^0 v_0 + v^1 v_1 + v^2 v_2 + v^3 v_3 = v^\mu v_\mu.$$

Here **lowering of an index** with the metric was used. Technically, a non-degenerate bilinear form provides a map between a vector space and its dual, in this context, the map is between the tangent spaces of M and the cotangent spaces of M. At a point in M, the tangent and cotangent spaces are dual. Just as an authentic inner product on a vector space with one argument fixed, by Riesz representation theorem, may be expressed as the action of a linear functional on the

vector space, the same holds for the Minkowski inner product of Minkowski space.

Thus if $v^\mu$ are the components of a vector in a tangent space, then $\eta_{\mu\nu} v^\mu = v_\nu$ are the components of a vector in the cotangent space (a linear functional). Due to the identification of vectors in tangent spaces with vectors in $M$ itself, this is mostly ignored, and vectors with lower indices are referred to as **covariant vectors**. In this latter interpretation, the covariant vectors are (almost always implicitly) identified with vectors (linear functionals) in the dual of Minkowski space. The ones with upper indices are **contravariant vectors**. In the same fashion, the inverse of the map from tangent to cotangent spaces, explicitly given by the inverse of $\eta$ in matrix representation, can be used to define **raising of an index**. The components of this inverse are denoted $\eta^{\mu\nu}$. It happens that $\eta^{\mu\nu} = \eta_{\mu\nu}$. These maps between a vector space and its dual can be denoted $\eta^\flat$ (eta-flat) and $\eta^\sharp$ (eta-sharp) by the musical analogy.[9]

The time-proven robustness of the formalism itself, sometimes referred to as index gymnastics, ensures that moving vectors around and changing from contravariant to covariant vectors and vice versa is mathematically sound. Incorrect expressions tend to reveal themselves quickly.

### 1.2.4    Geometry

## 1.3   Lorentz transformations and symmetry

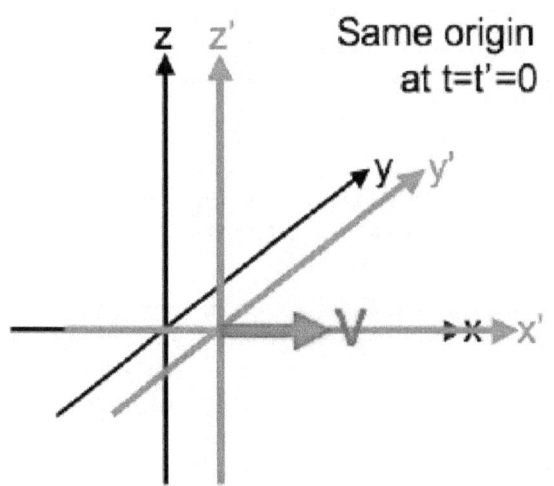

*Standard configuration of coordinate systems for Lorentz transformations.*

The Poincaré group is the group of all transformations preserving the interval. The interval is quite easily seen to be preserved by the translation group in 4 dimensions. The other transformations are those that preserve the interval and leave the origin fixed. Given the bilinear form associated with the Minkowski metric, the appropriate group follows directly from the theory (in particular the definition) of classical groups. In the linked article, one should identify $\eta$ (in its a matrix representation) with the matrix $\Phi$.

The appropriate group is O(3,1), in this context called the Lorentz group. Its elements are called (homogeneous) Lorentz transformations. For other methods of derivation, with a more physical twist, see derivations of the Lorentz transformations.

Among the simplest Lorentz transformations is a Lorentz boost. For reference, a boost in the $x$-direction is given by

$$
\begin{bmatrix} U_0' \\ U_1' \\ U_2' \\ U_3' \end{bmatrix} = \begin{bmatrix} \gamma & -\beta\gamma & 0 & 0 \\ -\beta\gamma & \gamma & 0 & 0 \\ 0 & 0 & 1 & 0 \\ 0 & 0 & 0 & 1 \end{bmatrix} \begin{bmatrix} U_0 \\ U_1 \\ U_2 \\ U_3 \end{bmatrix},
$$

where

$$
\gamma = \frac{1}{\sqrt{1 - \frac{v^2}{c^2}}}
$$

is the Lorentz factor, and

$$
\beta = \frac{v}{c}.
$$

Other Lorentz transformations are pure rotations, and hence elements of the SO(3) subgroup of O(3,1). A general homogeneous Lorentz transformation is a product of a pure boost and a pure rotation. An *inhomogeneous* Lorentz transformation is a homogeneous transformation followed by a translation in space and time. Special transformations are those that invert the space coordinates (P) and time coordinate (T) respectively, or both (PT).

All four-vectors in Minkowski space transform, by definition, according to the same formula under Lorentz transformations. Minkowski diagrams illustrate Lorentz transformations.

## 1.4 Causal structure

Main article: Causal structure

Vectors $v = (ct, x, y, z) = (ct, \mathbf{r})$ are classified according to the sign of $c^2t^2 - r^2$. A vector is **timelike** if $c^2t^2 > r^2$, **spacelike**

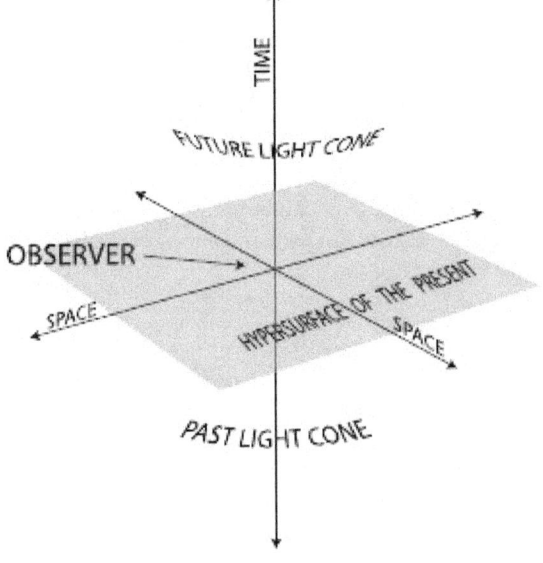

*Subdivision of Minkowski spacetime with respect to an event in four disjoint sets. The light cone, the **absolute future**, the **absolute past**, and **elsewhere**. The terminology is from Sard (1970).*

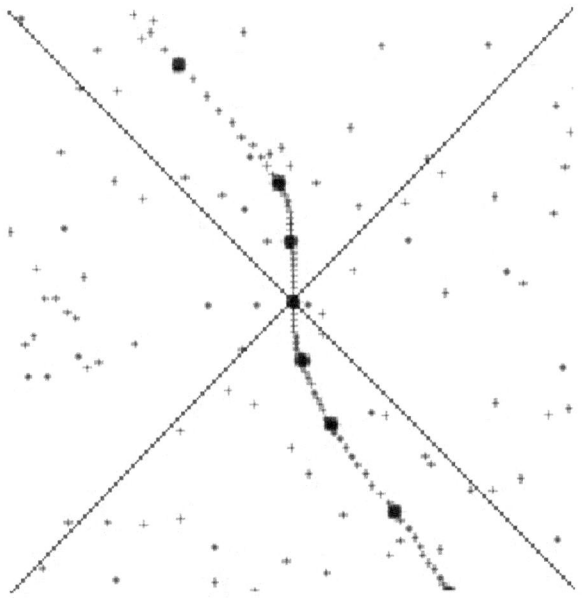

*The momentarily co-moving inertial frames along the trajectory ("world line") of a rapidly accelerating observer (center). The vertical direction indicates time, while the horizontal indicates distance, the dashed line is the spacetime of the observer. The small dots are specific events in spacetime. Note how the momentarily co-moving inertial frame changes when the observer accelerates.*

if $c^2t^2 < r^2$, and **null** or **lightlike** if $c^2t^2 = r^2$. This can be expressed in terms of the sign of $\eta(v,v)$ as well, but depends on the signature. The classification of any vector will be the same in all frames of reference, because of the invariance of the interval.

The set of all null vectors at an event[nb 5] of Minkowski space constitutes the light cone of that event. Given a timelike vector $v$, there is a worldline of constant velocity associated with it, represented by a straight line in a Minkowski diagram.

Once a direction of time is chosen,[nb 6] timelike and null vectors can be further decomposed into various classes. For timelike vectors one has

1. future-directed timelike vectors whose first component is positive, (tip of vector located in absolute future in figure) and

2. past-directed timelike vectors whose first component is negative (absolute past).

Null vectors fall into three classes:

1. the zero vector, whose components in any basis are (0,0,0,0) (origin),

2. future-directed null vectors whose first component is positive (upper light cone), and

3. past-directed null vectors whose first component is negative (lower light cone).

Spacelike vectors are in elsewhere. The terminology stems from the fact that spacelike separated events are connected by vectors requiring faster-than-light travel, and so cannot possibly influence each other. Together with spacelike and lightlike vectors there are 7 classes in all.

An orthonormal basis for Minkowski space necessarily consists of one timelike and three spacelike unit vectors. If one wishes to work with non-orthonormal bases it is possible to have other combinations of vectors. For example, one can easily construct a (non-orthonormal) basis consisting entirely of null vectors, called a **null basis**. Over the reals, if two null vectors are orthogonal (zero Minkowski tensor value), then they must be proportional. However, allowing complex numbers, one can obtain a null tetrad, which is a basis consisting of null vectors, some of which are orthogonal to each other.

Vector fields are called timelike, spacelike or null if the associated vectors are timelike, spacelike or null at each point where the field is defined.

### 1.4.1   Chronological and causality relations

Let $x, y \in M$. We say that

1. $x$ *chronologically precedes* $y$ if $y - x$ is future-directed timelike. This relation has the transitive property and so can be written x < y.

2. $x$ *causally precedes* $y$ if $y - x$ is future-directed null or future-directed timelike. It gives a partial ordering of space-time and so can be written x ≤ y.

### 1.4.2   Reversed triangle inequality

If $v$ and $w$ are both future-directed timelike four-vectors, then in the (+ - - -) sign convention for norm,

$$\|v + w\| \geq \|v\| + \|w\| .$$

## 1.5   Relationships to other formulations

### 1.5.1   Different number of dimensions

Strictly speaking, Minkowski space refers to a mathematical formulation in four dimensions. However, the mathematics can easily be extended or simplified to create an analogous "Minkowski space" in any number of dimensions. If $n \geq 2$, $n$-dimensional Minkowski space is a vector space of real dimension $n$ on which there is a constant Lorentz metric of signature $(n - 1, 1)$ or $(1, n - 1)$. These generalizations are used in theories where spacetime is assumed to have more or less than 4 dimensions. String theory and M-theory are two examples where $n > 4$. In string theory, there appears conformal field theories with 1 + 1 spacetime dimensions.

### 1.5.2   Flat versus curved space

As a *flat spacetime*, the three spatial components of Minkowski spacetime always obey the Pythagorean Theorem. Minkowski space is a suitable basis for special relativity, a good description of physical systems over finite distances in systems without significant gravitation. However, in order to take gravity into account, physicists use the theory of general relativity, which is formulated in the mathematics of a non-Euclidean geometry. When this geometry is used as a model of physical space, it is known as curved space.

Even in curved space, Minkowski space is still a good description in an infinitesimal region surrounding any point (barring gravitational singularities).[nb 7] More abstractly, we say that in the presence of gravity spacetime is described by a curved 4-dimensional manifold for which the tangent space to any point is a 4-dimensional Minkowski space. Thus, the structure of Minkowski space is still essential in the description of general relativity.

## 1.6  See also

- Causal structure
- Euclidean space
- Four vector
- Hyperboloid model
- Introduction to mathematics of general relativity
- Lorentzian manifold
- Metric tensor
- Minkowski diagram
- Minkowski plane
- Speed of light
- Super Minkowski space
- World line

## 1.7  Remarks

[1] This makes spacetime distance an invariant.

[2] Minkowski space can be formulated as an equivalent 4-D Euclidean space if you assume time is always an imaginary number. This is how the spacetime was first formulated, but since Minkowski reworked the structure, time is almost always required to be a real number.

[3] Consistent use of the term "Minkowski inner product" is intended for the bilinear form here, since it is in widespread use. It is by no means "standard" in the literature, but no such standard seems to exist.

[4] The Minkowski inner product is not an inner product, since it is not positive-definite, i.e. the quadratic form $\eta(v, v)$ need not be positive for nonzero $v$. The positive-definite condition has been replaced by the weaker condition of non-degeneracy. The bilinear form is said to be *indefinite*.

[5] Translate the coordinate system so that the event is the new origin.

[6] This corresponds to the time coordinate either increasing or decreasing when proper time for any particle increases. An application of T flips this direction.

[7] This similarity between flat and curved space at infinitesimally small distance scales is foundational to the definition of a manifold in general.

## 1.8  Notes

[1] Landau & Lifshitz 2002, p. 5

[2] Poincaré 1905–1906, pp. 129–176 Wikisource translation: On the Dynamics of the Electron

[3] Minkowski 1907–1908, pp. 53–111 *Wikisource translation: The Fundamental Equations for Electromagnetic Processes in Moving Bodies.

[4] Minkowski 1907–1909, pp. 75–88 Various English translations on Wikisource: Space and Time.

[5] Lee 2003, Proposition 3.8. The identification is routinely done in mathematics.

[6] Lee 2003, See Lee's discussion on geometric tangent vectors early in chapter 3.

[7] Sard 1970, p. 71

[8] Landau & Lifshitz 2002, p. 4

[9] Lee 2003, The tangent-cotangent isomorphism p. 282.

## 1.9  References

- Corry, L. (1997). "Hermann Minkowski and the postulate of relativity". *Arch. Hist. Exact Sci.* (Springer-Verlag) **51** (4): 273–314. doi:10.1007/BF00518231. ISSN 0003-9519. (subscription required (help)).

- Catoni, F.; et al. (2008). *Mathematics of Minkowski Space*. Frontiers in Mathematics. Basel: Birkhäuser Verlag. doi:10.1007/978-3-7643-8614-6. ISBN 978-3-7643-8613-9. ISSN 1660-8046.

- Galison, P. L. (1979). R McCormach; et al., eds. *Minkowski's Space-Time: from visual thinking to the absolute world*. Historical Studies in the Physical Sciences **10**. Johns Hopkins University Press. pp. 85–121. doi:10.2307/27757388. (subscription required (help)).

- Landau, L.D.; Lifshitz, E.M. (2002) [1939]. *The Classical Theory of Fields*. Course of Theoretical Physics **2** (4th ed.). Butterworth–Heinemann. ISBN 0 7506 2768 9.

- Lee, J. M. (2003). *Introduction to Smooth manifolds*. Springer Graduate Texts in Mathematics **218**. ISBN 0-387-95448-1.

- Minkowski, Hermann (1907–1908), "Die Grundgleichungen für die elektromagnetischen Vorgänge in bewegten Körpern" [The Fundamental Equations for Electromagnetic Processes in Moving Bodies],

*Nachrichten von der Gesellschaft der Wissenschaften zu Göttingen, Mathematisch-Physikalische Klasse*: 53–111 *Wikisource translation: The Fundamental Equations for Electromagnetic Processes in Moving Bodies

- Minkowski, Hermann (1907–1909), "Raum und Zeit" [Space and Time], *Physikalische Zeitschrift* **10**: 75–88 Various English translations on Wikisource: Space and Time

- Naber, G. L. (1992). *The Geometry of Minkowski Spacetime*. New York: Springer-Verlag. ISBN 0-387-97848-8.

- Penrose, Roger (2005). "18 Minkowskian geometry". *Road to Reality : A Complete Guide to the Laws of the Universe*. Alfred A. Knopf. ISBN 9780679454434.

- Poincaré, Henri (1905–1906), "Sur la dynamique de l'électron" [On the Dynamics of the Electron], *Rendiconti del Circolo matematico di Palermo* **21**: 129–176, doi:10.1007/BF03013466 Wikisource translation: On the Dynamics of the Electron

- Sard, R. D. (1970). *Relativistic Mechanics - Special Relativity and Classical Particle Dynamics*. New York: W. A. Benjamin. ISBN 978-0805384918.

- Shaw, R. (1982). "§ 6.6 Minkowski space, § 6.7,8 Canonical forms pp 221–242". *Linear Algebra and Group Representations*. Academic Press. ISBN 0-12-639201-3.

- Walter, Scott (1999). "Minkowski, Mathematicians, and the Mathematical Theory of Relativity". In Goenner, Hubert (ed.); et al. *The Expanding Worlds of General Relativity*. Boston: Birkhäuser. pp. 45–86. ISBN 0-8176-4060-6.

## 1.10   External links

Media related to Minkowski diagrams at Wikimedia Commons

- Animation clip on YouTube visualizing Minkowski space in the context of special relativity.

- The Geometry of Special Relativity: The Minkowski Space - Time Light Cone

# Chapter 2

# Spacetime

For other uses of this term, see Spacetime (disambiguation).

In physics, **spacetime** is any mathematical model that combines space and time into a single interwoven continuum. Since 300 BCE, the spacetime of our universe has historically been interpreted from a Euclidean space perspective, which regards space as consisting of three dimensions, and time as consisting of one dimension, the "fourth dimension". By combining space and time into a single manifold called Minkowski space in 1905, physicists have significantly simplified a large number of physical theories, as well as described in a more uniform way the workings of the universe at both the supergalactic and subatomic levels.

## 2.1  Explanation

In non-relativistic classical mechanics, the use of Euclidean space instead of spacetime is appropriate, because time is treated as universal with a constant rate of passage that is independent of the state of motion of an observer. In relativistic contexts, time cannot be separated from the three dimensions of space, because the observed rate at which time passes for an object depends on the object's velocity relative to the observer and also on the strength of gravitational fields, which can slow the passage of time for an object as seen by an observer outside the field.

In cosmology, the concept of spacetime combines space and time to a single abstract universe. Mathematically it is a manifold consisting of "events" which are described by some type of coordinate system. Typically **three spatial dimensions** (length, width, height), and one **temporal dimension** (time) are required. Dimensions are independent components of a coordinate grid needed to locate a point in a certain defined "space". For example, on the globe the latitude and longitude are two independent coordinates which together uniquely determine a location. In spacetime, a coordinate grid that spans the 3+1 dimensions locates events (rather than just points in space), i.e., time is added as another dimension to the coordinate grid. This way the coordinates specify *where* and *when* events occur. However, the unified nature of spacetime and the freedom of coordinate choice it allows imply that to express the temporal coordinate in one coordinate system requires both temporal and spatial coordinates in another coordinate system. Unlike in normal spatial coordinates, there are still restrictions for how measurements can be made spatially and temporally (see Spacetime intervals). These restrictions correspond roughly to a particular mathematical model which differs from Euclidean space in its manifest symmetry.

Until the beginning of the 20th century, time was believed to be independent of motion, progressing at a fixed rate in all reference frames; however, following its prediction by special relativity, later experiments confirmed that time slows at higher speeds of the reference frame relative to another reference frame. Such slowing, called time dilation, is explained in special relativity theory. Many experiments have confirmed time dilation, such as the relativistic decay of muons from cosmic ray showers and the slowing of atomic clocks aboard a Space Shuttle relative to synchronized Earth-bound inertial clocks.[1] The duration of time can therefore vary according to events and reference frames.

When dimensions are understood as mere components of the grid system, rather than physical attributes of space, it is easier to understand the alternate dimensional views as being simply the result of coordinate transformations.

The term *spacetime* has taken on a generalized meaning beyond treating spacetime events with the normal 3+1 dimensions. It is really the combination of space and time. Other proposed spacetime theories include additional dimensions—normally spatial but there exist some speculative theories that include additional temporal dimensions and even some that include dimensions that are neither temporal nor spatial (e.g., superspace). How many dimensions are needed to describe the universe is still an open question. Speculative theories such as string theory predict 10 or 26 dimensions (with M-theory predicting 11 dimensions: 10

9

spatial and 1 temporal), but the existence of more than four dimensions would only appear to make a difference at the subatomic level.[2]

## 2.2  Spacetime in literature

Incas regarded space and time as a single concept, referred to as *pacha* (Quechua: *pacha*, Aymara: *pacha*).[3][4] The peoples of the Andes maintain a similar understanding.[5]

The idea of a unified spacetime is stated by Edgar Allan Poe in his essay on cosmology titled *Eureka* (1848) that "Space and duration are one". In 1895, in his novel *The Time Machine*, H. G. Wells wrote, "There is no difference between time and any of the three dimensions of space except that our consciousness moves along it", and that "any real body must have extension in four directions: it must have Length, Breadth, Thickness, and Duration".

Marcel Proust, in his novel *Swann's Way* (published 1913), describes the village church of his childhood's Combray as "a building which occupied, so to speak, four dimensions of space—the name of the fourth being Time".

### 2.2.1  Mathematical concept

In Encyclopedie, published in 1754, under the term *dimension* Jean le Rond d'Alembert speculated that duration (time) might be considered a fourth dimension if the idea was not too novel.[6]

Another early venture was by Joseph Louis Lagrange in his *Theory of Analytic Functions* (1797, 1813). He said, "One may view mechanics as a geometry of four dimensions, and mechanical analysis as an extension of geometric analysis".[7]

The ancient idea of the cosmos gradually was described mathematically with differential equations, differential geometry, and abstract algebra. These mathematical articulations blossomed in the nineteenth century as electrical technology stimulated men like Michael Faraday and James Clerk Maxwell to describe the reciprocal relations of electric and magnetic fields. Daniel Siegel phrased Maxwell's role in relativity as follows:

> [...] the idea of the propagation of forces at the velocity of light through the electromagnetic field as described by Maxwell's equations—rather than instantaneously at a distance—formed the necessary basis for relativity theory.[8]

Maxwell used vortex models in his papers on On Physical

Lines of Force, but ultimately gave up on any substance but the electromagnetic field. Pierre Duhem wrote:

> [Maxwell] was not able to create the theory that he envisaged except by giving up the use of any model, and by extending by means of analogy the abstract system of electrodynamics to displacement currents.[9]

In Siegel's estimation, "this very abstract view of the electromagnetic fields, involving no visualizable picture of what is going on out there in the field, is Maxwell's legacy."[10] Describing the behaviour of electric fields and magnetic fields led Maxwell to view the combination as an electromagnetic field. These fields have a value at every point of spacetime. It is the intermingling of electric and magnetic manifestations, described by Maxwell's equations, that give spacetime its structure. In particular, the rate of motion of an observer determines the electric and magnetic profiles of the electromagnetic field. The propagation of the field is determined by the electromagnetic wave equation, which requires spacetime for description.

Spacetime was described as an affine space with quadratic form in Minkowski space of 1908.[11] In his 1914 textbook *The Theory of Relativity*, Ludwik Silberstein used biquaternions to represent events in Minkowski space. He also exhibited the Lorentz transformations between observers of differing velocities as biquaternion mappings. Biquaternions were described in 1853 by W. R. Hamilton, so while the physical interpretation was new, the mathematics was well known in English literature, making relativity an instance of applied mathematics.

The first inkling of general relativity in spacetime was articulated by W. K. Clifford. Description of the effect of gravitation on space and time was found to be most easily visualized as a "warp" or stretching in the geometrical fabric of space and time, in a smooth and continuous way that changed smoothly from point-to-point along the spacetime fabric. In 1947 James Jeans provided a concise summary of the development of spacetime theory in his book *The Growth of Physical Science*.[12]

## 2.3  Basic concepts

The basic elements of spacetime are events. In any given spacetime, an event is a unique position at a unique time. Because events are spacetime points, an example of an event in classical relativistic physics is $(x, y, z, t)$, the location of an elementary (point-like) particle at a particular time. A spacetime itself can be viewed as the union of all events in the same way that a line is the union of all of its points, formally organized into a manifold, a space which can be described at small scales using coordinate systems.

A spacetime is independent of any observer.[13] However, in describing physical phenomena (which occur at certain moments of time in a given region of space), each observer chooses a convenient metrical coordinate system. Events are specified by four real numbers in any such coordinate system. The trajectories of elementary (point-like) particles through space and time are thus a continuum of events called the world line of the particle. Extended or composite objects (consisting of many elementary particles) are thus a union of many world lines twisted together by virtue of their interactions through spacetime into a "world-braid".

However, in physics, it is common to treat an extended object as a "particle" or "field" with its own unique (e.g., center of mass) position at any given time, so that the world line of a particle or light beam is the path that this particle or beam takes in the spacetime and represents the history of the particle or beam. The world line of the orbit of the Earth (in such a description) is depicted in two spatial dimensions $x$ and $y$ (the plane of the Earth's orbit) and a time dimension orthogonal to $x$ and $y$. The orbit of the Earth is an ellipse in space alone, but its world line is a helix in spacetime.[14]

The unification of space and time is exemplified by the common practice of selecting a metric (the measure that specifies the interval between two events in spacetime) such that all four dimensions are measured in terms of units of distance: representing an event as $(x_0, x_1, x_2, x_3) = (ct, x, y, z)$ (in the Lorentz metric) or $(x_1, x_2, x_3, x_4) = (x, y, z, ict)$ (in the original Minkowski metric) where $c$ is the speed of light.[15] The metrical descriptions of Minkowski Space and spacelike, lightlike, and timelike intervals given below follow this convention, as do the conventional formulations of the Lorentz transformation.

### 2.3.1 Spacetime intervals in flat space

In a Euclidean space, the separation between two points is measured by the distance between the two points. The distance is purely spatial, and is always positive. In spacetime, the displacement four-vector $\Delta R$ is given by the space displacement vector $\Delta r$ and the time difference $\Delta t$ between the events. The *spacetime interval*, also called *invariant interval*, between the two events, $s^2$,[16] is defined as:

$$s^2 = \Delta r^2 - c^2 \Delta t^2 \text{ (spacetime interval)},$$

where $c$ is the speed of light. The choice of signs for $s^2$ above follows the space-like convention (−+++).[17] Spacetime intervals may be classified into three distinct types, based on whether the temporal separation ($c^2 \Delta t^2$) or the spatial separation ($\Delta r^2$) of the two events is greater: timelike, light-like or space-like.

Certain types of world lines are called geodesics of the spacetime – straight lines in the case of Minkowski space and their closest equivalent in the curved spacetime of general relativity. In the case of purely time-like paths, geodesics are (locally) the paths of greatest separation (spacetime interval) as measured along the path between two events, whereas in Euclidean space and Riemannian manifolds, geodesics are paths of shortest distance between two points.[18][19] The concept of geodesics becomes central in general relativity, since geodesic motion may be thought of as "pure motion" (inertial motion) in spacetime, that is, free from any external influences.

**Time-like interval**

$$c^2 \Delta t^2 > \Delta r^2$$
$$s^2 < 0$$

For two events separated by a time-like interval, enough time passes between them that there could be a cause–effect relationship between the two events. For a particle traveling through space at less than the speed of light, any two events which occur to or by the particle must be separated by a time-like interval. Event pairs with time-like separation define a negative spacetime interval ($s^2 < 0$) and may be said to occur in each other's future or past. There exists a reference frame such that the two events are observed to occur in the same spatial location, but there is no reference frame in which the two events can occur at the same time.

The measure of a time-like spacetime interval is described by the proper time interval, $\Delta \tau$:

$$\Delta \tau = \sqrt{\Delta t^2 - \frac{\Delta r^2}{c^2}} \text{ (proper time interval)}.$$

The proper time interval would be measured by an observer with a clock traveling between the two events in an inertial reference frame, when the observer's path intersects each event as that event occurs. (The proper time interval defines a real number, since the interior of the square root is positive.)

**Light-like interval**

$$c^2 \Delta t^2 = \Delta r^2$$
$$s^2 = 0$$

In a light-like interval, the spatial distance between two events is exactly balanced by the time between the two events. The events define a spacetime interval of zero (

$s^2 = 0$ ). Light-like intervals are also known as "null" intervals.

Events which occur to or are initiated by a photon along its path (i.e., while traveling at $c$ , the speed of light) all have light-like separation. Given one event, all those events which follow at light-like intervals define the propagation of a light cone, and all the events which preceded from a light-like interval define a second (graphically inverted, which is to say "*pastward*") light cone.

### Space-like interval

$$c^2 \Delta t^2 < \Delta r^2$$
$$s^2 > 0$$

When a space-like interval separates two events, not enough time passes between their occurrences for there to exist a causal relationship crossing the spatial distance between the two events at the speed of light or slower. Generally, the events are considered not to occur in each other's future or past. There exists a reference frame such that the two events are observed to occur at the same time, but there is no reference frame in which the two events can occur in the same spatial location.

For these space-like event pairs with a positive spacetime interval ( $s^2 > 0$ ), the measurement of space-like separation is the proper distance, $\Delta \sigma$ :

$$\Delta \sigma = \sqrt{s^2} = \sqrt{\Delta r^2 - c^2 \Delta t^2} \text{ (proper distance).}$$

Like the proper time of time-like intervals, the proper distance of space-like spacetime intervals is a real number value.

### 2.3.2   Interval as area

The interval has been presented as the area of an oriented rectangle formed by two events and isotropic lines through them. Time-like or space-like separations correspond to oppositely oriented rectangles, one type considered to have rectangles of negative area. The case of two events separated by light corresponds to the rectangle degenerating to the segment between the events and zero area.[20] The transformations leaving interval-length invariant are the area-preserving squeeze mappings.

The parameters traditionally used rely on quadrature of the hyperbola, which is the natural logarithm. This transcendental function is essential in mathematical analysis as its inverse unites circular functions and hyperbolic functions:

The exponential function, $e^t$, $t$ a real number, used in the hyperbola ($e^t$, $e^{-t}$ ), generates hyperbolic sectors and the hyperbolic angle parameter. The functions cosh and sinh, used with rapidity as hyperbolic angle, provide the common representation of squeeze in the form $\begin{pmatrix} \cosh \phi & \sinh \phi \\ \sinh \phi & \cosh \phi \end{pmatrix}$, or as the split-complex unit $e^{j\phi} = \cosh \phi + j \sinh \phi$.

## 2.4   Mathematics of spacetimes

For physical reasons, a spacetime continuum is mathematically defined as a four-dimensional, smooth, connected Lorentzian manifold $(M, g)$ . This means the smooth Lorentz metric $g$ has signature $(3, 1)$ . The metric determines the geometry of spacetime, as well as determining the geodesics of particles and light beams. About each point (event) on this manifold, coordinate charts are used to represent observers in reference frames. Usually, Cartesian coordinates $(x, y, z, t)$ are used. Moreover, for simplicity's sake, units of measurement are usually chosen such that the speed of light $c$ is equal to 1.

A reference frame (observer) can be identified with one of these coordinate charts; any such observer can describe any event $p$ . Another reference frame may be identified by a second coordinate chart about $p$ . Two observers (one in each reference frame) may describe the same event $p$ but obtain different descriptions.

Usually, many overlapping coordinate charts are needed to cover a manifold. Given two coordinate charts, one containing $p$ (representing an observer) and another containing $q$ (representing another observer), the intersection of the charts represents the region of spacetime in which both observers can measure physical quantities and hence compare results. The relation between the two sets of measurements is given by a non-singular coordinate transformation on this intersection. The idea of coordinate charts as local observers who can perform measurements in their vicinity also makes good physical sense, as this is how one actually collects physical data—locally.

For example, two observers, one of whom is on Earth, but the other one who is on a fast rocket to Jupiter, may observe a comet crashing into Jupiter (this is the event $p$ ). In general, they will disagree about the exact location and timing of this impact, i.e., they will have different 4-tuples $(x, y, z, t)$ (as they are using different coordinate systems). Although their kinematic descriptions will differ, dynamical (physical) laws, such as momentum conservation and the first law of thermodynamics, will still hold. In fact, relativity theory requires more than this in the sense that it stipulates these (and all other physical) laws must take the same form in all coordinate systems. This introduces tensors into relativity, by which all physical quantities are represented.

Geodesics are said to be time-like, null, or space-like if the tangent vector to one point of the geodesic is of this nature. Paths of particles and light beams in spacetime are represented by time-like and null (light-like) geodesics, respectively.

### 2.4.1   Topology

Main article: Spacetime topology

The assumptions contained in the definition of a spacetime are usually justified by the following considerations.

The connectedness assumption serves two main purposes. First, different observers making measurements (represented by coordinate charts) should be able to compare their observations on the non-empty intersection of the charts. If the connectedness assumption were dropped, this would not be possible. Second, for a manifold, the properties of connectedness and path-connectedness are equivalent, and one requires the existence of paths (in particular, geodesics) in the spacetime to represent the motion of particles and radiation.

Every spacetime is paracompact. This property, allied with the smoothness of the spacetime, gives rise to a smooth linear connection, an important structure in general relativity. Some important theorems on constructing spacetimes from compact and non-compact manifolds include the following:

- A compact manifold can be turned into a spacetime if, and only if, its Euler characteristic is 0. (Proof idea: the existence of a Lorentzian metric is shown to be equivalent to the existence of a nonvanishing vector field.)

- Any non-compact 4-manifold can be turned into a spacetime.[21]

### 2.4.2   Spacetime symmetries

Main article: Spacetime symmetries

Often in relativity, spacetimes that have some form of symmetry are studied. As well as helping to classify spacetimes, these symmetries usually serve as a simplifying assumption in specialized work. Some of the most popular ones include:

- Axisymmetric spacetimes

- Spherically symmetric spacetimes

- Static spacetimes

- Stationary spacetimes

### 2.4.3   Causal structure

Main article: Causal structure
See also: Causality (physics) and Causality

The causal structure of a spacetime describes causal relationships between pairs of points in the spacetime based on the existence of certain types of curves joining the points.

## 2.5   Spacetime in special relativity

Main article: Minkowski space

The geometry of spacetime in special relativity is described by the Minkowski metric on $\mathbb{R}^4$. This spacetime is called Minkowski space. The Minkowski metric is usually denoted by $\eta$ and can be written as a four-by-four matrix:

$$\eta_{ab} = \operatorname{diag}(1, -1, -1, -1)$$

where the Landau–Lifshitz time-like convention is being used. A basic assumption of relativity is that coordinate transformations must leave spacetime intervals invariant. Intervals are invariant under Lorentz transformations. This invariance property leads to the use of four-vectors (and other tensors) in describing physics.

Strictly speaking, one can also consider events in Newtonian physics as a single spacetime. This is Galilean–Newtonian relativity, and the coordinate systems are related by Galilean transformations. However, since these preserve spatial and temporal distances independently, such a spacetime can be decomposed into spatial coordinates plus temporal coordinates, which is not possible in the general case.

## 2.6   Spacetime in general relativity

In general relativity, it is assumed that spacetime is curved by the presence of matter (energy), this curvature being represented by the Riemann tensor. In special relativity, the Riemann tensor is identically zero, and so this concept of "non-curvedness" is sometimes expressed by the statement *Minkowski spacetime is flat.*

The earlier discussed notions of time-like, light-like and space-like intervals in special relativity can similarly be used to classify one-dimensional curves through curved spacetime. A time-like curve can be understood as one where the interval between any two infinitesimally close events on the curve is time-like, and likewise for light-like and space-like curves. Technically the three types of curves are usually defined in terms of whether the tangent vector at each point on the curve is time-like, light-like or space-like. The world line of a slower-than-light object will always be a time-like curve, the world line of a massless particle such as a photon will be a light-like curve, and a space-like curve could be the world line of a hypothetical tachyon. In the local neighborhood of any event, time-like curves that pass through the event will remain inside that event's past and future light cones, light-like curves that pass through the event will be on the surface of the light cones, and space-like curves that pass through the event will be outside the light cones. One can also define the notion of a three-dimensional "space-like hypersurface", a continuous three-dimensional "slice" through the four-dimensional property with the property that every curve that is contained entirely within this hypersurface is a space-like curve.[22]

Many spacetime continua have physical interpretations which most physicists would consider bizarre or unsettling. For example, a compact spacetime has closed timelike curves, which violate our usual ideas of causality (that is, future events could affect past ones). For this reason, mathematical physicists usually consider only restricted subsets of all the possible spacetimes. One way to do this is to study "realistic" solutions of the equations of general relativity. Another way is to add some additional "physically reasonable" but still fairly general geometric restrictions and try to prove interesting things about the resulting spacetimes. The latter approach has led to some important results, most notably the Penrose–Hawking singularity theorems.

## 2.7  Quantized spacetime

Main article: Quantum spacetime

In general relativity, spacetime is assumed to be smooth and continuous—and not just in the mathematical sense. In the theory of quantum mechanics, there is an inherent discreteness present in physics. In attempting to reconcile these two theories, it is sometimes postulated that spacetime should be quantized at the very smallest scales. Current theory is focused on the nature of spacetime at the Planck scale. Causal sets, loop quantum gravity, string theory, causal dynamical triangulation, and black hole thermodynamics all predict a quantized spacetime with agreement on the order of magnitude. Loop quantum gravity makes precise predic-

tions about the geometry of spacetime at the Planck scale.

Spin networks provide a language to describe quantum geometry of space. Spin foam does the same job on spacetime. A spin network is a one-dimensional graph, together with labels on its vertices and edges which encodes aspects of a spatial geometry.

## 2.8  See also

- Anthropic_principle § Applications of the principle §§ Spacetime
- Basic introduction to the mathematics of curved spacetime
- Four-vector
- Frame-dragging
- Global spacetime structure
- Hole argument
- List of mathematical topics in relativity
- Local spacetime structure
- Lorentz invariance
- Manifold
- Mathematics of general relativity
- Metric space
- Philosophy of space and time
- Relativity of simultaneity
- Strip photography
- World manifold

## 2.9  References

[1] Ashby, Neil (2003). "Relativity in the Global Positioning System" (PDF). Living Reviews in Relativity 6: 16. Bibcode:2003LRR.....6....1A. doi:10.12942/lrr-2003-1.

[2] Kopeikin, Sergei; Efroimsky, Michael; Kaplan, George (2011). Relativistic Celestial Mechanics of the Solar System. John Wiley & Sons. p. 157. ISBN 3527634576. Retrieved 2016-02-28. Extract of page 157

[3] Atuq Eusebio Manga Qespi, Instituto de lingüística y Cultura Amerindia de la Universidad de Valencia. Pacha: un concepto andino de espacio y tiempo. Revísta española de Antropología Americana, 24, p. 155–189. Edit. Complutense, Madrid. 1994

[4] Paul Richard Steele, Catherine J. Allen, *Handbook of Inca mythology*, p. 86, (ISBN 1-57607-354-8)

[5] Shirley Ardener, University of Oxford, *Women and space: ground rules and social maps*, p. 36 (ISBN 0-85496-728-1)

[6] Jean d'Alembert (1754) Dimension from ARTFL Encyclopedie project

[7] R.C. Archibald (1914) *Time as a fourth dimension Bulletin of the American Mathematical Society 20:409.*

[8] Daniel M. Siegel (2014) "Maxwell's contributions to electricity and magnetism", chapter 10 in *James Clerk Maxwell: Perspectives on his Life and Work*, Raymond Flood, Mark McCartney, Andrew Whitaker, editors, Oxford University Press ISBN 978-0-19-966437-5

[9] Pierre Duhem (1954) *The Aim and Structure of Physical Theory*, page 98, Princeton University Press

[10] Siegel 2014 p 191

[11] Minkowski, Hermann (1909), "Raum und Zeit", *Physikalische Zeitschrift* 10: 75–88

  • Various English translations on Wikisource: Space and Time.

[12] James Jeans (1947) The Growth of Physical Science, "Space-time", pp. 205–301, link from Internet Archive

[13] Matolcsi, Tamás (1994). *Spacetime Without Reference Frames*. Budapest: Akadémiai Kiadó.

[14] Ellis, G. F. R.; Williams, Ruth M. (2000). *Flat and curved space–times* (2nd ed.). Oxford University Press. p. 9. ISBN 0-19-850657-0.

[15] Petkov, Vesselin (2010). *Minkowski Spacetime: A Hundred Years Later*. Springer. p. 70. ISBN 90-481-3474-9. Retrieved 2016-02-28., Section 3.4, p. 70

[16] Note that the term *spacetime interval* is applied by several authors to the quantity $s^2$ and not to $s$. The reason that the quantity $s^2$ is used and not $s$ is that $s^2$ can be positive, zero or negative, and is a more generally convenient and useful quantity than the Minkowski norm with a timelike/null/spacelike distinguisher: the pair $(\sqrt{|s^2|}, \text{sgn}(s^2))$. Despite the notation, it should not be regarded as the square of a number, but as a symbol. The cost for this convenience is that this "interval" is quadratic in linear separation along a straight line.

[17] More generally the spacetime interval in flat space can be written as $s^2 = g_{\alpha\beta}\Delta x^\alpha \Delta x^\beta$ with metric tensor g independent of spacetime position.

[18] This characterization is not universal: both the arcs between two points of a great circle on a sphere are geodesics.

[19] Berry, Michael V. (1989). *Principles of Cosmology and Gravitation*. CRC Press. p. 58. ISBN 0-85274-037-9. Retrieved 2016-02-28. Extract of page 58, caption of Fig. 25

[20] I. M. Yaglom (1979) *A Simple Non-Euclidean Geometry and its Physical Basis*, page 178, Springer, ISBN 0387-90332-1, MR 520230

[21] Geroch, Robert; Horowitz, Gary T. (1979). "Chapter 5. Global structure of spacetimes". In Hawking, S.W.; Israel, W. *General Relativity: An Einstein Centenary Survey*. Cambridge University Press. p. 219. ISBN 0521299284.

[22] See "Quantum Spacetime and the Problem of Time in Quantum Gravity" by Leszek M. Sokolowski, where on this page he writes "Each of these hypersurfaces is spacelike, in the sense that every curve, which entirely lies on one of such hypersurfaces, is a spacelike curve." More commonly a spacelike hypersurface is defined technically as a surface such that the normal vector at every point is time-like, but the definition above may be somewhat more intuitive.

## 2.10 Further reading

• Albert Einstein on Space-Time 13th edition Encyclopedia Britannica Historical: Albert Einstein's 1926 article

• Ehrenfest, Paul (1920) "How do the fundamental laws of physics make manifest that Space has 3 dimensions?" *Annalen der Physik 366*: 440.

• George F. Ellis and Ruth M. Williams (1992) *Flat and curved space–times*. Oxford Univ. Press. ISBN 0-19-851164-7

• Encyclopedia of Space-time and gravitation Scholarpedia Expert articles

## 2.11 External links

• http://universaltheory.org

• Barrow, John D.; Tipler, Frank J. (1988). *The Anthropic Cosmological Principle*. Oxford University Press. ISBN 978-0-19-282147-8. LCCN 87028148.

• Isenberg, J. A. (1981). "Wheeler–Einstein–Mach spacetimes". *Phys. Rev. D* 24 (2): 251–256. Bibcode:1981PhRvD..24..251I. doi:10.1103/PhysRevD.24.251.

• Kant, Immanuel (1929) "Thoughts on the true estimation of living forces" in J. Handyside, trans., *Kant's Inaugural Dissertation and Early Writings on Space*. Univ. of Chicago Press.

• Lorentz, H. A., Einstein, Albert, Minkowski, Hermann, and Weyl, Hermann (1952) *The Principle of Relativity: A Collection of Original Memoirs*. Dover.

- Lucas, John Randolph (1973) *A Treatise on Time and Space*. London: Methuen.

- Penrose, Roger (2004). *The Road to Reality*. Oxford: Oxford University Press. ISBN 0-679-45443-8. Chpts. 17–18.

- Poe, Edgar A. (1848). *Eureka; An Essay on the Material and Spiritual Universe*. Hesperus Press Limited. ISBN 1-84391-009-8.

- Robb, A. A. (1936). *Geometry of Time and Space*. University Press.

- Erwin Schrödinger (1950) *Space–time structure*. Cambridge Univ. Press.

- Schutz, J. W. (1997). *Independent axioms for Minkowski Space–time*. Addison-Wesley Longman. ISBN 0-582-31760-6.

- Tangherlini, F. R. (1963). "Schwarzschild Field in n Dimensions and the Dimensionality of Space Problem". *Nuovo Cimento* **14** (27): 636.

- Taylor, E. F.; Wheeler, John A. (1963). *Spacetime Physics*. W. H. Freeman. ISBN 0-7167-2327-1.

- Wells, H.G. (2004). *The Time Machine*. New York: Pocket Books. ISBN 0-671-57554-6. (pp. 5–6)

- Stanford Encyclopedia of Philosophy: "Space and Time: Inertial Frames" by Robert DiSalle.

# Chapter 3

# Minkowski plane

This article is about the Benz plane. It is not to be confused with Minkowski space.

In mathematics, a **Minkowski plane** (named after Hermann Minkowski) is one of the Benz planes: Möbius plane, Laguerre plane and Minkowski plane.

## 3.1 Classical real Minkowski plane

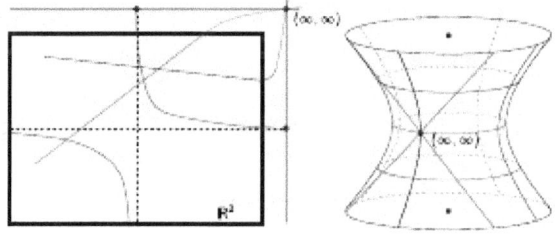

*classical Minkowski plane: 2d/3d-model*

Applying the pseudo-euclidean distance $d(P_1, P_2) = (x'_1 - x'_2)^2 - (y'_1 - y'_2)^2$ on two points $P_i = (x'_i, y'_i)$ (instead of the euclidean distance) we get the geometry of *hyperbolas*, because a pseudo-euclidean circle $\{P \in \mathbb{R}^2 \mid d(P, M) = r\}$ is a hyperbola with midpoint $M$.

By a transformation of coordinates $x_i = x'_i + y'_i$, $y_i = x'_i - y'_i$, the pseudo-euclidean distance can be rewritten as $d(P_1, P_2) = (x_1 - x_2)(y_1 - y_2)$. The hyperbolas then have asymptotes parallel to the non-primed coordinate axes.

The following completion (see Möbius and Laguerre planes) *homogenizes* the geometry of hyperbolas:

$$\mathcal{P} := (\mathbb{R} \cup \{\infty\})^2 = \mathbb{R}^2 \cup (\{\infty\} \times \mathbb{R}) \cup (\mathbb{R} \times \{\infty\}) \cup \{(\infty, \infty)\}, \ \infty \notin \mathbb{R}, \text{ the set of } \textbf{points},$$

$$\mathcal{Z} := \{\{(x, y) \in \mathbb{R}^2 \mid y = ax + b\} \cup \{(\infty, \infty)\}\} \mid a, b \in \mathbb{R}, a \neq 0\}$$

$$\cup \{\{(x, y) \in \mathbb{R}^2 \mid y = \frac{a}{x-b} + c, x \neq b\} \cup \{(b, \infty), (\infty, c)\} \mid a, b, c \in \mathbb{R}, a \neq 0\}, \text{ the set of } \textbf{cycles}.$$

The incidence structure $(\mathcal{P}, \mathcal{Z}, \in)$ is called the **classical real Minkowski plane**.

The set of points consists of $\mathbb{R}^2$, two copies of $\mathbb{R}$ and the point $(\infty, \infty)$.

Any line $y = ax + b, a \neq 0$ is completed by point $(\infty, \infty)$, any hyperbola $y = \frac{a}{x-b} + c, a \neq 0$ by the two points $(b, \infty), (\infty, c)$ (see figure).

Two points $(x_1, y_1) \neq (x_2, y_2)$ can not be connected by a cycle if and only if $x_1 = x_2$ or $y_1 = y_2$.

We define: Two points $P_1, P_2$ are (+)-**parallel** ( $P_1 \parallel_+ P_2$ ) if $x_1 = x_2$ and (−)-**parallel** ( $P_1 \parallel_- P_2$ ) if $y_1 = y_2$. Both these relations are equivalence relations on the set of points.

Two points $P_1, P_2$ are called **parallel** ( $P_1 \parallel P_2$ ) if $P_1 \parallel_+ P_2$ or $P_1 \parallel_- P_2$.

From the definition above we find:

**Lemma:**

- For any pair of non parallel points $A, B$ there is exactly one point $C$ with $A \parallel_+ C \parallel_- B$.

- For any point $P$ and any cycle $z$ there are exactly two points $A, B \in z$ with $A \parallel_+ P \parallel_- B$.

- For any three points $A$, $B$, $C$, pairwise non parallel, there is exactly one cycle $z$ that contains $A, B, C$.

- For any cycle $z$, any point $P \in z$ and any point $Q, P \nparallel Q$ and $Q \notin z$ there exists exactly one cycle $z'$ such that $z \cap z' = \{P\}$, i.e. $z$ **touches** $z'$ at point P.

Like the classical Möbius and Laguerre planes Minkowski planes can be described as the geometry of plane sections of a suitable quadric. But in this case the quadric lives in **projective** 3-space: The classical real Minkowski plane is isomorphic to the geometry of plane sections of a hyperboloid of one sheet (not degenerated quadric of index 2).

## 3.2 The axioms of a Minkowski plane

Let be $(\mathcal{P}, \mathcal{Z}; \|_+, \|_-, \in)$ an incidence structure with the set $\mathcal{P}$ of points, the set $\mathcal{Z}$ of cycles and two equivalence relations $\|_+$ ((+)-parallel) and $\|_-$ ((−)-parallel) on set $\mathcal{P}$. For $P \in \mathcal{P}$ we define: $\overline{P}_+ := \{Q \in \mathcal{P} \mid Q \|_+ P\}$ and $\overline{P}_- := \{Q \in \mathcal{P} \mid Q \|_- P\}$. An equivalence class $\overline{P}_+$ or $\overline{P}_-$ is called **(+)-generator** and **(−)-generator**, respectively. (For the space model of the classical Minkowski plane a generator is a line on the hyperboloid.)

Two points $A, B$ are called **parallel** ( $A \parallel B$ ) if $A \|_+ B$ or $A \|_- B$.

An incidence structure $\mathfrak{M} := (\mathcal{P}, \mathcal{Z}; \|_+, \|_-, \in)$ is called **Minkowski plane** if the following axioms hold:

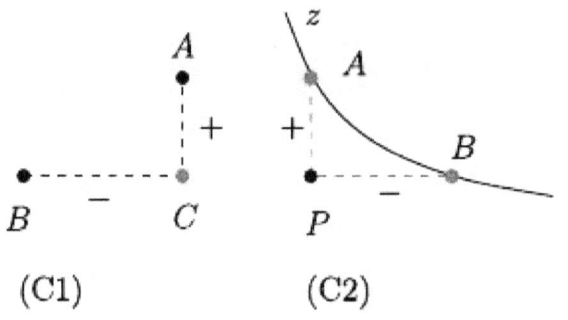

(C1)            (C2)

*Minkowski-axioms-c1-c2*

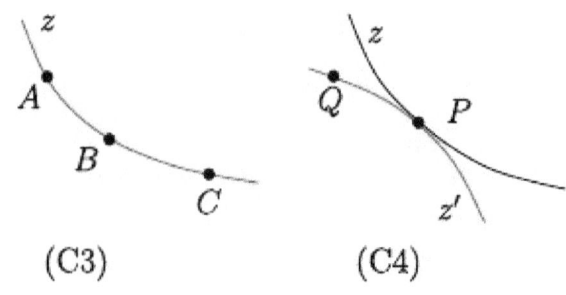

(C3)            (C4)

*Minkowski-axioms-c3-c4*

- **C1**: For any pair of non parallel points $A, B$ there is exactly one point $C$ with $A \|_+ C \|_- B$ .

- **C2**: For any point $P$ and any cycle $z$ there are exactly two points $A, B \in z$ with $A \|_+ P \|_- B$ .

- **C3**: For any three points $A, B, C$ , pairwise non parallel, there is exactly one cycle $z$ which contains $A, B, C$

- **C4**: For any cycle $z$ , any point $P \in z$ and any point $Q, P \not\parallel Q$ and $Q \notin z$ there exists exactly one cycle $z'$ such that $z \cap z' = \{P\}$ , i.e. $z$ **touches** $z'$ at point $P$ .

- **C5**: Any cycle contains at least 3 points. There is at least one cycle $z$ and a point $P$ not in $z$ .

For investigations the following statements on parallel classes (equivalent to C1, C2 respectively) are advantageous.

**C1′**:  For any two points $A, B$ we have $|\overline{A}_+ \cap \overline{B}_-| = 1$ .

**C2′**: For any point $P$ and any cycle $z$ we have: $|\overline{P}_+ \cap z| = 1 = |\overline{P}_- \cap z|$ .

First consequences of the axioms are

**Lemma:** For a Minkowski plane $\mathfrak{M}$ the following is true

- a) Any point is contained in at least one cycle.

- b) Any generator contains at least 3 points.

- c) Two points can be connected by a cycle if and only if they are non parallel.

Analogously to Möbius and Laguerre planes we get the connection to the linear geometry via the residues.

For a Minkowski plane $\mathfrak{M} = (\mathcal{P}, \mathcal{Z}; \|_+, \|_-, \in)$ and $P \in \mathcal{P}$ we define the local structure

$$\mathfrak{A}_P := (\mathcal{P} \backslash \overline{P}, \{$$

$$z \backslash \{\overline{P}\} \mid P \in z \in \mathcal{Z}\} \cup \{E \backslash \overline{P} \mid E \in \mathcal{E} \backslash \{\overline{P}_+, \overline{P}_-\}\}, \in)$$

and call it the **residue at point P**.

For the classical Minkowski plane $\mathfrak{A}_{(\infty, \infty)}$ is the real affine plane $\mathbb{R}^2$ .

An immediate consequence of axioms C1 to C4 and C1′, C2′ are the following two theorems.

**Theorem:** For a Minkowski plane $\mathfrak{M} = (\mathcal{P}, \mathcal{Z}; \|_+, \|, \in)$ any residue is an affine plane.

**Theorem:** Let be $\mathfrak{M} = (\mathcal{P}, \mathcal{Z}; \|_-, \in)$ an incidence structure with two equivalence relations $\|_+$ and $\|_-$ on the set $\mathcal{P}$ of points (see above).

$\mathfrak{M}$ is a Minkowski plane if and only if for any point $P$ the residue $\mathfrak{A}_P$ is an affine plane.

## 3.2.1 Minimal model

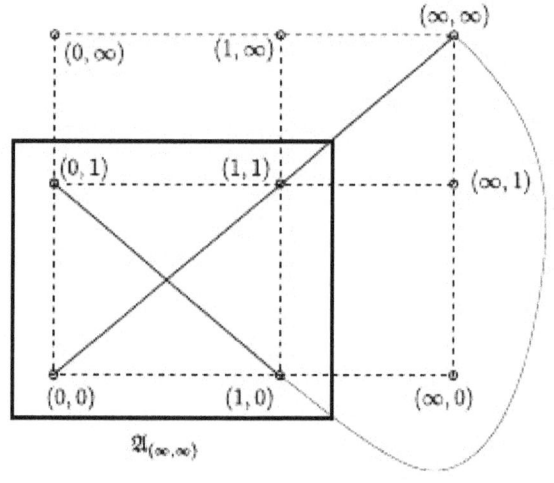

*Minkowski plane: minimal model*

The **minimal model** of a Minkowski plane can be established over the set $\overline{K} := \{0, 1, \infty\}$ of three elements:

$\mathcal{P} := \overline{K}^2$

$\mathcal{Z} := \{\{(a_1, b_1), (a_2, b_2), (a_3, b_3)\}|$

$|\{a_1, a_2, a_3\} = \{b_1, b_2, b_3\} = \overline{K}\} =$

$\{\{(0,0), (1,1), (\infty,\infty)\}, \quad \{(0,0), (1,\infty), (\infty,1)\},$
$\{(0,1), (1,0), (\infty,\infty)\}, \quad \{(0,1), (1,\infty), (\infty,0)\},$
$\{(0,\infty), (1,1), (\infty,0)\}, \{(0,\infty), (1,0), (\infty,1)\}\}$

Parallel points:

$(x_1, y_1) \parallel_+ (x_2, y_2)$ if and only if $x_1 = x_2$

$(x_1, y_1) \parallel_- (x_2, y_2)$ if and only if $y_1 = y_2$ .

Hence: $|\mathcal{P}| = 9$ and $|\mathcal{Z}| = 6$ .

## 3.2.2 Finite Minkowski-planes

For finite Minkowski-planes we get from C1', C2':

**Lemma:** Let be $\mathfrak{M} = (\mathcal{P}, \mathcal{Z}; \parallel_+, \parallel_-, \in)$ a finite Minkowski plane, i.e. $|\mathcal{P}| < \infty$ . For any pair of cycles $z_1, z_2$ and any pair of generators $e_1, e_2$ we have: $|z_1| = |z_2| = |e_1| = |e_2|$ .

This gives rise of the **definition**:
For a finite Minkowski plane $\mathfrak{M}$ and a cycle $z$ of $\mathfrak{M}$ we call the integer $n = |z| - 1$ the **order** of $\mathfrak{M}$ .

Simple combinatorial considerations yield

**Lemma:** For a finite Minkowski plane $\mathfrak{M} = (\mathcal{P}, \mathcal{Z}; \parallel_+, \parallel_-, \in)$ the following is true:

$n$

$|\mathcal{P}| = (n+1)^2$
$|\mathcal{Z}| = (n+1)n(n-1)$

## 3.3 Miquelian Minkowski planes

We get the most important examples of Minkowski planes by generalizing the classical real model: Just replace $\mathbb{R}$ by an arbitrary field $K$ then we get *in any case* a Minkowski plane $\mathfrak{M}(K) = (\mathcal{P}, \mathcal{Z}; \parallel_+, \parallel_-, \in)$ .

Analogously to Möbius and Laguerre planes the Theorem of Miquel is a characteristic property of a Minkowski plane $\mathfrak{M}(K)$ .

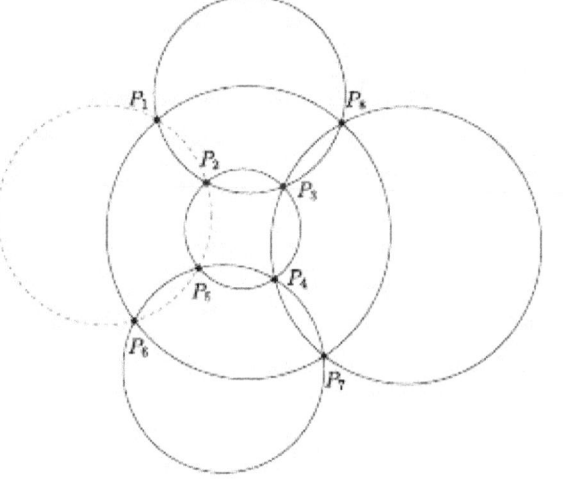

*Theorem of Miquel*

**Theorem (Miquel):** For the Minkowski plane $\mathfrak{M}(K)$ the following is true:

$P_1, ..., P_8$

(For a better overview in the figure there are circles drawn instead of hyperbolas.)

**Theorem (Chen):** Only a Minkowski plane $\mathfrak{M}(K)$ satisfies the theorem of Miquel.

Because of the last theorem $\mathfrak{M}(K)$ is called a **miquelian Minkowski plane**.

**Remark:** The **minimal model** of a Minkowski plane is miquelian.

It is isomorphic to the Minkowski plane $\mathfrak{M}(K)$ with $K = \text{GF}(2)$ (field $\{0, 1\}$ ).

An astonishing result is

**Theorem (Heise):** Any Minkowski plane of *even* order is miquelian.

**Remark:** A suitable stereographic projection shows: $\mathfrak{M}(K)$ is isomorphic to the geometry of the plane sections on a hyperboloid of one sheet (quadric of index 2) in projective 3-space over field $K$ .

**Remark:** There are a lot of Minkowski planes that are **not miquelian** (s. weblink below). But there are no "ovoidal Minkowski" planes, in difference to Möbius and Laguerre planes. Because any quadratic set of index 2 in projective 3-space is a quadric (see quadratic set).

## 3.4   See also

- Conformal geometry

## 3.5   References

- W. Benz, *Vorlesungen über Geomerie der Algebren*, Springer (1973)

- F. Buekenhout (ed.), *Handbook of Incidence Geometry*, Elsevier (1995) ISBN 0-444-88355-X

## 3.6   External links

- Benz plane at SpringerLink

- Lecture Note *Planar Circle Geometries*, an Introduction to Moebius-, Laguerre- and Minkowski Planes

# Chapter 4

# Four-dimensional space

*3D projection of a tesseract undergoing a simple rotation in four dimensional space.*

In mathematics, **four-dimensional space** ("4D") is a geometric space with four dimensions. It typically is more specifically four-dimensional Euclidean space, generalizing the rules of three-dimensional Euclidean space. It has been studied by mathematicians and philosophers for over two centuries, both for its own interest and for the insights it offered into mathematics and related fields.

Algebraically, it is generated by applying the rules of vectors and coordinate geometry to a space with four dimensions. In particular a vector with four elements (a 4-tuple) can be used to represent a position in four-dimensional space. The space is a Euclidean space, so has a metric and norm, and so all directions are treated as the same: the additional dimension is indistinguishable from the other three.

In modern physics, space and time are unified in a four-dimensional Minkowski continuum called spacetime, whose metric treats the time dimension differently from the three spatial dimensions (see below for the definition of the

Minkowski metric/pairing). Spacetime is *not* a Euclidean space.

## 4.1 History

See also: n-dimensional space § History

Lagrange wrote in his *Mécanique analytique* (published 1788, based on work done around 1755) that mechanics can be viewed as operating in a four-dimensional space — three of dimensions of space, and one of time.[1] In 1827 Möbius realized that a fourth dimension would allow a three-dimensional form to be rotated onto its mirror-image,[2] and by 1853 Ludwig Schläfli had discovered many polytopes in higher dimensions, although his work was not published until after his death.[3] Higher dimensions were soon put on firm footing by Bernhard Riemann's 1854 Habilitationsschrift, *Über die Hypothesen welche der Geometrie zu Grunde liegen*, in which he considered a "point" to be any sequence of coordinates $(x_1, ..., xn)$. The possibility of geometry in higher dimensions, including four dimensions in particular, was thus established.

An arithmetic of four dimensions called quaternions was defined by William Rowan Hamilton in 1843. This associative algebra was the source of the science of vector analysis in three dimensions as recounted in *A History of Vector Analysis*. Soon after tessarines and coquaternions were introduced as other four-dimensional algebras over **R**.

One of the first major expositors of the fourth dimension was Charles Howard Hinton, starting in 1880 with his essay *What is the Fourth Dimension?*; published in the Dublin University magazine.[4] He coined the terms *tesseract, ana* and *kata* in his book *A New Era of Thought*, and introduced a method for visualising the fourth dimension using cubes in the book *Fourth Dimension*.[5][6] In 1886 Victor Schlegel described[7] his method of visualizing four-dimensional objects with Schlegel diagrams.

In 1908, Hermann Minkowski presented a paper[8] consolidating the role of time as the fourth dimension of spacetime, the basis for Einstein's theories of special and general relativity.[9] But the geometry of spacetime, being non-Euclidean, is profoundly different from that popularised by Hinton. The study of Minkowski space required new mathematics quite different from that of four-dimensional Euclidean space, and so developed along quite different lines. This separation was less clear in the popular imagination, with works of fiction and philosophy blurring the distinction, so in 1973 H. S. M. Coxeter felt compelled to write:

> Little, if anything, is gained by representing the fourth Euclidean dimension as *time*. In fact, this idea, so attractively developed by H. G. Wells in *The Time Machine*, has led such authors as John William Dunne (*An Experiment with Time*) into a serious misconception of the theory of Relativity. Minkowski's geometry of space-time is *not* Euclidean, and consequently has no connection with the present investigation.
> — H. S. M. Coxeter, *Regular Polytopes*[10]

## 4.2  Vectors

Mathematically four-dimensional space is simply a space with four spatial dimensions, that is a space that needs four parameters to specify a point in it. For example, a general point might have position vector $\mathfrak{a}$, equal to

$$\mathfrak{a} = \begin{pmatrix} a_1 \\ a_2 \\ a_3 \\ a_4 \end{pmatrix}.$$

This can be written in terms of the four standard basis vectors ($\mathbf{e}_1, \mathbf{e}_2, \mathbf{e}_3, \mathbf{e}_4$), given by

$$\mathbf{e}_1 = \begin{pmatrix} 1 \\ 0 \\ 0 \\ 0 \end{pmatrix}; \mathbf{e}_2 = \begin{pmatrix} 0 \\ 1 \\ 0 \\ 0 \end{pmatrix}; \mathbf{e}_3 = \begin{pmatrix} 0 \\ 0 \\ 1 \\ 0 \end{pmatrix}; \mathbf{e}_4 = \begin{pmatrix} 0 \\ 0 \\ 0 \\ 1 \end{pmatrix},$$

so the general vector $\mathfrak{a}$ is

$$\mathfrak{a} = a_1\mathbf{e}_1 + a_2\mathbf{e}_2 + a_3\mathbf{e}_3 + a_4\mathbf{e}_4.$$

Vectors add, subtract and scale as in three dimensions.

The dot product of Euclidean three-dimensional space generalizes to four dimensions as

$$\mathfrak{a} \cdot \mathbf{b} = a_1b_1 + a_2b_2 + a_3b_3 + a_4b_4.$$

It can be used to calculate the norm or length of a vector,

$$|\mathfrak{a}| = \sqrt{\mathfrak{a} \cdot \mathfrak{a}} = \sqrt{a_1{}^2 + a_2{}^2 + a_3{}^2 + a_4{}^2},$$

and calculate or define the angle between two vectors as

$$\theta = \arccos \frac{\mathbf{a} \cdot \mathbf{b}}{|\mathbf{a}|\,|\mathbf{b}|}.$$

Minkowski spacetime is four-dimensional space with geometry defined by a nondegenerate pairing different from the dot product:

$$\mathfrak{a} \cdot \mathbf{b} = a_1b_1 + a_2b_2 + a_3b_3 - a_4b_4.$$

As an example, the distance squared between the points $(0,0,0,0)$ and $(1,1,1,0)$ is 3 in both the Euclidean and Minkowskian 4-spaces, while the distance squared between $(0,0,0,0)$ and $(1,1,1,1)$ is 4 in Euclidean space and 2 in Minkowski space; increasing $b_4$ actually decreases the metric distance. This leads to many of the well known apparent "paradoxes" of relativity.

The cross product is not defined in four dimensions. Instead the exterior product is used for some applications, and is defined as follows:

$$\mathfrak{a} \wedge \mathbf{b} = (a_1b_2 - a_2b_1)\mathbf{e}_{12} + (a_1b_3 - a_3b_1)\mathbf{e}_{13} + (a_1b_4$$
$$- a_4b_1)\ \mathbf{e}_{14} + (a_2b_3 - a_3b_2)\mathbf{e}_{23}$$
$$+ (a_2b_4 - a_4b_2)\mathbf{e}_{24} + (a_3b_4 - a_4b_3)\mathbf{e}_{34}.$$

This is bivector valued, with bivectors in four dimensions forming a six-dimensional linear space with basis ($\mathbf{e}_{12}, \mathbf{e}_{13}, \mathbf{e}_{14}, \mathbf{e}_{23}, \mathbf{e}_{24}, \mathbf{e}_{34}$). They can be used to generate rotations in four dimensions.

## 4.3  Orthogonality and vocabulary

In the familiar 3-dimensional space in which we live there are three coordinate axes — usually labeled $x$, $y$, and $z$ — with each axis orthogonal (i.e. perpendicular) to the other two. The six cardinal directions in this space can be called *up, down, east, west, north,* and *south*. Positions along these axes can be called *altitude, longitude,* and *latitude*. Lengths measured along these axes can be called *height*, *width*, and *depth*.

Comparatively, 4-dimensional space has an extra coordinate axis, orthogonal to the other three, which is usually labeled *w*. To describe the two additional cardinal directions, Charles Howard Hinton coined the terms *ana* and *kata*, from the Greek words meaning "up toward" and "down from", respectively. A position along the *w* axis can be called *spissitude*, as coined by Henry More.

## 4.4 Geometry

See also: Rotations in 4-dimensional Euclidean space

The geometry of 4-dimensional space is much more complex than that of 3-dimensional space, due to the extra degree of freedom.

Just as in 3 dimensions there are polyhedra made of two dimensional polygons, in 4 dimensions there are 4-polytopes made of polyhedra. In 3 dimensions there are 5 regular polyhedra known as the Platonic solids. In 4 dimensions there are 6 convex regular 4-polytopes, the analogues of the Platonic solids. Relaxing the conditions for regularity generates a further 58 convex uniform 4-polytopes, analogous to the 13 semi-regular Archimedean solids in three dimensions. Relaxing the conditions for convexity generates a further 10 nonconvex regular 4-polytopes.

In 3 dimensions, a circle may be extruded to form a cylinder. In 4 dimensions, there are several different cylinder-like objects. A sphere may be extruded to obtain a spherical cylinder (a cylinder with spherical "caps", known as a spherinder), and a cylinder may be extruded to obtain a cylindrical prism (a cubinder). The Cartesian product of two circles may be taken to obtain a duocylinder. All three can "roll" in 4-dimensional space, each with its own properties.

In 3 dimensions, curves can form knots but surfaces cannot (unless they are self-intersecting). In 4 dimensions, however, knots made using curves can be trivially untied by displacing them in the fourth direction, but 2-dimensional surfaces can form non-trivial, non-self-intersecting knots in 4-dimensional space.[11] Because these surfaces are 2-dimensional, they can form much more complex knots than strings in 3-dimensional space can. The Klein bottle is an example of such a knotted surface . Another such surface is the real projective plane.

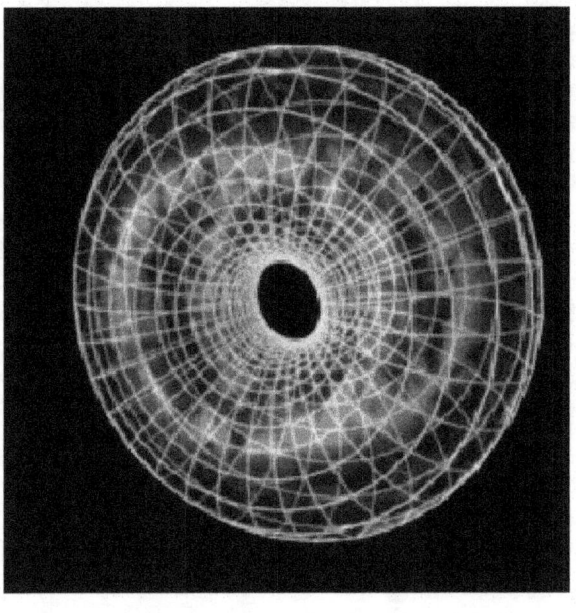

*Stereographic projection of a Clifford torus: the set of points (cos(a), sin(a), cos(b), sin(b)), which is a subset of the 3-sphere.*

### 4.4.1 Hypersphere

Main article: Hypersphere

The set of points in Euclidean 4-space having the same distance R from a fixed point $P_0$ forms a hypersurface known as a 3-sphere. The hyper-volume of the enclosed space is:

$$\mathbf{V} = \tfrac{1}{2}\pi^2 R^4$$

This is part of the Friedmann–Lemaître–Robertson–Walker metric in General relativity where *R* is substituted by function *R(t)* with *t* meaning the cosmological age of the universe. Growing or shrinking *R* with time means expanding or collapsing universe, depending on the mass density inside.[12]

## 4.5 Cognition

Research using virtual reality finds that humans in spite of living in a three-dimensional world can without special practice make spatial judgments based on the length of, and angle between, line segments embedded in four-dimensional space.[13] The researchers noted that "the participants in our study had minimal practice in these tasks, and it remains an open question whether it is possible to obtain more sustainable, definitive, and richer 4D representations with increased perceptual experience in 4D virtual environments."[13] In another study,[14] the ability of humans to orient themselves in 2D, 3D and 4D mazes has been tested. Each maze consisted of four path segments of random length and connected with orthogonal random bends, but without branches or loops (i.e. actually labyrinths). The graphical interface was based on John McIntosh's free 4D Maze game.[15] The participating persons had to navigate through the path and finally estimate the linear direction back to the starting point. The researchers found that some of the participants were able to mentally integrate their path after some practice in 4D (the lower-dimensional cases were for comparison and for the participants to learn the method).

## 4.6   Dimensional analogy

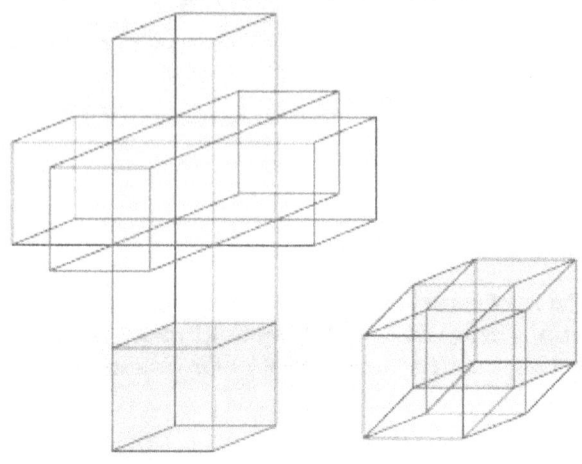

A *net of a tesseract*

To understand the nature of four-dimensional space, a device called *dimensional analogy* is commonly employed. Dimensional analogy is the study of how $(n-1)$ dimensions relate to $n$ dimensions, and then inferring how $n$ dimensions would relate to $(n + 1)$ dimensions.[16]

Dimensional analogy was used by Edwin Abbott Abbott in the book *Flatland*, which narrates a story about a square that lives in a two-dimensional world, like the surface of a piece of paper. From the perspective of this square, a three-dimensional being has seemingly god-like powers, such as ability to remove objects from a safe without breaking it open (by moving them across the third dimension), to see everything that from the two-dimensional perspective is enclosed behind walls, and to remain completely invisible by standing a few inches away in the third dimension.

By applying dimensional analogy, one can infer that a four-dimensional being would be capable of similar feats from our three-dimensional perspective. Rudy Rucker illustrates this in his novel *Spaceland*, in which the protagonist encounters four-dimensional beings who demonstrate such powers.

### 4.6.1   Cross-sections

As a three-dimensional object passes through a two-dimensional plane, a two-dimensional being would only see a cross-section of the three-dimensional object. For example, if a spherical balloon passed through a sheet of paper, a being on the paper would see first a single point, then a circle gradually growing larger, then smaller again until it shrank to a point and then disappeared. Similarly, if a four-dimensional object passed through three dimensions, we would see a three-dimensional cross-section of the four-dimensional object—for example, a hypersphere would appear first as a point, then as a growing sphere, with the sphere then shrinking to a single point and then disappearing.[17] This means of visualizing aspects of the fourth dimension was used in the novel Flatland and also in several works of Charles Howard Hinton.[18]

### 4.6.2   Projections

A useful application of dimensional analogy in visualizing the fourth dimension is in projection. A projection is a way for representing an $n$-dimensional object in $n-1$ dimensions. For instance, computer screens are two-dimensional, and all the photographs of three-dimensional people, places and things are represented in two dimensions by projecting the objects onto a flat surface. When this is done, depth is removed and replaced with indirect information. The retina of the eye is also a two-dimensional array of receptors but the brain is able to perceive the nature of three-dimensional objects by inference from indirect information (such as shading, foreshortening, binocular vision, etc.). Artists often use perspective to give an illusion of three-dimensional depth to two-dimensional pictures.

Similarly, objects in the fourth dimension can be mathematically projected to the familiar 3 dimensions, where they can be more conveniently examined. In this case, the 'retina' of the four-dimensional eye is a three-dimensional array of receptors. A hypothetical being with such an eye would perceive the nature of four-dimensional objects by inferring four-dimensional depth from indirect information in the three-dimensional images in its retina.

The perspective projection of three-dimensional objects into the retina of the eye introduces artifacts such as foreshortening, which the brain interprets as depth in the third dimension. In the same way, perspective projection from four dimensions produces similar foreshortening effects. By applying dimensional analogy, one may infer four-dimensional "depth" from these effects.

As an illustration of this principle, the following sequence of images compares various views of the 3-dimensional cube with analogous projections of the 4-dimensional tesseract into three-dimensional space.

### 4.6.3 Shadows

A concept closely related to projection is the casting of shadows.

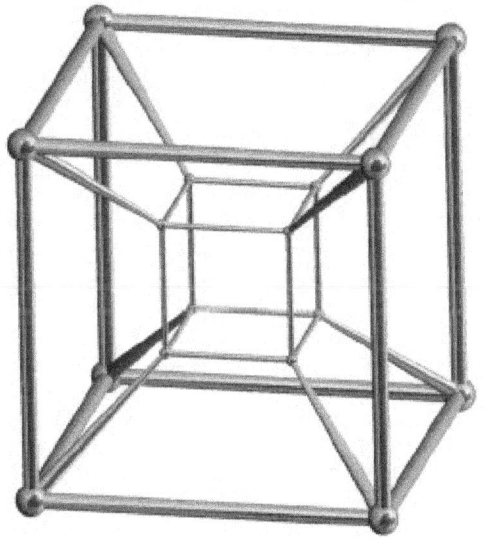

If a light is shone on a three dimensional object, a two-dimensional shadow is cast. By dimensional analogy, light shone on a two-dimensional object in a two-dimensional world would cast a one-dimensional shadow, and light on a one-dimensional object in a one-dimensional world would cast a zero-dimensional shadow, that is, a point of non-light. Going the other way, one may infer that light shone on a four-dimensional object in a four-dimensional world would cast a three-dimensional shadow.

If the wireframe of a cube is lit from above, the resulting shadow is a square within a square with the corresponding corners connected. Similarly, if the wireframe of a tesseract were lit from "above" (in the fourth dimension), its shadow would be that of a three-dimensional cube within another three-dimensional cube. (Note that, technically, the visual representation shown here is actually a two-dimensional image of the three-dimensional shadow of the four-dimensional wireframe figure.)

### 4.6.4 Bounding volumes

Dimensional analogy also helps in inferring basic properties of objects in higher dimensions. For example, two-dimensional objects are bounded by one-dimensional boundaries: a square is bounded by four edges. Three-dimensional objects are bounded by two-dimensional surfaces: a cube is bounded by 6 square faces. By applying dimensional analogy, one may infer that a four-dimensional cube, known as a tesseract, is bounded by three-dimensional volumes. And indeed, this is the case: mathematics shows that the tesseract is bounded by 8 cubes. Knowing this is key to understanding how to interpret a three-dimensional projection of the tesseract. The boundaries of the tesseract project to *volumes* in the image, not merely two-dimensional surfaces.

### 4.6.5 Visual scope

Being three-dimensional, we are only able to see the world with our eyes in two dimensions. A four-dimensional being would be able to see the world in three dimensions. For example, it would be able to see all six sides of an opaque box simultaneously, and in fact, what is inside the box at the same time, just as we can see the interior of a square on a piece of paper. It would be able to see all points in 3-dimensional space simultaneously, including the inner structure of solid objects and things obscured from our three-dimensional viewpoint. Our brains receive images in the second dimension and use reasoning to help us "picture" three-dimensional objects.

### 4.6.6 Limitations

Reasoning by analogy from familiar lower dimensions can be an excellent intuitive guide, but care must be exercised not to accept results that are not more rigorously tested. For example, consider the formulas for the circumference of a circle $C = 2\pi r$ and the surface area of a sphere: $A = 4\pi r^2$. One might be tempted to suppose that the surface volume of a hypersphere is $V = 6\pi r^3$, or perhaps $V = 8\pi r^3$, but either of these would be wrong. The correct formula is $V = 2\pi^2 r^3$. [10]

## 4.7 See also

- Euclidean space

- Euclidean geometry

- 4-manifold

- Exotic $\mathbf{R}^4$

- Fourth dimension in art

- Dimension

- Four-dimensionalism

- Fifth dimension

- Sixth dimension

- 4-polytope

- Polytope

- List of geometry topics

- Block Theory of the Universe

- *Flatland*, a book by Edwin A. Abbott about two- and three-dimensional spaces, to understand the concept of four dimensions

- *Sphereland*, an unofficial sequel to *Flatland*

- The Planiverse, a book by A. K. Dewdney about a two-dimensional being, in a logically consistent physical world, in contact with University students in our 3d world

- Charles Howard Hinton

- *Dimensions*, a set of films about two-, three- and four-dimensional polytopes

- List of four-dimensional games

## 4.8 References

[1] Bell, E.T. (1937). *Men of Mathematics*, Simon and Schuster, p. 154.

[2] Coxeter, H. S. M. (1973). *Regular Polytopes*, Dover Publications, Inc., p. 141.

[3] Coxeter, H. S. M. (1973). *Regular Polytopes*, Dover Publications, Inc., pp. 142–143.

[4] Rudolf v.B. Rucker, editor *Speculations on the Fourth Dimension: Selected Writings of Charles H. Hinton*, p. vii, Dover Publications Inc., 1980 ISBN 0-486-23916-0

[5] Hinton, Charles Howard (1904). *Fourth Dimension*. ISBN 1-5645-9708-3.

[6] Gardner, Martin (1975). *Mathematical Carnival*. Knopf Publishing. pp. 42, 52–53. ISBN 0-394-49406-7.

[7] Victor Schlegel (1886) *Ueber Projectionsmodelle der regelmässigen vier-dimensionalen Körper*, Waren

[8] Minkowski, Hermann (1909), "Raum und Zeit", *Physikalische Zeitschrift* 10: 75–88

  - Various English translations on Wikisource: Space and Time

[9] C Møller (1952). *The Theory of Relativity*. Oxford UK: Clarendon Press. p. 93. ISBN 0-19-851256-2.

[10] Coxeter, H. S. M. (1973). Regular Polytopes, Dover Publications, Inc., p. 119.

[11] J. Scott Carter, Masahico Saito Knotted Surfaces and Their Diagrams

[12] Ray d'Inverno (1992), *Introducing Einstein's Relativity*, Clarendon Press, chp. 22.8 *Geometry of 3-spaces of constant curvature*, p.319ff, ISBN 0-19-859653-7

[13] Ambinder MS, Wang RF, Crowell JA, Francis GK, Brinkmann P. (2009). Human four-dimensional spatial intuition in virtual reality. Psychon Bull Rev. 16(5):818-23. doi:10.3758/PBR.16.5.818 PMID 19815783 online supplementary material

[14] Aflalo TN, Graziano MS (2008). Four-Dimensional Spatial Reasoning in Humans. *Journal of Experimental Psychology*: Human Perception and Performance 34(5):1066-1077. doi:10.1037/0096-1523.34.5.1066 Preprint

[15] John McIntosh's four dimensional maze game. Free software

[16] Michio Kaku (1994). *Hyperspace: A Scientific Odyssey Through Parallel Universes, Time Warps, and the Tenth Dimension*, Part I, chapter 3, *The Man Who "Saw" the Fourth Dimension (about tesseracts in years 1870–1910)*. ISBN 0-19-286189-1.

[17] Rucker, Rudy (1984), *The Fourth Dimension /A Guided Tour of the Higher Universes*, Houghton Mifflin, p. 18, ISBN 0-395-39388-4

[18] In particular, Hinton, Charles Howard (1904). *Fourth Dimension*. pp. 11–14. ISBN 1-5645-9708-3.

## 4.9 Further reading

- Andrew Forsyth (1930) Geometry of Four Dimensions, link from Internet Archive.

- Gamow, George (1988). *One Two Three . . . Infinity: Facts and Speculations of Science* (3rd ed.). Courier Dover Publications. p. 68. ISBN 0-486-25664-2. Extract of page 68

- E. H. Neville (1921) The Fourth Dimension, Cambridge University Press, link from University of Michigan Historical Math Collection.

## 4.10 External links

- "Dimensions" videos, showing several different ways to visualize four dimensional objects

- Science News article summarizing the "Dimensions" videos, with clips

- Garrett Jones' tetraspace page

- Flatland: a Romance of Many Dimensions (second edition)

- TeV scale gravity, mirror universe, and ... dinosaurs Article from Acta Physica Polonica B by Z.K. Silagadze.

- Exploring Hyperspace with the Geometric Product

- 4D Euclidean space

- 4D Building Blocks - Interactive game to explore 4D space

- 4DNav - A small tool to view a 4D space as four 3D space uses ADSODA algorithm

- MagicCube 4D A 4-dimensional analog of traditional Rubik's Cube.

- Frame-by-frame animations of 4D - 3D analogies

# Chapter 5

# Minkowski diagram

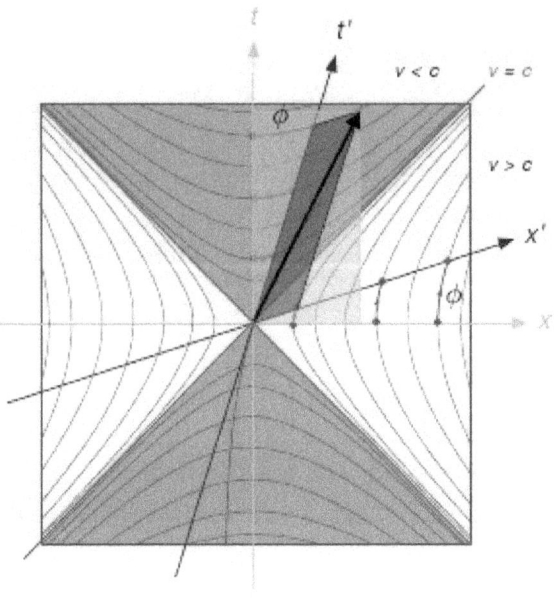

*Minkowski diagram with resting frame (x, t), moving frame (x', t'), light cone, and hyperbolas marking out time and space with respect to the origin.*

The **Minkowski diagram**, also known as a **spacetime diagram**, was developed in 1908 by Hermann Minkowski and provides an illustration of the properties of space and time in the special theory of relativity. It allows a qualitative understanding of the corresponding phenomena like time dilation and length contraction without mathematical equations.

The term Minkowski diagram is used in both a generic and particular sense. In general, a Minkowski diagram is a graphic depiction of a portion of Minkowski space, often where space has been curtailed to a single dimension. These two-dimensional diagrams portray worldlines as curves in a plane that correspond to motion along the spatial axis. The vertical axis is usually temporal, and the units of measurement are taken such that the light cone at an event consists of the lines of slope plus or minus one through that event.[1]

A particular Minkowski diagram illustrates the result of a Lorentz transformation. The horizontal corresponds to the usual notion of *simultaneous events*, for a stationary observer at the origin. The Lorentz transformation relates two inertial frames of reference, where an observer makes a change of velocity at the event (0, 0). The new time axis of the observer forms an angle $\alpha$ with the previous time axis, with $\alpha < \pi/4$. After the Lorentz transformation the new simultaneous events lie on a line inclined by $\alpha$ to the previous line of simultaneity. Whatever the magnitude of $\alpha$, the line $t = x$ forms the universal[2] bisector.

## 5.1 Basics

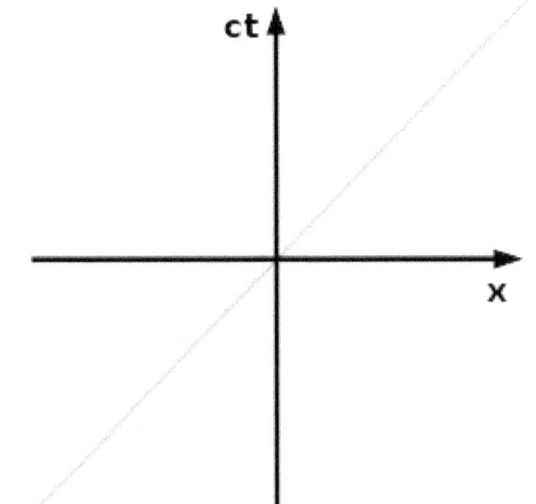

*A photon moving right at the origin corresponds to the yellow track of events, a straight line with a slope of 45°.*

For simplification in Minkowski diagrams, usually only events in a universe of one space dimension and one time dimension are considered. Unlike common distance-time diagrams, the distance will be displayed on the horizontal axis and the time on the vertical axis. In this manner the events happening in the one dimension of space can be transferred easily to a horizontal line in the diagram. Objects plotted on the diagram can be thought of as moving from bottom to top as time passes. In this way each object, like an observer or a vehicle, follows in the diagram a certain curve which is called its world line.

Each point in the diagram represents a certain position in space and time. Such a position is called an **event** whether or not anything happens at that position. The space and time units of measurement on the axes may, for example, be taken as one of the following pairs:

- units of 30 centimetres length and nanoseconds, or

- astronomical units and intervals of 8 minutes and 20 seconds (500 seconds), or

- light years and years.

This way light paths are represented by lines bisecting the axes.

## 5.2 Path-time diagram in Newtonian physics

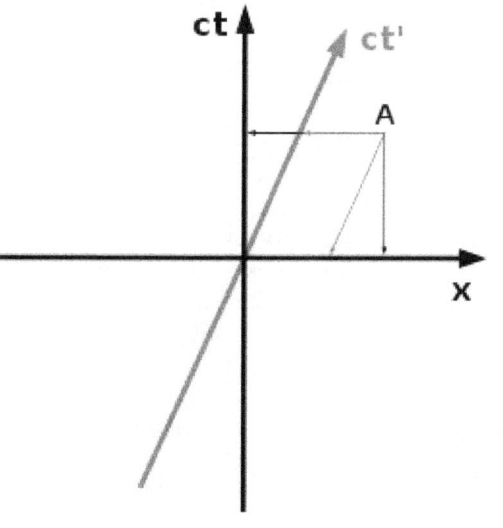

*In Newtonian physics for both observers the event at A is assigned to the same point in time.*

The black axes labelled $x$ and $ct$ on the adjoining diagram are the coordinate system of an observer which we will refer to as 'at rest', and who is positioned at $x = 0$. His world line is identical with the time axis. Each parallel line to this axis would correspond also to an object at rest but at another position. The blue line, however, describes an object moving with constant speed $v$ to the right, such as a moving observer.

This blue line labelled $ct'$ may be interpreted as the time axis for the second observer. Together with the path axis (labeled $x$, which is identical for both observers) it represents his coordinate system. Both observers agree on the location of the origin of their coordinate systems. The axes for the moving observer are not perpendicular to each other and the scale on his time axis is stretched. To determine the coordinates of a certain event, two lines, each parallel to one of the two axes, must be constructed passing through the event, and their intersections with the axes read off.

Determining position and time of the event A as an example in the diagram leads to the same time for both observers, as expected. Only for the position different values result, because the moving observer has approached the position of the event A since $t = 0$. Generally stated, all events on a line parallel to the path axis (x axis) happen simultaneously for both observers. There is only one universal time $t = t'$ which corresponds with the existence of only one common path axis. On the other hand, due to two different time axes the observers usually measure different path coordinates for the same event. This graphical translation from $x$ and $t$ to $x'$ and $t'$ and vice versa is described mathematically by the so-called Galilean transformation.

## 5.3 Minkowski diagram in special relativity

Albert Einstein (1905) discovered that the Newtonian description is not correct,[3] with Hermann Minkowski (1908) providing the graphical representation.[4] Space and time have properties which lead to different rules for the translation of coordinates in case of moving observers. In particular, events which are estimated to happen simultaneously from the viewpoint of one observer, happen at different times for the other.

In the Minkowski diagram this relativity of simultaneity corresponds with the introduction of a separate path axis for the moving observer. Following the rule described above each observer interprets all events on a line parallel to his path axis as simultaneous. The sequence of events from the viewpoint of an observer can be illustrated graphically by shifting this line in the diagram from bottom to top.

If $ct$ instead of $t$ is assigned on the time axes, the angle $\alpha$ between both path axes will be identical with that between

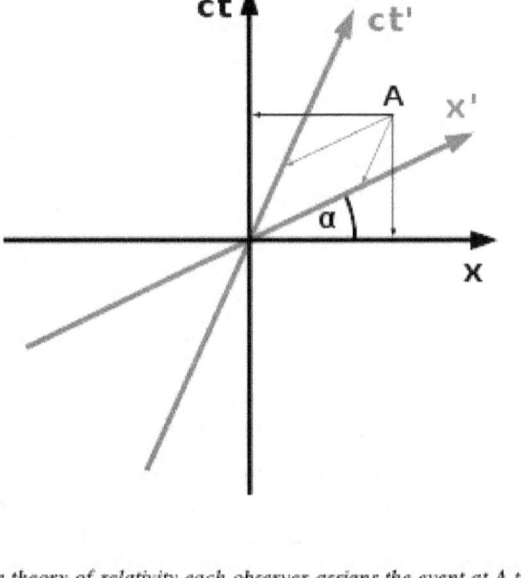

*In the theory of relativity each observer assigns the event at A to a different time and location.*

both time axes. This follows from the second postulate of the special relativity, saying that the speed of light is the same for all observers, regardless of their relative motion (see below). $\alpha$ is given by

$$\tan(\alpha) = \frac{v}{c} = \beta$$

The corresponding translation from $x$ and $t$ to $x'$ and $t'$

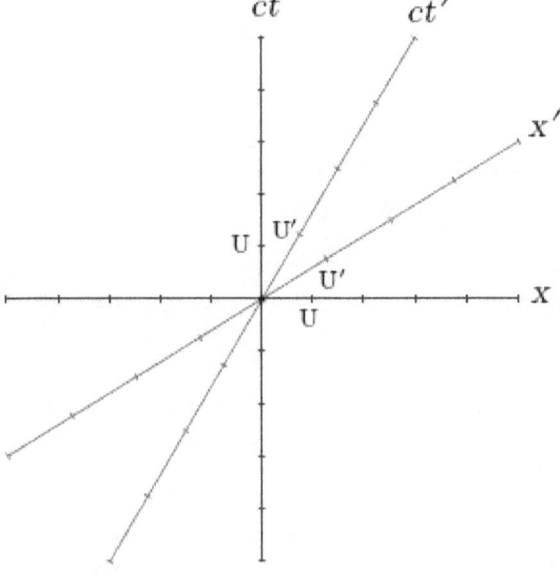

*Different scales on the axes.*

and vice versa is described mathematically by the so-called

Lorentz transformation. Whatever space and time axes arise through such transformation, in a Minkowski diagram they correspond to conjugate diameters of a pair of hyperbolas. The scales on the axes are given as follows: If $U$ is the unit length on the axes of $ct$ and $x$ respectively, the unit length on the axes of $ct'$ and $x'$ is:[5]

$$U' = U \cdot \sqrt{\frac{1 + \beta^2}{1 - \beta^2}}$$

The $ct$-axis represents the worldline of a clock resting in $S$, with $U$ representing the duration between two events happening on this worldline, also called the proper time between these events. Length $U$ upon the $x$-axis represents the rest length or proper length of a rod resting in $S$. The same interpretation can also be applied to distance $U'$ upon the $ct'$- and $x'$-axis for clocks and rods resting in $S'$.

### History

In Minkowski's 1908 paper there were three diagrams, first to illustrate the Lorentz transformation, then the partition of the plane by the light-cone, and finally illustration of worldlines.[4] The first diagram used a branch of the unit hyperbola $t^2 - x^2 = 1$ to show the locus of a unit of proper time depending on velocity, thus illustrating time dilation. The second diagram showed the conjugate hyperbola to calibrate space, where a similar stretching leaves the impression of FitzGerald contraction. In 1914 Ludwik Silberstein[6] included a diagram of "Minkowski's representation of the Lorentz transformation". This diagram included the unit hyperbola, its conjugate, and a pair of conjugate diameters. Since the 1960s a version of this more complete configuration has been referred to as The Minkowski Diagram, and used as a standard illustration of the transformation geometry of special relativity. E. T. Whittaker has pointed out that the principle of relativity is tantamount to the arbitrariness of what hyperbola radius is selected for time in the Minkowski diagram. In 1912 Gilbert N. Lewis and Edwin B. Wilson applied the methods of synthetic geometry to develop the properties of the non-Euclidean plane that has Minkowski diagrams.[7][8]

## 5.4   Loedel diagram

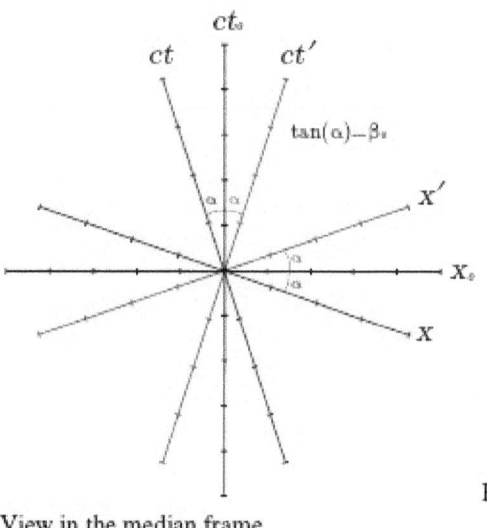

Fig.     1:

View in the median frame

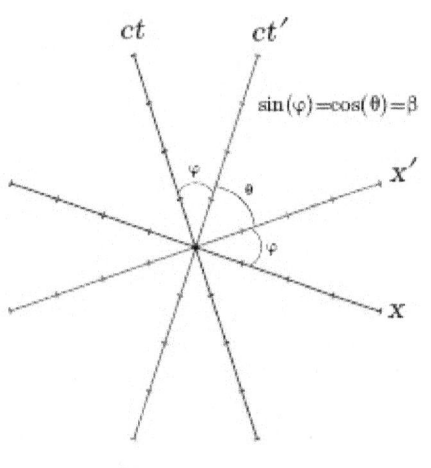

Fig.     2:

Symmetric diagram

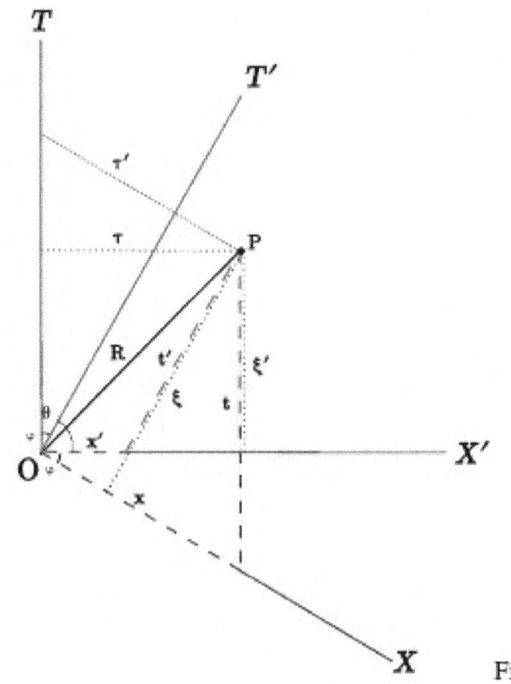

Fig.     3:

Covariant and contravariant components.

While the rest frame has space and time axes at right angles, the moving frame has primed axes which form an acute angle. Since the frames are meant to be equivalent, the asymmetry may be disturbing. However, several authors showed that there is a frame of reference between the resting and moving ones where their symmetry would be apparent ("median frame").[9] In this frame, the two other frames are moving in opposite directions with equal speed. Using such coordinates makes the units of length and time the same for both axes. If $\beta = v/c$ and $\gamma=1/\sqrt{1-\beta^2}$ is given between S and S', then these expressions are connected with the values in their median frame $S_0$ as follows:[9][10]

$$(1) \quad \beta = \frac{2\beta_0}{1+\beta_0^2},$$

$$(2) \quad \beta_0 = \frac{\gamma-1}{\beta\gamma}.$$

For instance, if $\beta = 0.5$ between S and S', then by (2) they are moving in their median frame $S_0$ with approximately $\pm0.268c$ each in opposite directions. On the other hand, if $\beta_0 = 0.5$ in $S_0$, then by (1) the relative velocity between S and S' in their own rest frames is 0.8c. The construction of the axes of S and S' is done in accordance with the ordinary method using $\tan\alpha = \beta_0$ with respect to the orthogonal axes of the median frame (Fig. 1).

However, it turns out that, when drawing such a symmetric diagram, it is possible to derive the diagram's relations even without mentioning the median frame and $\beta_0$ at all. Instead, the relative velocity $\beta = v/c$ between S and S' can directly be used in the following construction, providing the same result:[11] If $\varphi$ is the angle between the axes of $ct'$ and $ct$ (or between $x$ and $x'$), and $\theta$ between the axes of $x'$ and $ct'$, it is given:[11][12][13][14]

$$\sin \varphi = \cos \theta = \beta,$$
$$\cos \varphi = \sin \theta = 1/\gamma,$$
$$\tan \varphi = \cot \theta = \beta \cdot \gamma.$$

Two methods of construction are obvious from Fig. 2: (a) The $x$-axis is drawn perpendicular to the $ct'$-axis, the $x'$ and $ct$-axes are added at angle $\varphi$ ; (b) the $x'$-axis is drawn at angle $\theta$ with respect to the $ct'$-axis, the $x$-axis is added perpendicular to the $ct'$-axis and the $ct$-axis perpendicular to the $x'$-axis.

Also the components of a vector can be vividly demonstrated by such diagrams (Fig. 3): The parallel projections $(x, t; x', t')$ of vector $R$ are its contravariant components, $(\xi, \tau; \xi', \tau')$ its covariant components.[12][13]

### History

- Max Born (1920) drew Minkowski diagrams by placing the $ct'$-axis almost perpendicular to the x-axis, as well as the $ct$-axis to the $x'$-axis, in order to demonstrate length contraction and time dilation in the symmetric case of two rods and two clocks moving in opposite direction.[15]

- Dmitry Mirimanoff (1921) showed that there is always a median frame with respect to two relatively moving frames, and derived the relations between them from the Lorentz transformation. However, he didn't give a graphical representation in a diagram.[9]

- Symmetric diagrams were systematically developed by Paul Gruner in collaboration with Josef Sauter in two papers in 1921. Relativistic effects such as length contraction and time dilation and some relations to covariant and contravariant vectors were demonstrated by them.[12][13] Gruner extended this method in subsequent papers (1922-1924), and gave credit to Mirimanoff's treatment as well.[16][17][18][19][20][21]

- The construction of symmetric Minkowski diagrams was later independently rediscovered by several authors.   For instance, starting in 1948, Enrique Loedel Palumbo published a series of papers in Spanish language, presenting the details of such an approach.[22][23] In 1955, Henri Amar also published a paper presenting such relations, and gave credit to Loedel in a subsequent paper in 1957.[24][25] Some authors of textbooks use symmetric Minkowski diagrams, denoting as *Loedel diagrams*.[11][14]

## 5.5   Time dilation

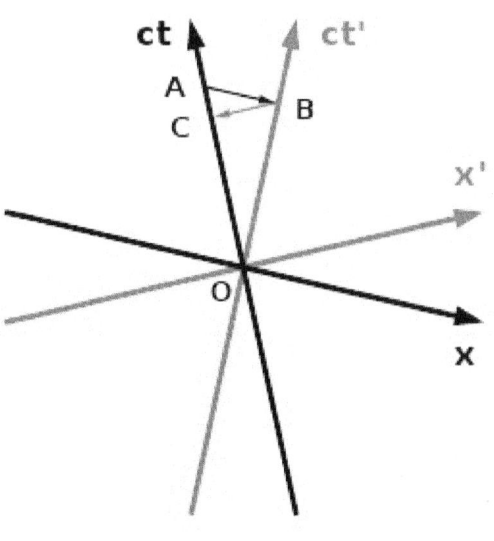

*Time dilation: Both observers consider the clock of the other as running slower.*

Relativistic time dilation means that a clock (indicating its proper time) that moves relative to an observer is observed to run slower. In fact, time itself in the frame of the moving clock is observed to run slower. This can be read immediately from the adjoining Loedel diagram quite straightforwardly because unit lengths in the two system of axes are identical. Thus, in order to compare reading between the two systems, we can simply compare lengths as they appear on the page: we do not need to consider the fact that unit lengths on each axis are warped by the factor $(1+\beta^2)^{1/2}(1-\beta^2)^{-1/2}$ , which we would have to account for in the corresponding Minkowski diagram.

The observer whose reference frame is given by the black axes is assumed to move from the origin O towards A. The moving clock has the reference frame given by the blue axes and moves from O to B. For the black observer, all events happening simultaneously with the event at A are located on a straight line parallel to its space axis. This line passes through A and B, so A and B are simultaneous from the reference frame of the observer with black axes. However, the clock that is moving relative to the black observer marks off time along the blue time axis. This is represented by the distance from O to B. Therefore, the observer at A with the black axes notices his or her clock as reading the distance from O to A while he or she observes the clock moving relative him or her to read the distance from O to B. Due to the distance from O to B being smaller than the distance from O to A, he or she concludes that the time passed on the clock moving relative to him or her is smaller than that passed on his or her own clock.

A second observer, having moved together with the clock from O to B, will argue that the other clock has reached only C until this moment and therefore this clock runs slower. The reason for these apparently paradoxical statements is the different determination of the events happening synchronously at different locations. Due to the principle of relativity, the question of "Who is right?" has no answer and does not make sense.

## 5.6 Length contraction

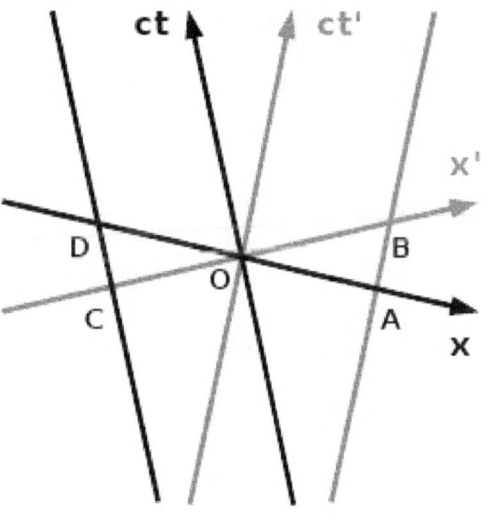

*Length contraction: Both observers consider objects moving with the other observer as being shorter.*

Relativistic length contraction means that the proper length of an object moving relative to an observer is decreased and finally also the space itself is contracted in this system. The observer is assumed again to move along the $ct$-axis. The world lines of the endpoints of an object moving relative to him are assumed to move along the $ct'$-axis and the parallel line passing through A and B. For this observer the endpoints of the object at $t = 0$ are O and A. For a second observer moving together with the object, so that for him the object is at rest, it has the proper length OB at $t' = 0$. Due to OA < OB. the object is contracted for the first observer.

The second observer will argue that the first observer has evaluated the endpoints of the object at O and A respectively and therefore at different times, leading to a wrong result due to his motion in the meantime. If the second

observer investigates the length of another object with endpoints moving along the $ct$-axis and a parallel line passing through C and D he concludes the same way this object to be contracted from OD to OC. Each observer estimates objects moving with the other observer to be contracted. This apparently paradoxical situation is again a consequence of the relativity of simultaneity as demonstrated by the analysis via Minkowski diagram.

For all these considerations it was assumed, that both observers take into account the speed of light and their distance to all events they see in order to determine the actual times at which these events happen from their point of view.

## 5.7 Constancy of the speed of light

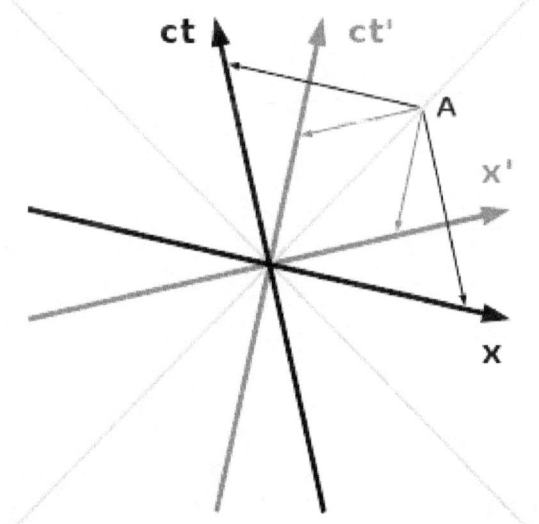

*For the speed of a photon passing A both observers measure the same value even though they move relative to each other.*

Another postulate of special relativity is the constancy of the speed of light. It says that any observer in an inertial reference frame measuring the vacuum speed of light relative to himself obtains the same value regardless of his own motion and that of the light source. This statement seems to be paradoxical, but it follows immediately from the differential equation yielding this, and the Minkowski diagram agrees. It explains also the result of the Michelson–Morley experiment which was considered to be a mystery before the theory of relativity was discovered, when photons were thought to be waves through an undetectable medium.

For world lines of photons passing the origin in different directions $x = ct$ and $x = -ct$ holds. That means any position on such a world line corresponds with steps on $x$- and $ct$-axis of equal absolute value. From the rule for reading off coordinates in coordinate system with tilted axes follows that the two world lines are the angle bisectors of the $x$- and $ct$-axis. The Minkowski diagram shows, that they are angle bisectors of the $x'$- and $ct'$-axis as well. That means both observers measure the same speed $c$ for both photons.

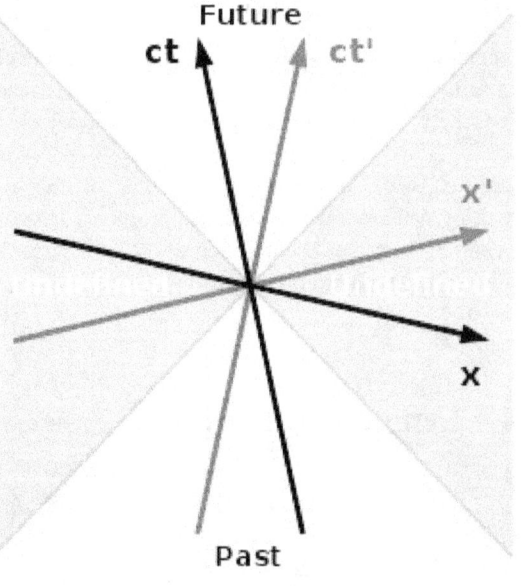

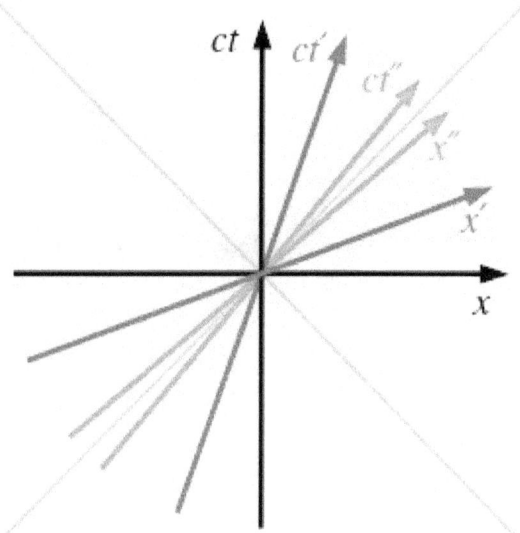

*Minkowski diagram for 3 coordinate systems. For the speeds relative to the system in black $v' = 0.4c$ and $v'' = 0.8c$ holds.*

Further coordinate systems corresponding to observers with arbitrary velocities can be added to this Minkowski diagram. For all these systems both photon world lines represent the angle bisectors of the axes. The more the relative speed approaches the speed of light the more the axes approach the corresponding angle bisector. The path axis is always more flat and the time axis more steep than the photon world lines. The scales on both axes are always identical, but usually different from those of the other coordinate systems.

## 5.8  Speed of light and causality

Straight lines passing the origin which are steeper than both photon world lines correspond with objects moving more slowly than the speed of light. If this applies to an object, then it applies from the viewpoint of all observers, because the world lines of these photons are the angle bisectors for any inertial reference frame. Therefore, any point above the origin and between the world lines of both photons can be reached with a speed smaller than that of the light and can have a cause-effect-relationship with the origin. This

*Past and future relative to the origin. For the grey areas a corresponding temporal classification is not possible.*

area is the absolute future, because any event there happens later compared to the event represented by the origin regardless of the observer, which is obvious graphically from the Minkowski diagram.

Following the same argument the range below the origin and between the photon world lines is the absolute past relative to the origin. Any event there belongs definitely to the past and can be the cause of an effect at the origin.

The relationship between any such pairs of event is called *timelike*, because they have a time distance greater than zero for all observers. A straight line connecting these two events is always the time axis of a possible observer for whom they happen at the same place. Two events which can be connected just with the speed of light are called *lightlike*.

In principle a further dimension of space can be added to the Minkowski diagram leading to a three-dimensional representation. In this case the ranges of future and past become cones with apexes touching each other at the origin. They are called light cones.

## 5.9  The speed of light as a limit

Following the same argument, all straight lines passing through the origin and which are more nearly horizontal than the photon world lines, would correspond to objects or signals moving faster than light regardless of the speed of the observer. Therefore, no event outside the light cones

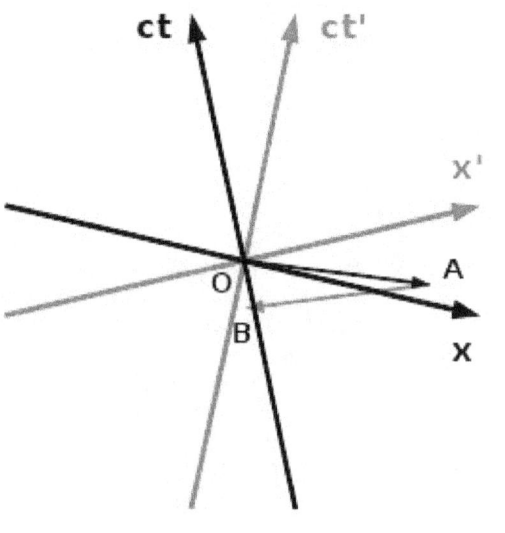

*Sending a message at superluminal speed from O via A to B into the past. Both observers consider the temporal order of the pairs of events O and A as well as A and B different.*

can be reached from the origin, even by a light-signal, nor by any object or signal moving with less than the speed of light. Such pairs of events are called *spacelike* because they have a finite spatial distance different from zero for all observers. On the other hand, a straight line connecting such events is always the space coordinate axis of a possible observer for whom they happen at the same time. By a slight variation of the velocity of this coordinate system in both directions it is always possible to find two inertial reference frames whose observers estimate the chronological order of these events to be different.

Therefore, an object moving faster than light, say from O to A in the adjoining diagram, would imply that, for any observer watching the object moving from O to A, there can be found another observer (moving at less than the speed of light with respect to the first) for whom the object moves from A to O. The question of which observer is right has no unique answer, and therefore makes no physical sense. Any such moving object or signal would violate the principle of causality.

Also, any general technical means of sending signals faster than light would permit information to be sent into the originator's own past. In the diagram, an observer at O in the *x-ct*-system sends a message moving faster than light to A. At A it is received by another observer, moving so as to be in the *x'-ct'*-system, who sends it back, again faster than light by the same technology, arriving at B. But B is in the past relative to O. The absurdity of this process becomes obvious when both observers subsequently confirm that they

received no message at all but all messages were directed towards the other observer as can be seen graphically in the Minkowski diagram. Indeed, if it were possible to accelerate an observer to the speed of light, then the space and time axes would coincide with their angle bisector. The coordinate system would collapse.

These considerations show that the speed of light as a limit is a consequence of the properties of spacetime, and not of the properties of objects such as technologically imperfect space ships. The prohibition of faster-than-light motion actually has nothing in particular to do with electromagnetic waves or light, but depends on the structure of spacetime.

## 5.10 Eponym

When Taylor and Wheeler composed *Spacetime Physics* (1966), they did *not* use the term "Minkowski diagram" for their spacetime geometry. Instead they included an acknowledgement of Minkowski's contribution to philosophy by the totality of his innovation of 1908.[26]

As an eponym, the term *Minkowski diagram* is subject to Stigler's law of eponymy, namely that Minkowski is wrongly designated as originator. The earlier works of Alexander Macfarlane contain algebra and diagrams that correspond well with the Minkowski diagram. See for instance the plate of figures in *Proceedings of the Royal Society in Edinburgh* for 1900. Macfarlane was building on what one sees in William Kingdon Clifford's *Elements of Dynamic* (1878), page 90.

When abstracted to a line drawing, then any figure showing conjugate hyperbolas, with a selection of conjugate diameters, falls into this category. Students making drawings to accompany the exercises in George Salmon's *A Treatise on Conic Sections* (1900) at pages 165–71 (on conjugate diameters) will be making Minkowski diagrams.

## 5.11 Spacetime diagram of an accelerating observer in special relativity

The momentarily co-moving inertial frames along the world line of a rapidly accelerating observer (center). The vertical direction indicates time, while the horizontal indicates distance, the dashed line is the spacetime trajectory ("world line") of the observer. The small dots are specific events in spacetime. If one imagines these events to be the flashing of a light, then the events that pass the two diagonal lines in the bottom half of the image (the past light cone of the observer

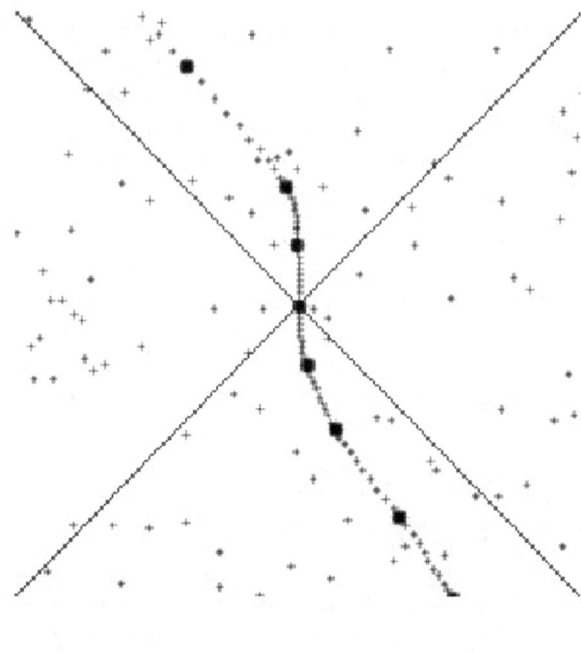

in the origin) are the events visible to the observer. The slope of the world line (deviation from being vertical) gives the relative velocity to the observer. Note how the momentarily co-moving inertial frame changes when the observer accelerates.

## 5.12  See also

- Minkowski space

- Penrose diagram

- Rapidity

## 5.13  References

[1]  Mermin (1968) Chapter 17

[2]  See Vladimir Karapetoff

[3]  Einstein, Albert (1905a), "Zur Elektrodynamik bewegter Körper" (PDF), *Annalen der Physik* **322** (10): 891–921, Bibcode:1905AnP...322..891E, doi:10.1002/andp.19053221004. See also: English translation.

[4]  Minkowski, Hermann (1909), "Raum und Zeit", *Physikalische Zeitschrift* 10: 75–88

   - Various English translations on Wikisource: Space and Time

[5]  Jürgen Freund (2008). *Special Relativity for Beginners: A Textbook for Undergraduates*. World Scientific. p. 49. ISBN 981277159X.

[6]  Silberstein (1914) *The Theory of Relativity*, page 131

[7]  Edwin B. Wilson & Gilbert N. Lewis (1912). "The Spacetime Manifold of Relativity. The Non-Euclidean Geometry of Mechanics and Electromagnetics", Proceedings of the American Academy of Arts and Sciences 48:387–507

[8]  Synthetic Spacetime, a digest of the axioms used, and theorems proved, by Wilson and Lewis. Archived by WebCite

[9]  Mirimanoff, Dmitry (1921). "La transformation de Lorentz-Einstein et le temps universel de M. Ed. Guillaume". *Archives des sciences physiques et naturelles (supplement)*. 5 3: 46–48. (Translation: The Lorentz-Einstein transformation and the universal time of Ed. Guillaume)

[10]  Albert Shadowitz (2012). *The Electromagnetic Field* (Reprint of 1975 ed.). Courier Dover Publications. p. 460. ISBN 0486132013. See Google books, p. 460

[11]  Leo Sartori (1996). *Understanding Relativity: a simplified approach to Einstein's theories*. University of California Press. pp. 151ff. ISBN 0-520-20029-2.

[12]  Gruner, Paul & Sauter, Josef (1921). "Représentation géométrique élémentaire des formules de la théorie de la relativité". *Archives des sciences physiques et naturelles*. 5 3: 295–296. (Translation: Elementary geometric representation of the formulas of the special theory of relativity)

[13]  Gruner, Paul (1921). "Eine elementare geometrische Darstellung der Transformationsformeln der speziellen Relativitätstheorie". *Physikalische Zeitschrift* **22**: 384–385. (Translation: An elementary geometrical representation of the transformation formulas of the special theory of relativity)

[14]  Albert Shadowitz (1988). *Special relativity* (Reprint of 1968 ed.). Courier Dover Publications. pp. 20–22. ISBN 0-486-65743-4.

[15]  Born, Max (1920). *Die Relativitätstheorie Einsteins* (First ed.). Springer. pp. 177–180. See also Reprint (2013) of third edition (1922) at Google books, p. 187

[16]  Gruner, Paul (1922). *Elemente der Relativitätstheorie* [*Elements of the theory of relativity*]. Bern: P. Haupt.

[17]  Gruner, Paul (1922). "Graphische Darstellung der speziellen Relativitätstheorie in der vierdimensionalen Raum-Zeit-Welt I" [Graphical representation of the special theory of relativity in the four-dimensional spacetime-world I]. *Zeitschrift für Physik* **10** (1): 22–37. Bibcode:1922ZPhy...10...22G. doi:10.1007/BF01332542.

[18]  Gruner, Paul (1922). "Graphische Darstellung der speziellen Relativitätstheorie in der vierdimensionalen Raum-Zeit-Welt II" [Graphical representation of the special theory of relativity in the four-dimensional spacetime-world II]. *Zeitschrift für Physik* **10** (1): 227–235. Bibcode:1922ZPhy...10..227G. doi:10.1007/BF01332563.

[19] Gruner, Paul (1921). "a) Représentation graphique de l'univers espace-temps à quatre dimensions. b) Représentation graphique du temps universel dans la théorie de la relativité". *Archives des sciences physiques et naturelles*. 5 4: 234–236. (Translation: Graphical representation of the four-dimensional space-time universe)

[20] Gruner, Paul (1922). "Die Bedeutung "reduzierter" orthogonaler Koordinatensysteme für die Tensoranalysis und die spezielle Relativitätstheorie" [The importance of "reduced" orthogonal coordinate-systems for tensor analysis and the special theory of relativity]. *Zeitschrift für Physik* 10 (1): 236–242. Bibcode:1922ZPhy...10..236G. doi:10.1007/BF01332564.

[21] Gruner, Paul (1924). "Geometrische Darstellungen der speziellen Relativitätstheorie, insbesondere des elektromagnetischen Feldes bewegter Körper" [Geometrich representations of the special theory of relativity, in particular the electromagnetic field of moving bodies]. *Zeitschrift für Physik* 21 (1): 366–371. Bibcode:1924ZPhy...21..366G. doi:10.1007/BF01328285.

[22] Loedel, Enrique (1948). "Aberracion y Relatividad". *Anales soc. cient. argentina* 145: 3–13.

[23] *Fisica relativista*, Kapelusz Editorial, Buenos Aires, Argentina (1955).

[24] Amar, Henri (1955). "New Geometric Representation of the Lorentz Transformation". *American Journal of Physics* 23 (8): 487–489. Bibcode:1955AmJPh..23..487A. doi:10.1119/1.1934074.

[25] Amar, Henri & Loedel, Enrique (1957). "Geometric Representation of the Lorentz Transformation". *American Journal of Physics* 25 (5): 326–327. Bibcode:1957AmJPh..25..326A. doi:10.1119/1.1934453.

[26] Taylor/Wheeler (1966) page 37: "Minkowski's insight is central to the understanding of the physical world. It focuses attention on those quantities, such as interval, which are the same in all frames of reference. It brings out the relative character of quantities, such as velocity, energy, time, distance, which depend on the frame of reference."

• Anthony French (1968) *Special Relativity*, pages 82 & 83, New York: W W Norton & Company.

• E.N. Glass (1975) "Lorentz boosts and Minkowski diagrams" American Journal of Physics 43:1013,4.

• N. David Mermin (1968) *Space and Time in Special Relativity*, Chapter 17 Minkowski diagrams: The Geometry of Spacetime, pages 155–99 McGraw-Hill.

• Rindler, Wolfgang (2001). *Relativity: Special, General and Cosmological*. Oxford University Press. ISBN 0-19-850836-0.

• W.G.V. Rosser (1964) *An Introduction to the Theory of Relativity*, page 256, Figure 6.4, London: Butterworths.

• Edwin F. Taylor and John Archibald Wheeler (1963) *Spacetime Physics*, pages 27 to 38, New York: W. H. Freeman and Company, Second edition (1992).

• Walter, Scott (1999), "The non-Euclidean style of Minkowskian relativity" (PDF), in J. Gray, *The Symbolic Universe: Geometry and Physics*, Oxford University Press, pp. 91–127 (see page 10 of e-link)

## 5.14 External links

Media related to Minkowski diagrams at Wikimedia Commons

# Chapter 6

# Minkowski space (number field)

For Minkowski space-time, see Minkowski space.

In mathematics, in the field of algebraic number theory, a **Minkowski space** is a Euclidean space associated with an algebraic number field.

If $K$ is a number field of degree $d$ then there are $d$ distinct embeddings of $K$ into $\mathbf{C}$. We let $K\mathbf{C}$ be the image of $K$ in the product $\mathbf{C}^d$, considered as equipped with the usual Hermitian inner product. If $c$ denotes complex conjugation, let $K\mathbf{R}$ denote the subspace of $K\mathbf{C}$ fixed by $c$, equipped with a scalar product. This is the Minkowski space of $K$.

## 6.1 References

- Neukirch, Jürgen (1999). *Algebraic Number Theory*. Grundlehren der Mathematischen Wissenschaften **322**. Springer-Verlag. ISBN 978-3-540-65399-8. MR 1697859. Zbl 0956.11021.

# Chapter 7

# Poincaré group

*Henri Poincaré*

## 7.1 Overview

A Minkowski spacetime isometry has the property that the interval between events is left invariant. For example, if everything was postponed by two hours, including the two events and the path you took to go from one to the other, then the time interval between the events recorded by a stop-watch you carried with you would be the same. Or if everything was shifted five miles to the west, or turned 60 degrees to the right, you would also see no change in the interval. It turns out that the proper length of an object is also unaffected by such a shift. A time or space reversal (a reflection) is also an isometry of this group.

In Minkowski space (i.e. ignoring the effects of gravity), there are ten degrees of freedom of the isometries, which may be thought of as translation through time or space (four degrees, one per dimension); reflection through a plane (three degrees, the freedom in orientation of this plane); or a "boost" in any of the three spatial directions (three degrees). Composition of transformations is the operator of the Poincaré group, with proper rotations being produced as the composition of an even number of reflections.

In classical physics, the Galilean group is a comparable ten-parameter group that acts on absolute time and space. Instead of boosts, it features shear mappings to relate co-moving frames of reference.

## 7.2 Details

For the Poincaré group (fundamental group) of a topological space, see Fundamental group.

The **Poincaré group**, named after Henri Poincaré (1906),[1] was first defined by Minkowski (1908) being the group of Minkowski spacetime isometries.[2][3] It is a ten-generator non-abelian Lie group of fundamental importance in physics.

The Poincaré group is the group of Minkowski spacetime isometries. It is a ten-dimensional noncompact Lie group. The abelian group of translations is a normal subgroup, while the Lorentz group is also a subgroup, the stabilizer of the origin. The Poincaré group itself is the minimal subgroup of the affine group which includes all translations and Lorentz transformations. More precisely, it is a semidirect product of the translations and the Lorentz group,

$\mathbf{R}^{1,3} \rtimes SO(1,3)$ .

Another way of putting this is that the Poincaré group is a group extension of the Lorentz group by a vector representation of it; it is sometimes dubbed, informally, as the *"inhomogeneous Lorentz group"*. In turn, it can also be obtained as a group contraction of the de Sitter group $SO(4,1) \sim Sp(2,2)$, as the de Sitter radius goes to infinity.

Its positive energy unitary irreducible representations are indexed by mass (nonnegative number) and spin (integer or half integer) and are associated with particles in quantum mechanics (see Wigner's classification).

In accordance with the Erlangen program, the geometry of Minkowski space is defined by the Poincaré group: Minkowski space is considered as a homogeneous space for the group.

The **Poincaré algebra** is the Lie algebra of the Poincaré group. It is a Lie algebra extension of the Lie algebra of the Lorentz group. More specifically, the proper ($\det\Lambda = 1$), orthochronous ($\Lambda^0{}_0 \geq 1$) part of the Lorentz subgroup (its identity component), $SO^+(1, 3)$, is connected to the identity and is thus provided by the exponentiation $\exp(ia_\mu P^\mu) \exp(i\omega_{\mu\nu} M^{\mu\nu}/2)$ of this Lie algebra. In component form, the Poincaré algebra is given by the commutation relations:[4][5]

where P is the generator of translations, M is the generator of Lorentz transformations, and $\eta$ is the (+,−,−,−) Minkowski metric (see Sign convention).

The bottom commutation relation is the ("homogeneous") Lorentz group, consisting of rotations, $J_i = -\epsilon_{imn}M^{mn}/2$, and boosts, $K_i = M_{i0}$. In this notation, the entire Poincaré algebra is expressible in noncovariant (but more practical) language as

$$[J_m, P_n] = i\epsilon_{mnk}P_k ,$$

$$[J_i, P_0] = 0 ,$$

$$[K_i, P_k] = i\eta_{ik}P_0 ,$$

$$[K_i, P_0] = -iP_i ,$$

$$[J_m, J_n] = i\epsilon_{mnk}J_k ,$$

$$[J_m, K_n] = i\epsilon_{mnk}K_k ,$$

$$[K_m, K_n] = -i\epsilon_{mnk}J_k ,$$

where the bottom line commutator of two boosts is often referred to as a "Wigner rotation". Note the important simplification $[Jm+i\,Km , Jn-i\,Kn] = 0$, which permits reduction of the Lorentz subalgebra to $\mathbf{su}(2) \oplus \mathbf{su}(2)$ and efficient treatment of its associated representations.

The Casimir invariants of this algebra are $P\mu P^\mu$ and $W\mu\,W^\mu$ where $W\mu$ is the Pauli–Lubanski pseudovector; they serve as labels for the representations of the group.

The Poincaré group is the full symmetry group of any relativistic field theory. As a result, all elementary particles fall in representations of this group. These are usually specified by the *four-momentum* squared of each particle (i.e. its mass squared) and the intrinsic quantum numbers $J^{PC}$, where J is the spin quantum number, P is the parity and C is the charge-conjugation quantum number. In practice, charge conjugation and parity are violated by many quantum field theories; where this occurs, P and C are forfeited. Since CPT symmetry is invariant in quantum field theory, a time-reversal quantum number may be constructed from those given.

As a topological space, the group has four connected components: the component of the identity; the time reversed component; the spatial inversion component; and the component which is both time-reversed and spatially inverted.

## 7.3   Poincaré symmetry

**Poincaré symmetry** is the full symmetry of special relativity. It includes:

- *translations* (displacements) in time and space ($P$), forming the abelian Lie group of translations on space-time;

- *rotations* in space, forming the non-Abelian Lie group of three-dimensional rotations ($J$);

- *boosts*, transformations connecting two uniformly moving bodies ($K$).

The last two symmetries, $J$ and $K$, together make the Lorentz group (see also Lorentz invariance); the semi-direct product of the translations group and the Lorentz group then produce the Poincaré group. Objects which are invariant under this group are then said to possess **Poincaré invariance** or **relativistic invariance**.

## 7.4   See also

- Euclidean group

- Representation theory of the Poincaré group

- Wigner's classification

- Symmetry in quantum mechanics

- Center of mass (relativistic)

- Pauli–Lubanski pseudovector

- Particle physics and representation theory

## 7.5 Notes

[1] Poincaré, Henri, "Sur la dynamique de l'électron", *Rendiconti del Circolo matematico di Palermo* 21: 129–176, doi:10.1007/bf03013466 (Wikisource translation: On the Dynamics of the Electron). The group defined in this paper would now be described as the homogeneous Lorentz group with scalar multipliers.

[2] Minkowski, Hermann, "Die Grundgleichungen für die elektromagnetischen Vorgänge in bewegten Körpern", *Nachrichten von der Gesellschaft der Wissenschaften zu Göttingen, Mathematisch-Physikalische Klasse*: 53–111 (Wikisource translation: The Fundamental Equations for Electromagnetic Processes in Moving Bodies).

[3] Minkowski, Hermann, "Raum und Zeit", *Physikalische Zeitschrift* 10: 75–88

[4] N.N. Bogolubov (1989). *General Principles of Quantum Field Theory* (2nd ed.). Springer. p. 272. ISBN 0-7923-0540-X.

[5] T. Ohlsson (2011). *Relativistic Quantum Physics: From Advanced Quantum Mechanics to Introductory Quantum Field Theory*. Cambridge University Press. p. 10. ISBN 1-13950-4320.

## 7.6 References

- Wu-Ki Tung (1985). *Group Theory in Physics*. World Scientific Publishing. ISBN 9971-966-57-3.

- Weinberg, Steven (1995). *The Quantum Theory of Fields* 1. Cambridge: Cambridge University press. ISBN 978-0-521-55001-7.

- L.H. Ryder (1996). *Quantum Field Theory* (2nd ed.). Cambridge University Press. p. 62. ISBN 0-52147-8146.

# Chapter 8

# Euclidean space

This article is about Euclidean spaces of all dimensions. For 3-dimensional Euclidean space, see 3-dimensional space.

In geometry, **Euclidean space** encompasses the two-

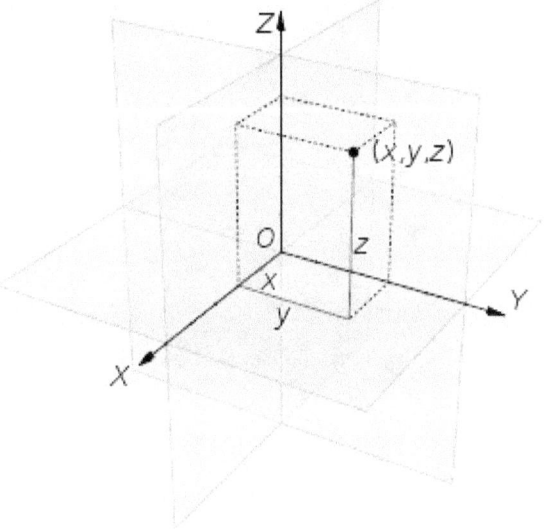

*Every point in three-dimensional Euclidean space is determined by three coordinates.*

dimensional Euclidean plane, the three-dimensional space of Euclidean geometry, and certain other spaces. It is named after the Ancient Greek mathematician Euclid of Alexandria.[1] The term "Euclidean" distinguishes these spaces from other types of spaces considered in modern geometry. Euclidean spaces also generalize to higher dimensions.

Classical Greek geometry defined the Euclidean plane and Euclidean three-dimensional space using certain postulates, while the other properties of these spaces were deduced as theorems. Geometric constructions are also used to define rational numbers. When algebra and mathematical analysis became developed enough, this relation reversed and now it is more common to define Euclidean space using Cartesian coordinates and the ideas of analytic geometry. It means

that points of the space are specified with collections of real numbers, and geometric shapes are defined as equations and inequalities. This approach brings the tools of algebra and calculus to bear on questions of geometry and has the advantage that it generalizes easily to Euclidean spaces of more than three dimensions.

From the modern viewpoint, there is essentially only one Euclidean space of each dimension. With Cartesian coordinates it is modelled by the real coordinate space ($\mathbf{R}^n$) of the same dimension. In one dimension, this is the real line; in two dimensions, it is the Cartesian plane; and in higher dimensions it is a coordinate space with three or more real number coordinates. Mathematicians denote the n-dimensional Euclidean space by $\mathbf{E}^n$ if they wish to emphasize its Euclidean nature, but $\mathbf{R}^n$ is used as well since the latter is assumed to have the standard Euclidean structure, and these two structures are not always distinguished. Euclidean spaces have finite dimension.[2]

## 8.1 Intuitive overview

One way to think of the Euclidean plane is as a set of points satisfying certain relationships, expressible in terms of distance and angle. For example, there are two fundamental operations (referred to as motions) on the plane. One is translation, which means a shifting of the plane so that every point is shifted in the same direction and by the same distance. The other is rotation about a fixed point in the plane, in which every point in the plane turns about that fixed point through the same angle. One of the basic tenets of Euclidean geometry is that two figures (usually considered as subsets) of the plane should be considered equivalent (congruent) if one can be transformed into the other by some sequence of translations, rotations and reflections (see below).

In order to make all of this mathematically precise, the theory must clearly define the notions of distance, angle, translation, and rotation for a mathematically described space.

Even when used in physical theories, Euclidean space is an abstraction detached from actual physical locations, specific reference frames, measurement instruments, and so on. A purely mathematical definition of Euclidean space also ignores questions of units of length and other physical dimensions: the distance in a "mathematical" space is a number, not something expressed in inches or metres. The standard way to define such space, as carried out in the remainder of this article, is to define the Euclidean plane as a two-dimensional real vector space equipped with an inner product.[2] The reason for working with arbitrary vector spaces instead of $\mathbf{R}^n$ is that it is often preferable to work in a *coordinate-free* manner (that is, without choosing a preferred basis). For then:

- the vectors in the vector space correspond to the points of the Euclidean plane,

- the addition operation in the vector space corresponds to translation, and

- the inner product implies notions of angle and distance, which can be used to define rotation.

Once the Euclidean plane has been described in this language, it is actually a simple matter to extend its concept to arbitrary dimensions. For the most part, the vocabulary, formulae, and calculations are not made any more difficult by the presence of more dimensions. (However, rotations are more subtle in high dimensions, and visualizing high-dimensional spaces remains difficult, even for experienced mathematicians.)

A Euclidean space is not technically a vector space but rather an affine space, on which a vector space acts by translations, or, conversely, a Euclidean vector is the difference (displacement) in an ordered pair of points, not a single point. Intuitively, the distinction says merely that there is no canonical choice of where the origin should go in the space, because it can be translated anywhere. When a certain point is chosen, it can be declared the origin and subsequent calculations may ignore the difference between a point and its coordinate vector, as said above. See point–vector distinction for details.

## 8.2 Euclidean structure

These are distances between points and the angles between lines or vectors, which satisfy certain conditions (see below), which makes a set of points a Euclidean space. The natural way to obtain these quantities is by introducing and using the standard inner product (also known as the dot product) on $\mathbf{R}^n$.[2] The inner product of any two real n-vectors $\mathbf{x}$ and $\mathbf{y}$ is defined by

$$\mathbf{x} \cdot \mathbf{y} = \sum_{i=1}^{n} x_i y_i = x_1 y_1 + x_2 y_2 + \cdots + x_n y_n,$$

where $x_i$ and $y_i$ are ith coordinates of vectors $\mathbf{x}$ and $\mathbf{y}$ respectively. The result is always a real number.

### 8.2.1 Distance

Main article: Euclidean distance

The inner product of $\mathbf{x}$ with itself is always non-negative. This product allows us to define the "length" of a vector $\mathbf{x}$ through square root:

$$\|\mathbf{x}\| = \sqrt{\mathbf{x} \cdot \mathbf{x}} = \sqrt{\sum_{i=1}^{n} (x_i)^2}.$$

This length function satisfies the required properties of a norm and is called the **Euclidean norm** on $\mathbf{R}^n$.

Finally, one can use the norm to define a metric (or distance function) on $\mathbf{R}^n$ by

$$d(\mathbf{x}, \mathbf{y}) = \|\mathbf{x} - \mathbf{y}\| = \sqrt{\sum_{i=1}^{n} (x_i - y_i)^2}.$$

This distance function is called the Euclidean metric. This formula expresses a special case of the Pythagorean theorem.

This distance function (which makes a metric space) is sufficient to define all Euclidean geometry, including the dot product. Thus, a real coordinate space together with this Euclidean structure is called **Euclidean space**. Its vectors form an inner product space (in fact a Hilbert space), and a normed vector space.

The metric space structure is the main reason behind the use of real numbers $\mathbf{R}$, not some other ordered field, as the mathematical foundation of Euclidean (and many other) spaces. Euclidean space is a complete metric space, a property which is impossible to achieve operating over rational numbers, for example.

### 8.2.2 Angle

Main article: Angle
The (**non-reflex**) **angle** $\theta$ ($0° \leq \theta \leq 180°$) between vectors $\mathbf{x}$ and $\mathbf{y}$ is then given by

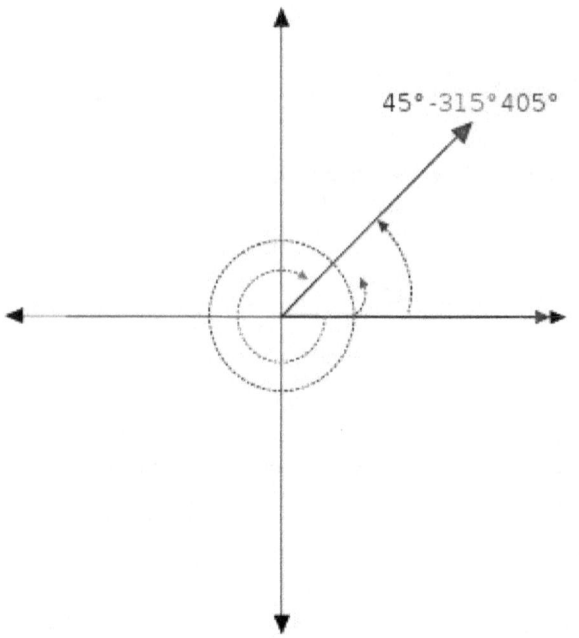

*Positive and negative angles on the oriented plane*

$$\theta = \arccos\left(\frac{\mathbf{x} \cdot \mathbf{y}}{\|\mathbf{x}\|\|\mathbf{y}\|}\right)$$

where arccos is the arccosine function. It is useful only for $n > 1$,[footnote 1] and the case $n = 2$ is somewhat special. Namely, on an oriented Euclidean plane one can define an angle between two vectors as a number defined modulo 1 turn (usually denoted as either $2\pi$ or $360°$), such that $\angle \mathbf{y} \, \mathbf{x} = -\angle \mathbf{x} \, \mathbf{y}$. This oriented angle is equal either to the angle $\theta$ from the formula above or to $-\theta$. If one non-zero vector is fixed (such as the first basis vector), then each non-zero vector is uniquely defined by its magnitude and angle.

The angle does not change if vectors $\mathbf{x}$ and $\mathbf{y}$ are multiplied by positive numbers.

Unlike the aforementioned situation with distance, the scale of angles is the same in pure mathematics, physics, and computing. It does not depend on the scale of distances; all distances may be multiplied by some fixed factor, and all angles will be preserved. Usually, the angle is considered a dimensionless quantity, but there are different units of measurement, such as radian (preferred in pure mathematics and theoretical physics) and degree (°) (preferred in most applications).

## 8.2.3  Rotations and reflections

Main articles: Rotation (mathematics), Reflection (mathematics) and Orthogonal group

See also: rotational symmetry and reflection symmetry

Symmetries of a Euclidean space are transformations which preserve the Euclidean metric (called *isometries*). Although aforementioned translations are most obvious of them, they have the same structure for any affine space and do not show a distinctive character of Euclidean geometry. Another family of symmetries leave one point fixed, which may be seen as the origin without loss of generality. All transformations, which preserves the origin and the Euclidean metric, are linear maps. Such transformations Q must, for any $\mathbf{x}$ and $\mathbf{y}$, satisfy:

$$Q\mathbf{x} \cdot Q\mathbf{y} = \mathbf{x} \cdot \mathbf{y} \text{ (explain the notation),}$$
$$|Q\mathbf{x}| = |\mathbf{x}|.$$

Such transforms constitute a group called the *orthogonal group* O($n$). Its elements Q are exactly solutions of a matrix equation

$$Q^{\mathsf{T}}Q = QQ^{\mathsf{T}} = I,$$

where $Q^{\mathsf{T}}$ is the transpose of Q and $I$ is the identity matrix.

But a Euclidean space is orientable.[footnote 2] Each of these transformations either preserves or reverses orientation depending on whether its determinant is +1 or −1 respectively. Only transformations which preserve orientation, which form the *special orthogonal* group SO($n$), are considered (proper) rotations. This group has, as a Lie group, the same dimension $n(n - 1)/2$ and is the identity component of O($n$).

Groups SO($n$) are well-studied for $n \leq 4$. There are no non-trivial rotations in 0- and 1-spaces. Rotations of a Euclidean plane ($n = 2$) are parametrized by the angle (modulo 1 turn). Rotations of a 3-space are parametrized with axis and angle, whereas a rotation of a 4-space is a superposition of two 2-dimensional rotations around perpendicular planes.

Among linear transforms in O($n$) which reverse the orientation are hyperplane reflections. This is the only possible case for $n \leq 2$, but starting from three dimensions, such isometry in the general position is a rotoreflection.

## 8.2.4  Euclidean group

Main article: Euclidean group

The Euclidean group $E(n)$, also referred to as the group of all isometries ISO($n$), treats translations, rotations, and reflections in a uniform way, considering them as group actions in the context of group theory, and especially in Lie group theory. These group actions preserve the Euclidean structure.

As the group of all isometries, ISO($n$), the Euclidean group is important because it makes Euclidean geometry a case of Klein geometry, a theoretical framework including many alternative geometries.

The structure of Euclidean spaces – distances, lines, vectors, angles (up to sign), and so on – is invariant under the transformations of their associated Euclidean group. For instance, translations form a commutative subgroup that acts freely and transitively on $\mathbf{E}^n$, while the stabilizer of any point there is the aforementioned O($n$).

Along with translations, rotations, reflections, as well as the identity transformation, Euclidean motions comprise also glide reflections (for $n \geq 2$), screw operations and rotoreflections (for $n \geq 3$), and even more complex combinations of primitive transformations for $n \geq 4$.

The group structure determines which conditions a metric space needs to satisfy to be a Euclidean space:

1. Firstly, a metric space must be translationally invariant with respect to some (finite-dimensional) real vector space. This means that the space itself is an affine space, that the space is *flat*, not curved, and points do not have different properties, and so any point can be translated to any other point.

2. Secondly, the metric must correspond in the aforementioned way to some positive-defined quadratic form on this vector space, because point stabilizers have to be isomorphic to O($n$).

## 8.3   Non-Cartesian coordinates

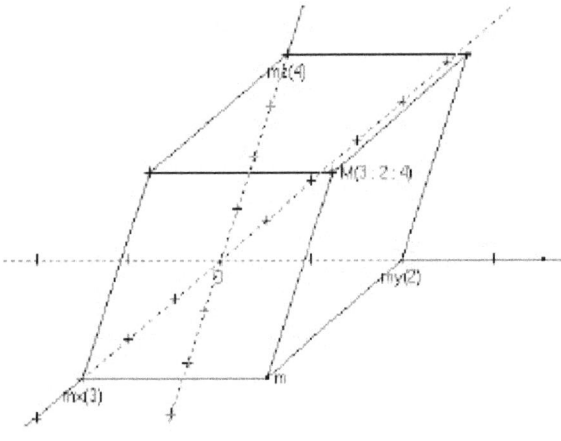

*3-dimensional skew coordinates*

Main article: Coordinate system

Cartesian coordinates are arguably the standard, but not the only possible option for a Euclidean space. Skew coordinates are compatible with the affine structure of $\mathbf{E}^n$, but make formulae for angles and distances more complicated.

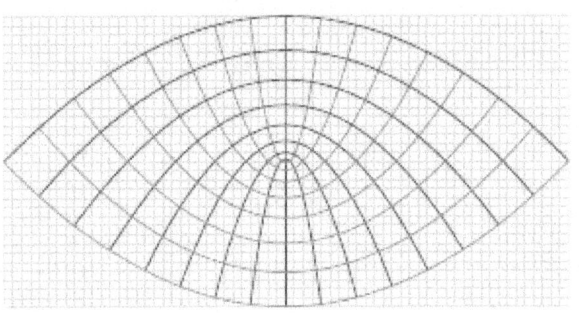

*Parabolic coordinates*

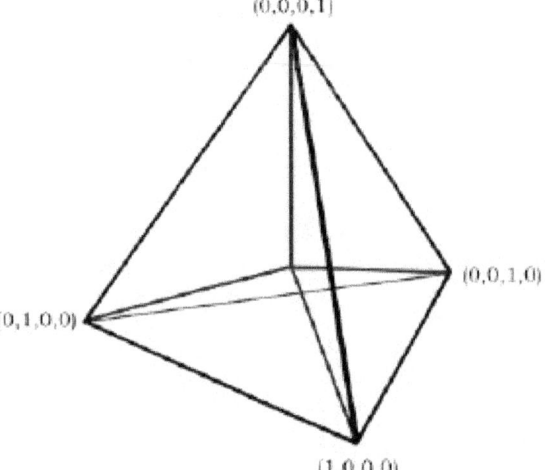

*Barycentric coordinates in 3-dimensional space: four coordinates are related with one linear equation*

Another approach, which goes in line with ideas of differential geometry and conformal geometry, is orthogonal coordinates, where coordinate hypersurfaces of different coordinates are orthogonal, although curved. Examples include the polar coordinate system on Euclidean plane, the second important plane coordinate system.

See below about expression of the Euclidean structure in curvilinear coordinates.

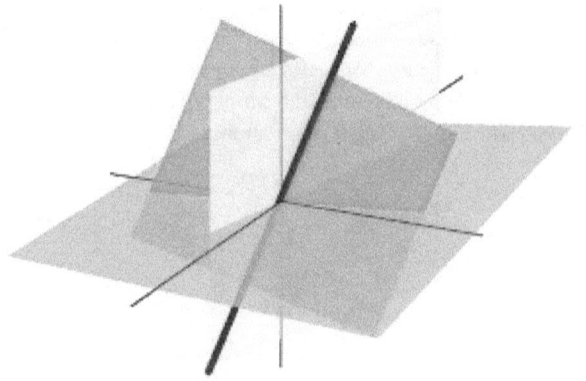

*Three mutually transversal planes in the 3-dimensional space and their intersections, three lines*

## 8.4   Geometric shapes

See also: List of mathematical shapes

### 8.4.1   Lines, planes, and other subspaces

Main article: Flat (geometry)

The simplest (after points) objects in Euclidean space are flats, or Euclidean *subspaces* of lesser dimension. Points are 0-dimensional flats, 1-dimensional flats are called *(straight) lines*, and 2-dimensional flats are *planes*. $(n-1)$-dimensional flats are called *hyperplanes*.

Any two distinct points lie on exactly one line. Any line and a point outside it lie on exactly one plane. More generally, the properties of flats and their incidence of Euclidean space are shared with affine geometry, whereas the affine geometry is devoid of distances and angles.

### 8.4.2   Line segments and triangles

Main articles: Line segment and Triangle geometry

This is not only a line which a pair $(A, B)$ of distinct points defines. Points on the line which lie between A and B, together with A and B themselves, constitute a line segment *A B*. Any line segment has the length, which equals to distance between A and B. If $A = B$, then the segment is degenerate and its length equals to 0, otherwise the length is positive.

A (non-degenerate) triangle is defined by three points not lying on the same line. Any triangle lies on one plane. The concept of triangle is not specific to Euclidean spaces, but

Euclidean triangles have numerous special properties and define many derived objects.

A triangle can be thought of as a 3-gon on a plane, a special (and the first meaningful in Euclidean geometry) case of a polygon.

### 8.4.3   Polytopes and root systems

Main articles: Polytope and Root system
See also: List of polygons, polyhedra and polytopes and List of regular polytopes

Polytope is a concept that generalizes polygons on a plane and polyhedra in 3-dimensional space (which are among the earliest studied geometrical objects). A simplex is a generalization of a line segment (1-simplex) and a triangle (2-simplex). A tetrahedron is a 3-simplex.

The concept of a polytope belongs to affine geometry, which is more general than Euclidean. But Euclidean geometry distinguish *regular polytopes*. For example, affine geometry does not see the difference between an equilateral triangle and a right triangle, but in Euclidean space the former is regular and the latter is not.

Root systems are special sets of Euclidean vectors. A root system is often identical to the set of vertices of a regular polytope.

### 8.4.4   Curves

Main article: Euclidean geometry of curves
See also: List of curves

### 8.4.5   Balls, spheres, and hypersurfaces

Main articles: Ball (mathematics) and Hypersurface
See also: n-sphere and List of surfaces

## 8.5   Topology

Main article: Real coordinate space § Topological properties

Since Euclidean space is a metric space, it is also a topological space with the natural topology induced by the metric. The metric topology on $\mathbf{E}^n$ is called the **Euclidean topology**, and it is identical to the standard topology on

$\mathbf{R}^n$. A set is open if and only if it contains an open ball around each of its points; in other words, open balls form a base of the topology. The topological dimension of the Euclidean n-space equals n, which implies that spaces of different dimension are not homeomorphic. A finer result is the invariance of domain, which proves that any subset of n-space, that is (with its subspace topology) homeomorphic to an open subset of n-space, is itself open.

## 8.6 Applications

Aside from countless uses in fundamental mathematics, a Euclidean model of the physical space can be used to solve many practical problems with sufficient precision. Two usual approaches are a fixed, or *stationary* reference frame (i.e. the description of a motion of objects as their positions that change continuously with time), and the use of Galilean space-time symmetry (such as in Newtonian mechanics). To both of them the modern Euclidean geometry provides a convenient formalism; for example, the space of Galilean velocities is itself a Euclidean space (see relative velocity for details).

Topographical maps and technical drawings are planar Euclidean. An idea behind them is the scale invariance of Euclidean geometry, that permits to represent large objects in a small sheet of paper, or a screen.

## 8.7 Alternatives and generalizations

Although Euclidean spaces are no longer considered to be the only possible setting for a geometry, they act as prototypes for other geometric objects. Ideas and terminology from Euclidean geometry (both traditional and analytic) are pervasive in modern mathematics, where other geometric objects share many similarities with Euclidean spaces, share part of their structure, or embed Euclidean spaces.

### 8.7.1 Curved spaces

Main article: Riemannian geometry

A smooth manifold is a Hausdorff topological space that is locally diffeomorphic to Euclidean space. Diffeomorphism does not respect distance and angle, but if one additionally prescribes a smoothly varying inner product on the manifold's tangent spaces, then the result is what is called a Riemannian manifold. Put differently, a Riemannian manifold is a space constructed by deforming and patching together Euclidean spaces. Such a space enjoys notions of dis-

tance and angle, but they behave in a curved, non-Euclidean manner. The simplest Riemannian manifold, consisting of $\mathbf{R}^n$ with a constant inner product, is essentially identical to Euclidean n-space itself. Less trivial examples are n-sphere and hyperbolic spaces. Discovery of the latter in the 19th century was branded as the non-Euclidean geometry.

Also, the concept of a Riemannian manifold permits an expression of the Euclidean structure in any smooth coordinate system, via metric tensor. From this tensor one can compute the Riemann curvature tensor. Where the latter equals to zero, the metric structure is locally Euclidean (it means that at least some open set in the coordinate space is isometric to a piece of Euclidean space), no matter whether coordinates are affine or curvilinear.

### 8.7.2 Indefinite quadratic form

See also: Sylvester's law of inertia

If one replaces the inner product of a Euclidean space with an indefinite quadratic form, the result is a pseudo-Euclidean space. Smooth manifolds built from such spaces are called pseudo-Riemannian manifolds. Perhaps their most famous application is the theory of relativity, where flat spacetime is a pseudo-Euclidean space called Minkowski space, where rotations correspond to motions of hyperbolic spaces mentioned above. Further generalization to curved spacetimes form pseudo-Riemannian manifolds, such as in general relativity.

### 8.7.3 Other number fields

Another line of generalization is to consider other number fields than one of real numbers. Over complex numbers, a Hilbert space can be seen as a generalization of Euclidean dot product structure, although the definition of the inner product becomes a sesquilinear form for compatibility with metric structure.

### 8.7.4 Infinite dimensions

Main articles: inner product space and Hilbert space

## 8.8 See also

- Function of several real variables, a coordinate presentation of a function on a Euclidean space

- Geometric algebra, an alternative algebraic formalism

- Vector calculus, a standard algebraic formalism

## 8.9 Footnotes

[1] On the real line ($n = 1$) any two non-zero vectors are either parallel or antiparallel depending on whether their signs match or oppose. There are no angles between 0 and 180°.

[2] It is $\mathbf{R}^n$ which is oriented because of the ordering of elements of the standard basis. Although an orientation is not an attribute of the Euclidean structure, there are only two possible orientations, and any linear automorphism either keeps orientation or reverses (swaps the two).

## 8.10 References

[1] Ball, W.W. Rouse (1960) [1908]. *A Short Account of the History of Mathematics* (4th ed.). Dover Publications. pp. 50–62. ISBN 0-486-20630-0.

[2] E.D. Solomentsev (7 February 2011). "Euclidean space.". *Encyclopedia of Mathematics*. Springer. Retrieved 1 May 2014.

## 8.11 External links

- Hazewinkel, Michiel, ed. (2001), "Euclidean space", *Encyclopedia of Mathematics*, Springer, ISBN 978-1-55608-010-4

# Chapter 9

# Euclidean group

In mathematics, the **Euclidean group** E($n$), also known as ISO($n$) or similar, is the symmetry group of $n$-dimensional Euclidean space. Its elements, the isometries associated with the Euclidean metric, are called **Euclidean motions**.

These groups are among the oldest and most studied, at least in the cases of dimension 2 and 3 – implicitly, long before the concept of group was invented.

## 9.1 Overview

### 9.1.1 Dimensionality

The number of degrees of freedom for E($n$) is $n(n + 1)/2$, which gives 3 in case $n = 2$, and 6 for $n = 3$. Of these, $n$ can be attributed to available translational symmetry, and the remaining $n(n - 1)/2$ to rotational symmetry.

### 9.1.2 Direct and indirect isometries

There is a subgroup E$^+$($n$) of the **direct isometries**, i.e., isometries preserving orientation, also called **rigid motions**; they are the moves of a rigid body in $n$-dimensional space. These include the translations, and the rotations, which together generate E$^+$($n$). E$^+$($n$) is also called a **special Euclidean group**, and denoted SE($n$).

The others are the **indirect isometries**, also called **opposite isometries**. The subgroup E$^+$($n$) is of index 2. In other words, the indirect isometries form a single coset of E$^+$($n$). Given any indirect isometry, for example a given reflection $R$ that reverses orientation, all indirect isometries are given as $DR$, where $D$ is a direct isometry.

The Euclidean group for SE(3) is used for the kinematics of a rigid body, in classical mechanics. A *rigid body motion* is in effect the same as a curve in the Euclidean group. Starting with a body $B$ oriented in a certain way at time $t = 0$, its orientation at any other time is related to the starting orientation by a Euclidean motion, say $f(t)$. Setting $t = 0$,

we have $f(0) = I$, the identity transformation. This means that the curve will always lie inside E$^+$(3), in fact: starting at the identity transformation $I$, such a continuous curve can certainly never reach anything other than a direct isometry. This is for simple topological reasons: the determinant of the transformation cannot jump from +1 to −1.

The Euclidean groups are not only topological groups, they are Lie groups, so that calculus notions can be adapted immediately to this setting.

### 9.1.3 Relation to the affine group

The Euclidean group E($n$) is a subgroup of the affine group for $n$ dimensions, and in such a way as to respect the semidirect product structure of both groups. This gives, *a fortiori*, two ways of writing down elements in an explicit notation. These are:

1. by a pair ($A$, $b$), with $A$ an $n \times n$ orthogonal matrix, and $b$ a real column vector of size $n$; or

2. by a single square matrix of size $n + 1$, as explained for the affine group.

Details for the first representation are given in the next section.

In the terms of Felix Klein's Erlangen programme, we read off from this that Euclidean geometry, the geometry of the Euclidean group of symmetries, is therefore a specialisation of affine geometry. All affine theorems apply. The extra factor in Euclidean geometry is the notion of distance, from which angle can then be deduced.

## 9.2 Detailed discussion

### 9.2.1  Subgroup structure, matrix and vector representation

The Euclidean group is a subgroup of the group of affine transformations.

It has as subgroups the translational group T($n$), and the orthogonal group O($n$). Any element of E($n$) is a translation followed by an orthogonal transformation (the linear part of the isometry), in a unique way:

$$x \mapsto A(x + b)$$

where $A$ is an orthogonal matrix

or an orthogonal transformation followed by a translation:

$$x \mapsto Ax + b$$

T($n$) is a normal subgroup of E($n$): for any translation $t$ and any isometry $u$, we have

$$u^{-1}tu$$

again a translation (one can say, through a displacement that is $u$ acting on the displacement of $t$; a translation does not affect a displacement, so equivalently, the displacement is the result of the linear part of the isometry acting on $t$).

Together, these facts imply that E($n$) is the semidirect product of O($n$) extended by T($n$). In other words, O($n$) is (in the natural way) also the quotient group of E($n$) by T($n$):

$$\mathrm{O}(n) \cong \mathrm{E}(n) \,/\, \mathrm{T}(n).$$

Now SO($n$), the special orthogonal group, is a subgroup of O($n$), of index two. Therefore E($n$) has a subgroup E$^+$($n$), also of index two, consisting of *direct* isometries. In these cases the determinant of $A$ is 1.

They are represented as a translation followed by a rotation, rather than a translation followed by some kind of reflection (in dimensions 2 and 3, these are the familiar reflections in a mirror line or plane, which may be taken to include the origin, or in 3D, a rotoreflection).

We have:

$$\mathrm{SO}(n) \cong \mathrm{E}^+(n) \,/\, \mathrm{T}(n).$$

### 9.2.2  Subgroups

Types of subgroups of E($n$):

- Finite groups. They always have a fixed point. In 3D, for every point there are for every orientation two which are maximal (with respect to inclusion) among the finite groups: O$h$ and I$h$. The groups I$h$ are even maximal among the groups including the next category.

- Countably infinite groups without arbitrarily small translations, rotations, or combinations, i.e., for every point the set of images under the isometries is topologically discrete. E.g. for $1 \le m \le n$ a group generated by $m$ translations in independent directions, and possibly a finite point group. This includes lattices. Examples more general than those are the discrete space groups.

- Countably infinite groups with arbitrarily small translations, rotations, or combinations. In this case there are points for which the set of images under the isometries is not closed. Examples of such groups are, in 1D, the group generated by a translation of 1 and one of $\sqrt{2}$, and, in 2D, the group generated by a rotation about the origin by 1 radian.

- Non-countable groups, where there are points for which the set of images under the isometries is not closed. E.g. in 2D all translations in one direction, and all translations by rational distances in another direction.

- Non-countable groups, where for all points the set of images under the isometries is closed. E.g.

  - all direct isometries that keep the origin fixed, or more generally, some point (in 3D called the rotation group)
  - all isometries that keep the origin fixed, or more generally, some point (the orthogonal group)
  - all direct isometries E$^+$($n$)
  - the whole Euclidean group E($n$)
  - one of these groups in an $m$-dimensional subspace combined with a discrete group of isometries in the orthogonal ($n-m$)-dimensional space
  - one of these groups in an $m$-dimensional subspace combined with another one in the orthogonal ($n-m$)-dimensional space

Examples in 3D of combinations:

- all rotations about one fixed axis

- ditto combined with reflection in planes through the axis and/or a plane perpendicular to the axis

- ditto combined with discrete translation along the axis or with all isometries along the axis

- a discrete point group, frieze group, or wallpaper group in a plane, combined with any symmetry group in the perpendicular direction

- all isometries which are a combination of a rotation about some axis and a proportional translation along the axis; in general this is combined with $k$-fold rotational isometries about the same axis ($k \geq 1$); the set of images of a point under the isometries is a $k$-fold helix; in addition there may be a 2-fold rotation about a perpendicularly intersecting axis, and hence a $k$-fold helix of such axes.

- for any point group: the group of all isometries which are a combination of an isometry in the point group and a translation; for example, in the case of the group generated by inversion in the origin: the group of all translations and inversion in all points; this is the generalized dihedral group of $R^3$, $\mathrm{Dih}(R^3)$.

### 9.2.3 Overview of isometries in up to three dimensions

E(1), E(2), and E(3) can be categorized as follows, with degrees of freedom:

See also: Euclidean plane isometry

Chasles' theorem asserts that any element of $E^+(3)$ is a screw displacement.

See also 3D isometries that leave the origin fixed, space group, involution.

### 9.2.4 Commuting isometries

For some isometry pairs composition does not depend on order:

- two translations

- two rotations or screws about the same axis

- reflection with respect to a plane, and a translation in that plane, a rotation about an axis perpendicular to the plane, or a reflection with respect to a perpendicular plane

- glide reflection with respect to a plane, and a translation in that plane

- inversion in a point and any isometry keeping the point fixed

- rotation by 180° about an axis and reflection in a plane through that axis

- rotation by 180° about an axis and rotation by 180° about a perpendicular axis (results in rotation by 180° about the axis perpendicular to both)

- two rotoreflections about the same axis, with respect to the same plane

- two glide reflections with respect to the same plane

### 9.2.5 Conjugacy classes

The translations by a given distance in any direction form a conjugacy class; the translation group is the union of those for all distances.

In 1D, all reflections are in the same class.

In 2D, rotations by the same angle in either direction are in the same class. Glide reflections with translation by the same distance are in the same class.

In 3D:

- Inversions with respect to all points are in the same class.

- Rotations by the same angle are in the same class.

- Rotations about an axis combined with translation along that axis are in the same class if the angle is the same and the translation distance is the same.

- Reflections in a plane are in the same class

- Reflections in a plane combined with translation in that plane by the same distance are in the same class.

- Rotations about an axis by the same angle not equal to 180°, combined with reflection in a plane perpendicular to that axis, are in the same class.

## 9.3 See also

- Fixed points of isometry groups in Euclidean space

- Euclidean plane isometry

- Poincaré group

- Coordinate rotations and reflections

- Reflection through the origin

- Plane of rotation

## 9.4   References

- Cederberg, Judith N. (2001). *A Course in Modern Geometries.* pp. 136–164. ISBN 978-0-387-98972-3.

- William Thurston. *Three-dimensional geometry and topology. Vol. 1.* Edited by Silvio Levy. Princeton Mathematical Series, 35. Princeton University Press, Princeton, NJ, 1997. x+311 pp. ISBN 0-691-08304-5

# Chapter 10

# Pseudo-Euclidean space

In mathematics and theoretical physics, a **pseudo-Euclidean space** is a finite-dimensional real n-space together with a non-degenerate indefinite quadratic form $q$. Such a quadratic form can, given a suitable choice of basis $(e_1, ..., en)$, be applied to a vector $x = x_1 e_1 + ... + xnen$, giving (with $1 \leq k < n$)

$$q(x) = \left(x_1^2 + \cdots + x_k^2\right) - \left(x_{k+1}^2 + \cdots + x_n^2\right)$$
which is called the *magnitude* of the vector $x$.

For true Euclidean spaces, $k = n$, implying that the quadratic form is positive-definite rather than indefinite. Otherwise $q$ is an isotropic quadratic form. Note that if $i \leq k$ and $j > k$, then $q(ei + ej) = 0$, so that $ei + ej$ is a null vector. In a pseudo-Euclidean space, unlike in a Euclidean space, there exist vectors with negative magnitude.

As with the term *Euclidean space*, *pseudo-Euclidean space* may refer to either an affine space or a vector space (see point–vector distinction) over real numbers.

## 10.1 Geometry

The geometry of a pseudo-Euclidean space is consistent in spite of a breakdown of the some properties of Euclidean space; most notably that it is not a metric space as explained below. Though, its affine structure provides that concepts of line, plane and, generally, of an affine subspace (flat), can be used without modifications, as well as line segments.

### 10.1.1 Positive, zero, and negative magnitudes

A null vector is a vector whose magnitude is zero. Unlike in a Euclidean space, it can be non-zero, in which case it is perpendicular to itself. Every pseudo-Euclidean space has a linear cone of null vectors given by $\{x : q(x) = 0 \}$. When the pseudo-Euclidean space provides a model for

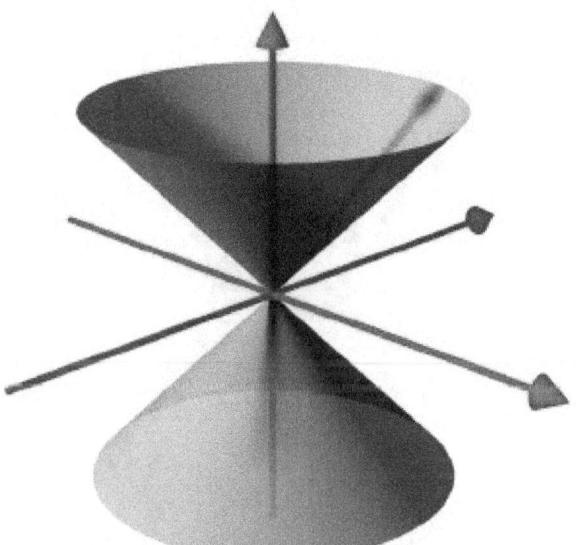

n = 3, k is either 1 or 2 depending on the choice of sign of q

spacetime (see below), the null cone is called the light cone of the origin.

The null cone separates two open sets[1] of positive-magnitude and negative-magnitude vectors. If $k > 1$, then the set of positive-magnitude vectors is connected. If $k = 1$, which means the quadratic form has the only $x_1^2$ square term with positive sign, then it consists of two disjoint parts, one with $x_1 > 0$ and another with $x_1 < 0$. Similar statements can be made for negative-magnitude vectors if k is replaced with $n - k$.

### 10.1.2 Distance

The magnitude q corresponds to the square of a vector (or its norm) in Euclidean case. To define the vector norm (and distance) in an invariant manner, one has to get square roots of magnitudes, which leads to possibly imaginary distances; see square root of negative numbers. But even for a triangle with positive magnitudes of all three sides (whose square roots are real and positive), the triangle inequality is not necessarily true.

That's why terms *norm* and *distance* are avoided in pseudo-Euclidean geometry, replaced with *magnitude* and *interval* respectively.

Though, for a curve whose tangent vectors all have the same sign of magnitude, the arc length is defined. It has important applications: see proper time, for example.

### 10.1.3   Rotations and spheres

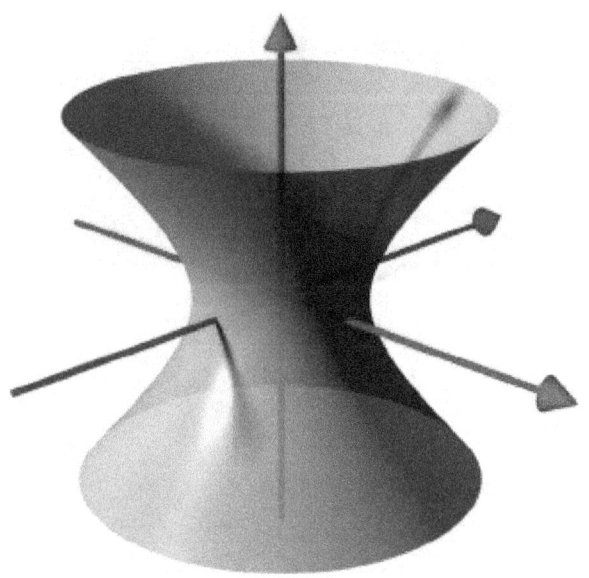

The rotations group of such space is indefinite orthogonal group O($q$), also denoted as O($k$, $n-k$) without a reference to particular quadratic form.[2] Such "rotations" preserve the form q and, hence, the magnitude of each vector whether is it positive, zero, or negative.

Whereas Euclidean space has a unit sphere, pseudo-Euclidean space has the hypersurfaces $\{x : q(x) = 1\}$ and $\{x : q(x) = -1\}$. Such a hypersurface called a hyperboloid or unit quasi-sphere is preserved by the appropriate indefinite orthogonal group.

### 10.1.4   Symmetric bilinear form

The quadratic form q gives rise to a symmetric bilinear form defined as follows:

$$\langle x, y \rangle = \frac{1}{2}[q(x+y) - q(x) - q(y)] = (x_1 y_1 + \cdots + x_k y_k) - (x_{k+1} y_{k+1} + \cdots + x_n y_n).$$

The quadratic form can be expressed in terms of the bilinear form: $\langle x, x \rangle = q(x)$.

When $\langle x, y \rangle = 0$, then x and y are orthogonal elements of the pseudo-Euclidean space. Some authors use the terms "inner product" or "dot product" for the bilinear form, but it does not define an inner product space and its properties do not match to dot product of Euclidean vectors, although these terms are seldom used to refer to this bilinear form.

The standard basis of the real n-space is orthogonal. There are no ortho*normal* bases in a pseudo-Euclidean space because there is no vector norm.

### 10.1.5   Subspaces and orthogonality

For a (positive-dimensional) subspace[3] U of a pseudo-Euclidean space, when the magnitude form q is restricted to U, following three cases are possible:

1. $q|U$ is either positively or negatively definite. Then, U is essentially Euclidean (up to sign of q).

2. $q|U$ is indefinite, but non-degenerate. Then, U is itself pseudo-Euclidean. It is possible only if dim $U \geq 2$; if dim $U = 2$, which means than U is a plane, then it is called a hyperbolic plane.

3. $q|U$ is degenerate.

One of most jarring properties (for a Euclidean intuition) of pseudo-Euclidean vectors and flats is their orthogonality. When two non-zero Euclidean vectors are perpendicular, they are certainly not collinear. Any Euclidean linear subspace intersects with its orthogonal complement only by the $\{0\}$ subspace. But the definition from the previous subsection immediately implies that any vector v of zero magnitude is perpendicular to itself. Hence, for the 1-subspace $N = \langle v \rangle$ generated by such non-zero vector, its orthogonal complement $N^\perp$ will be a superspace of N.

The formal definition of the orthogonal complement of a vector subspace in a pseudo-Euclidean space gives a perfectly well-defined result which satisfies the equality dim $U$ + dim $U^\perp = n$ due to the magnitude form's non-degeneracy. It is just the condition

$$U \cap U^\perp = \{0\} \text{ or, equivalently, } U + U^\perp = \text{all space}^{[4]}$$

which can be broken if the subspace U contains a null direction.[5] While subspaces form a distributive lattice, as in any vector space, they do not form a Boolean algebra with this $\perp$ operation, as in inner product spaces.

For a subspace N composed *entirely* of null vectors (which means that the magnitude q, restricted to N, equals to 0), always holds:

$N \subset N^{\perp}$ or, equivalently, $N \cap N^{\perp} = N$.

Such subspaces can have up to $\min(k, n - k)$ dimensions.[6]

For a (positive) Euclidean k-subspace its orthogonal complement is a $(n - k)$-dimensional negative "Euclidean" subspace, and vice versa. Generally, for a $(d_+ + d_- + d_0)$-dimensional subspace U consisting of $d_+$ positive and $d_-$ negative dimensions (see Sylvester's law of inertia for clarification), its orthogonal "complement" $U^{\perp}$ has $(k - d_+ - d_0)$ positive and $(n - k - d_- - d_0)$ negative dimensions, while the rest $d_0$ ones are degenerate and form the $U \cap U^{\perp}$ intersection.

### 10.1.6 Parallelogram law and Pythagorean theorem

The parallelogram law takes the form

$$q(x) + q(y) = \frac{1}{2}(q(x + y) + q(x - y)).$$

Using the square of the sum identity, for an arbitrary triangle one can express the magnitude of the third side from magnitudes of two sides and their bilinear form product:

$$q(x + y) = q(x) + q(y) + 2\langle x, y \rangle.$$

This demonstrates that, for perpendicular vectors, a pseudo-Euclidean analog of the Pythagorean theorem holds:

$$\langle x, y \rangle = 0 \Rightarrow q(x) + q(y) = q(x + y).$$

### 10.1.7 Angle

Generally, absolute value $|\langle x, y \rangle|$ of the bilinear form on two vectors may be greater than $\sqrt{|q(x)q(y)|}$, equal to it, or less. This causes similar problems with definition of angle (see dot product#Geometric definition) as appeared above for distances. If $k = 1$ (only one positive term in q), then for positive-magnitude vectors:

$$|\langle x, y \rangle| \geq \sqrt{q(x)q(y)},$$

which permits to define hyperbolic angle, an analog of angle between these vectors through inverse hyperbolic cosine:

$$\operatorname{arcosh} \frac{|\langle x, y \rangle|}{\sqrt{q(x)q(y)}} \, . \quad [7]$$

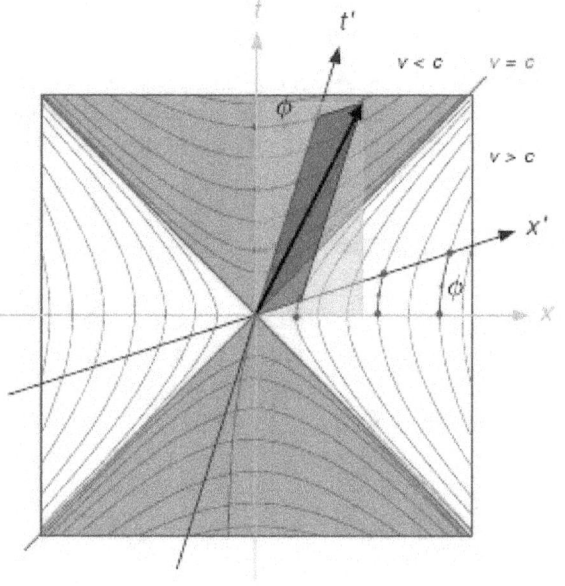

It corresponds to the distance on a $(n - 1)$-dimensional hyperbolic space. This is known as rapidity in the context of theory of relativity discussed below. Unlike Euclidean angle, it takes values from $[0, +\infty)$ and equals to 0 for antiparallel vectors.

There is no reasonable definition of the angle between a null vector and another vector (either null or non-null).

## 10.2 Algebra and tensor calculus

Like Euclidean spaces, a pseudo-Euclidean space possesses geometric algebra. Unlike properties above, where replacement of q to $-q$ changed numbers but not geometry, the sign reversal of the magnitude form actually *alters* $C\ell$, so for example $C\ell_{1,2}(\mathbf{R})$ and $C\ell_{2,1}(\mathbf{R})$ are not isomorphic.

Just like over any vector space, there are pseudo-Euclidean tensors. Like with a Euclidean structure, there are raising and lowering indices operators but, unlike the case with Euclidean tensors, there is no bases where these operations do not change values of components. If there is a vector $v^{\beta}$, the corresponding covariant vector is:

$$v_{\alpha} = q_{\alpha\beta}v^{\beta},$$

and with the standard-form

$$q_{\alpha\beta} = \begin{pmatrix} I_{k \times k} & 0 \\ 0 & -I_{(n-k) \times (n-k)} \end{pmatrix}$$

the first k components of $v\alpha$ are numerically the same as ones of $v^\beta$, but the rest $n - k$ have opposite signs.

The correspondence between contravariant and covariant tensors makes a tensor calculus on pseudo-Riemannian manifolds analogous to one on Riemannian manifolds.

## 10.3  Examples

A very important pseudo-Euclidean space is Minkowski space, which is the mathematical setting in which Albert Einstein's theory of special relativity is conveniently formulated. For Minkowski space, $n = 4$ and $k = 3$[8] so that

$$q(x) = x_1^2 + x_2^2 + x_3^2 - x_4^2,$$

The geometry associated with this pseudo-metric was investigated by Poincaré. Its rotation group is the Lorentz group. The Poincaré group includes also translations and plays the same role as Euclidean groups of ordinary Euclidean spaces.

Another pseudo-Euclidean space is the plane $z = x + y j$ consisting of split-complex numbers, equipped with the quadratic form

$$\|z\| = zz^* = z^*z = x^2 - y^2.$$

This is the simplest case of a pseudo-Euclidean space ($n = 2$, $k = 1$) and the only one where the null cone dissects the space to *four* open sets. The group $SO^+(1, 1)$ consists of so named hyperbolic rotations.

## 10.4  See also

- Hyperbolic equation
- Hyperboloid model
- Paravector

## 10.5  Footnotes

[1] The standard topology on $\mathbf{R}^n$ is assumed.

[2] What is the "rotations group" depends on exact definition of a rotation. "O" groups contain improper rotations. Transforms which preserves orientation form the group $SO(q)$, or $SO(k, n - k)$, but it also is not connected if both k and $n - k$ are positive. The group $SO^+(q)$, which preserves orientation on positive- and negative-magnitude parts separately,

is a (connected) analog of Euclidean rotations group $SO(n)$. Indeed, all these groups are Lie groups of dimension $n(n - 1)/2$ .

[3] A linear subspace is assumed, but same conclusions are true for an affine flat with the only complication that the magnitude form is always defined on vectors, not points.

[4] Violation of this equality makes the term "orthogonal complement" itself an oxymoron.

[5] Actually, $U \cap U^\perp$ is not zero if and only if the magnitude form q restricted to U is degenerate.

[6] Thomas E. Cecil (1992) *Lie Sphere Geometry*, page 24, Universitext Springer ISBN 0-387-97747-3

[7] Note that $\cos(i \cdot \operatorname{arcosh} s) = s$, so for $s > 0$ these can be understood as imaginary angles.

[8] Another well-established representation uses $k = 1$ and coordinate indices starting from 0 (thence $q(x) = x_0^2 - x_1^2 - x_2^2 - x_3^2$), but they are equivalent up to sign of q. See sign convention#Metric signature.

## 10.6  References

- Werner Greub (1963) *Linear Algebra*, 2nd edition, §12.4 Pseudo-Euclidean Spaces, pp. 237–49, Springer-Verlag.

- Walter Noll (1964) "Euclidean geometry and Minkowskian chronometry", American Mathematical Monthly 71:129–44.

- Novikov, S. P.; Fomenko, A.T.; [translated from the Russian by M. Tsaplina] (1990). *Basic elements of differential geometry and topology*. Dordrecht; Boston: Kluwer Academic Publishers. ISBN 0-7923-1009-8.

- Poincaré, *Science and Hypothesis* 1906 referred to in the book B.A. Rosenfeld, *A History of Non-Euclidean Geometry* Springer 1988 (English translation) p. 266.

- Szekeres, Peter (2004). *A course in modern mathematical physics: groups, Hilbert space, and differential geometry*. Cambridge University Press. ISBN 0-521-82960-7.

- Shafarevich, I. R.; A. O. Remizov (2012). *Linear Algebra and Geometry*. Springer. ISBN 978-3-642-30993-9.

## 10.7  External links

- D.D. Sokolov (originator), Pseudo-Euclidean space, Encyclopedia of Mathematics

# Chapter 11

# Time

For other uses of "Time", see Time (disambiguation).

**Time** is the indefinite continued progression of existence and events that occur in apparently irreversible succession from the past through the present to the future.[1][2][3] Time is a component quantity of various measurements used to sequence events, to compare the duration of events or the intervals between them, and to quantify rates of change of quantities in material reality or in the conscious experience.[4][5][6][7] Time is often referred to as the **fourth dimension**, along with the three spatial dimensions.[8]

Time has long been a major subject of study in religion, philosophy, and science, but defining it in a manner applicable to all fields without circularity has consistently eluded scholars.[2][6][7][9][10][11] Nevertheless, diverse fields such as business, industry, sports, the sciences, and the performing arts all incorporate some notion of time into their respective measuring systems.[12][13][14] Some simple definitions of time include "time is what clocks measure",[6][15] which is a problematically vague and self-referential definition that utilizes the device used to measure the subject as the definition of the subject, and "time is what keeps everything from happening at once", which is without substantive meaning in the absence of the definition of simultaneity in the context of the limitations of human sensation, observation of events, and the perception of such events.[16][17][18][19]

Two contrasting viewpoints on time divide many prominent philosophers. One view is that time is part of the fundamental structure of the universe—a dimension independent of events, in which events occur in sequence. Sir Isaac Newton subscribed to this realist view, and hence it is sometimes referred to as Newtonian time.[20][21] The opposing view is that *time* does not refer to any kind of "container" that events and objects "move through", nor to any entity that "flows", but that it is instead part of a fundamental intellectual structure (together with space and number) within which humans sequence and compare events. This second view, in the tradition of Gottfried Leibniz[15] and Immanuel Kant,[22][23] holds that *time* is neither an event nor a thing, and thus is not itself measurable nor can it be travelled.

Time is one of the seven fundamental physical quantities in both the International System of Units and International System of Quantities. Time is used to define other quantities—such as velocity—so defining time in terms of such quantities would result in circularity of definition.[24] An operational definition of time, wherein one says that observing a certain number of repetitions of one or another standard cyclical event (such as the passage of a free-swinging pendulum) constitutes one standard unit such as the second, is highly useful in the conduct of both advanced experiments and everyday affairs of life. The operational definition leaves aside the question whether there is something called time, apart from the counting activity just mentioned, that flows and that can be measured. Investigations of a single continuum called spacetime bring questions about space into questions about time, questions that have their roots in the works of early students of natural philosophy.

Furthermore, it may be that there is a subjective component to time, but whether or not time itself is "felt", as a sensation, or is a judgment, is a matter of debate.[2][6][7][25][26]

Temporal measurement has occupied scientists and technologists, and was a prime motivation in navigation and astronomy. Periodic events and periodic motion have long served as standards for units of time. Examples include the apparent motion of the sun across the sky, the phases of the moon, the swing of a pendulum, and the beat of a heart. Currently, the international unit of time, the second, is defined by measuring the electronic transition frequency of caesium atoms (see below). Time is also of significant social importance, having economic value ("time is money") as well as personal value, due to an awareness of the limited time in each day and in human life spans.

*The flow of sand in an hourglass can be used to keep track of elapsed time. It also concretely represents the present as being between the past and the future.*

# 11.1 Temporal measurement and history

Temporal measurement, chronometry, takes two distinct period forms: the calendar, a mathematical tool for organizing intervals of time,[27] and the clock, a physical mechanism that counts the passage of time. In day-to-day life, the clock is consulted for periods less than a day whereas the calendar is consulted for periods longer than a day. Increasingly, personal electronic devices display both calendars and clocks simultaneously. The number (as on a clock dial or calendar) that marks the occurrence of a specified event as to hour or date is obtained by counting from a fiducial epoch—a central reference point.

## 11.1.1 History of the calendar

Main article: Calendar

Artifacts from the Paleolithic suggest that the moon was used to reckon time as early as 6,000 years ago.[28] Lunar calendars were among the first to appear, either 12 or 13 lunar months (either 354 or 384 days). Without intercalation to add days or months to some years, seasons quickly drift in a calendar based solely on twelve lunar months. Lunisolar calendars have a thirteenth month added to some years to make up for the difference between a full year (now known to be about 365.24 days) and a year of just twelve lunar months. The numbers twelve and thirteen came to feature prominently in many cultures, at least partly due to this relationship of months to years. Other early forms of calendars originated in Mesoamerica, particularly in ancient Mayan civilization. These calendars were religiously and astronomically based, with 18 months in a year and 20 days in a month.[29]

The reforms of Julius Caesar in 45 BC put the Roman world on a solar calendar. This Julian calendar was faulty in that its intercalation still allowed the astronomical solstices and equinoxes to advance against it by about 11 minutes per year. Pope Gregory XIII introduced a correction in 1582; the Gregorian calendar was only slowly adopted by different nations over a period of centuries, but it is now the most commonly used calendar around the world, by far.

During the French Revolution, a new clock and calendar were invented in attempt to de-Christianize time and create a more rational system in order to replace the Gregorian Calendar. The French Republican Calendar's days consisted of ten hours of a hundred minutes of a hundred seconds, which marked a deviation from the duodecimal system used in many other devices by many cultures. The system was later abolished in 1806.[30]

## 11.1.2 History of time measurement devices

Main article: History of timekeeping devices
See also: Clock

A large variety of devices has been invented to measure time. The study of these devices is called horology.

An Egyptian device that dates to c.1500 BC, similar in shape to a bent T-square, measured the passage of time

*Horizontal sundial in Taganrog*

*A contemporary quartz watch, 2007*

from the shadow cast by its crossbar on a nonlinear rule. The T was oriented eastward in the mornings. At noon, the device was turned around so that it could cast its shadow in the evening direction.[31]

A sundial uses a gnomon to cast a shadow on a set of markings calibrated to the hour. The position of the shadow marks the hour in local time. The invention of the idea to separate the day into smaller parts is credited to Egyptians because of their sundials, which operated on a duodecimal system based in the number 12. The importance of the number 12 is due the number of lunar cycles in a year and the number of stars used to count the passage of night.[32]

The most precise timekeeping device of the ancient world was the water clock, or *clepsydra*, one of which was found in the tomb of Egyptian pharaoh Amenhotep I (1525–1504 BC). They could be used to measure the hours even at night, but required manual upkeep to replenish the flow of water. The Ancient Greeks and the people from Chaldea (southeastern Mesopotamia) regularly maintained timekeeping records as an essential part of their astronomical observations. Arab inventors and engineers in particular made improvements on the use of water clocks up to the Middle Ages.[33] In the 11th century, Chinese inventors and engineers invented the first mechanical clocks driven by an escapement mechanism.

The hourglass uses the flow of sand to measure the flow of time. They were used in navigation. Ferdinand Magellan used 18 glasses on each ship for his circumnavigation of the globe (1522).[34] Incense sticks and candles were, and are, commonly used to measure time in temples and churches across the globe. Waterclocks, and later, mechanical clocks, were used to mark the events of the abbeys and monasteries of the Middle Ages. Richard of Wallingford (1292–1336), abbot of St. Alban's abbey, famously built a mechanical clock as an astronomical orrery about 1330.[35][36] Great advances in accurate time-keeping were

made by Galileo Galilei and especially Christiaan Huygens with the invention of pendulum driven clocks along with the invention of the minute hand by Jost Burgi.[37]

The English word clock probably comes from the Middle Dutch word *klocke* which, in turn, derives from the medieval Latin word *clocca*, which ultimately derives from Celtic and is cognate with French, Latin, and German words that mean bell. The passage of the hours at sea were marked by bells, and denoted the time (see ship's bell). The hours were marked by bells in abbeys as well as at sea.

*Chip-scale atomic clocks, such as this one unveiled in 2004, are expected to greatly improve GPS location.*[38]

Clocks can range from watches, to more exotic varieties such as the Clock of the Long Now. They can be driven by a variety of means, including gravity, springs, and various forms of electrical power, and regulated by a variety of means such as a pendulum.

Alarm clocks first appeared in Ancient Greece around 250 B.C. with a water clock that would set off a whistle. This idea was later mechanized by Levi Hutchins and Seth E. Thomas.[37]

A chronometer is a portable timekeeper that meets certain precision standards. Initially, the term was used to refer to the marine chronometer, a timepiece used to determine longitude by means of celestial navigation, a precision firstly achieved by John Harrison. More recently, the term has also been applied to the chronometer watch, a watch that meets precision standards set by the Swiss agency COSC.

*The 555 timer IC is an integrated circuit (chip) used in a variety of timer, pulse generator and oscillator applications.*

The most accurate timekeeping devices are atomic clocks, which are accurate to seconds in many millions of years,[39] and are used to calibrate other clocks and timekeeping instruments. Atomic clocks use the frequency of electronic transitions in certain atoms to measure the second. One of the most common atoms used is caesium, most modern atomic clocks probe caesium with microwaves to determine the frequency of these electron vibrations.[40] Since 1967, the International System of Measurements bases its unit of time, the second, on the properties of caesium atoms. SI defines the second as 9,192,631,770 cycles of the radiation that corresponds to the transition between two electron spin energy levels of the ground state of the $^{133}$Cs atom.

Today, the Global Positioning System in coordination with the Network Time Protocol can be used to synchronize timekeeping systems across the globe.

In medieval philosophical writings, the **atom** was a unit of time referred to as the smallest possible division of time. The earliest known occurrence in English is in Byrhtferth's *Enchiridion* (a science text) of 1010–1012,[41] where it was defined as 1/564 of a *momentum* (1½ minutes),[42] and thus equal to 15/94 of a second. It was used in the *computus*, the

process of calculating the date of Easter.

As of May 2010, the smallest time interval uncertainty in direct measurements is on the order of 12 attoseconds (1.2 × 10$^{-17}$ seconds), about 3.7 × 10$^{26}$ Planck times.[43]

### 11.1.3   List of units

Main article: Time (Orders of magnitude)

## 11.2   Definitions and standards

The SI base unit for time is the SI second. The International System of Quantities, which incorporates the SI, also defines larger units of time equal to fixed integer multiples of one second (1 s), such as the minute, hour and day. These are not part of the SI, but may be used alongside the SI. Other units of time such as the month and the year are not equal to fixed multiples of 1 s, and instead exhibit significant variations in duration.[50]

The official SI definition of the second is as follows:[50][51]

> The second is the duration of 9,192,631,770 periods of the radiation corresponding to the transition between the two hyperfine levels of the ground state of the caesium 133 atom.

At its 1997 meeting, the CIPM affirmed that this definition refers to a caesium atom in its ground state at a temperature of 0 K.[50] Previous to 1967, the second was defined as:

> the fraction 1/31,556,925.9747 of the tropical year for 1900 January 0 at 12 hours ephemeris time.

The current definition of the second, coupled with the current definition of the metre, is based on the special theory of relativity, which affirms our spacetime to be a Minkowski space.

### 11.2.1   World time

Time-keeping is so critical to the functioning of modern societies that it is coordinated at an international level. The basis for scientific time is a continuous count of seconds based on atomic clocks around the world, known as the International Atomic Time (TAI). Other scientific time standards include Terrestrial Time and Barycentric Dynamical Time.

Coordinated Universal Time (UTC) is the basis for modern civil time. Since 1 January 1972, it has been defined to follow TAI with an exact offset of an integer number of seconds, changing only when a leap second is added to keep clock time synchronized with the rotation of the Earth. In TAI and UTC systems, the duration of a second is constant, as it is defined by the unchanging transition period of the caesium atom.

Greenwich Mean Time (GMT) is an older standard, adopted starting with British railways in 1847. Using telescopes instead of atomic clocks, GMT was calibrated to the mean solar time at the Royal Observatory, Greenwich in the UK. Universal Time (UT) is the modern term for the international telescope-based system, adopted to replace "Greenwich Mean Time" in 1928 by the International Astronomical Union. Observations at the Greenwich Observatory itself ceased in 1954, though the location is still used as the basis for the coordinate system. Because the rotational period of Earth is not perfectly constant, the duration of a second would vary if calibrated to a telescope-based standard like GMT or UT—in which a second was defined as a fraction of a day or year. The terms "GMT" and "Greenwich Mean Time" are sometimes used informally to refer to UT or UTC.

The Global Positioning System also broadcasts a very precise time signal worldwide, along with instructions for converting GPS time to UTC.

Earth is split up into a number of time zones. Most time zones are exactly one hour apart, and by convention compute their local time as an offset from UTC or GMT. In many locations these offsets vary twice yearly due to daylight saving time transitions.

### 11.2.2 Time conversions

These conversions are accurate at the millisecond level for time systems involving earth rotation (UT1 & TT). Conversions between atomic time systems (TAI, GPS, and UTC) are accurate at the microsecond level.

Definitions:

1. LS = TAI - UTC = Leap Seconds from http://maia. usno.navy.mil/ser7/tai-utc.dat

2. DUT1 = UT1 - UTC from http://maia.usno.navy.mil/ ser7/ser7.dat or http://maia.usno.navy.mil/search/ search.html

### 11.2.3 Sidereal time

Sidereal time is the measurement of time relative to a distant star (instead of solar time that is relative to the sun). It is used in astronomy to predict when a star will be overhead. Due to the orbit of the earth around the sun a sidereal day is about 4 minutes (1/366th) less than a solar day.

### 11.2.4 Chronology

Main article: Chronology

Another form of time measurement consists of studying the past. Events in the past can be ordered in a sequence (creating a chronology), and can be put into chronological groups (periodization). One of the most important systems of periodization is the geologic time scale, which is a system of periodizing the events that shaped the Earth and its life. Chronology, periodization, and interpretation of the past are together known as the study of history.

### 11.2.5 Time-like concepts: terminology

The term "time" is generally used for many close but different concepts, including:

- instant[52] as an object—one point on the time axes. Being an object, it has no value;

- time interval[53] as an object—part of the time axes limited by two instants. Being an object, it has no value;

- date[54] as a quantity characterizing an instant. As a quantity, it has a value which may be expressed in a variety of ways, for example "2014-04-26T09:42:36,75" in ISO standard format, or more colloquially such as "today, 9:42 a.m.";

- duration[55] as a quantity characterizing a time interval.[56] As a quantity, it has a value, such as a number of minutes, or may be described in terms of the quantities (such as times and dates) of its beginning and end.

## 11.3 Religion

Further information: Time and fate deities

### 11.3.1 Linear and cyclical time

See also: Time cycles and Wheel of time

Ancient cultures such as Incan, Mayan, Hopi, and other Native American Tribes - plus the Babylonians, Ancient Greeks, Hinduism, Buddhism, Jainism, and others - have a concept of a wheel of time: they regard time as cyclical and quantic, consisting of repeating ages that happen to every being of the Universe between birth and extinction.

In general, the Islamic and Judeo-Christian world-view regards time as linear[57] and directional,[58] beginning with the act of creation by God. The traditional Christian view sees time ending, teleologically,[59] with the eschatological end of the present order of things, the "end time".

In the Old Testament book Ecclesiastes, traditionally ascribed to Solomon (970–928 BC), time (as the Hebrew word עדן ,זמן *'iddan(time) zĕman(season)* is often translated) was traditionally regarded as a medium for the passage of predestined events. (Another word, "زمان" *zamān*, meant *time fit for an event*, and is used as the modern Arabic, Persian, and Hebrew equivalent to the English word "time".)

### 11.3.2   Time in Greek mythology

The Greek language denotes two distinct principles, Chronos and Kairos. The former refers to numeric, or chronological, time. The latter, literally "the right or opportune moment", relates specifically to metaphysical or Divine time. In theology, Kairos is qualitative, as opposed to quantitative.

In Greek mythology, Chronos (Ancient Greek: Χρόνος) is identified as the Personification of Time. His name in Greek means "time" and is alternatively spelled Chronus (Latin spelling) or Khronos. Chronos is usually portrayed as an old, wise man with a long, gray beard, such as "Father Time". Some English words whose etymological root is khronos/chronos include *chronology, chronometer, chronic, anachronism, synchronize,* and *chronicle.*

### 11.3.3   Time in Kabbalah

According to Kabbalists, "time" is a paradox[60] and an illusion.[61] Both the future and the past are recognized to be combined and simultaneously present.

## 11.4   Philosophy

Main articles: Philosophy of space and time and Temporal finitism

Two distinct viewpoints on time divide many prominent philosophers. One view is that time is part of the fundamental structure of the universe, a dimension in which events occur in sequence. Sir Isaac Newton subscribed to this realist view, and hence it is sometimes referred to as Newtonian time.[21] An opposing view is that *time* does not refer to any kind of actually existing dimension that events and objects "move through", nor to any entity that "flows", but that it is instead an intellectual concept (together with space and number) that enables humans to sequence and compare events.[62] This second view, in the tradition of Gottfried Leibniz[15] and Immanuel Kant,[22][23] holds that space and time "do not exist in and of themselves, but ... are the product of the way we represent things", because we can know objects only as they appear to us.

The *Vedas*, the earliest texts on Indian philosophy and Hindu philosophy dating back to the late 2nd millennium BC, describe ancient Hindu cosmology, in which the universe goes through repeated cycles of creation, destruction and rebirth, with each cycle lasting 4,320 million years.[63] Ancient Greek philosophers, including Parmenides and Heraclitus, wrote essays on the nature of time.[64] Plato, in the *Timaeus*, identified time with the period of motion of the heavenly bodies. Aristotle, in Book IV of his *Physica* defined time as 'number of movement in respect of the before and after'.[65]

In Book 11 of his *Confessions*, St. Augustine of Hippo ruminates on the nature of time, asking, "What then is time? If no one asks me, I know: if I wish to explain it to one that asketh, I know not." He begins to define time by what it is not rather than what it is,[66] an approach similar to that taken in other negative definitions. However, Augustine ends up calling time a "distention" of the mind (Confessions 11.26) by which we simultaneously grasp the past in memory, the present by attention, and the future by expectation.

In contrast to ancient Greek philosophers who believed that the universe had an infinite past with no beginning, medieval philosophers and theologians developed the concept of the universe having a finite past with a beginning. This view is shared by Abrahamic faiths as they believe time started by creation, therefore the only thing being infinite is God and everything else, including time, is finite.

Isaac Newton believed in absolute space and absolute time; Leibniz believed that time and space are relational.[67] The differences between Leibniz's and Newton's interpretations came to a head in the famous Leibniz–Clarke correspondence.

Time is not an empirical concept. For neither co-existence nor succession would be perceived by us, if the representation of time did not exist as a foundation *a priori*. Without this presupposition we could not represent to ourselves that things exist together at one and the same time, or at different times, that is, contemporaneously, or in succession.

"

"

Immanuel Kant, *Critique of Pure Reason* (1781), trans. Vasilis Politis (London: Dent., 1991), p.54.

Immanuel Kant, in the *Critique of Pure Reason*, described time as an *a priori* intuition that allows us (together with the other *a priori* intuition, space) to comprehend sense experience.[68] With Kant, neither space nor time are conceived as substances, but rather both are elements of a systematic mental framework that necessarily structures the experiences of any rational agent, or observing subject. Kant thought of time as a fundamental part of an abstract conceptual framework, together with space and number, within which we sequence events, quantify their duration, and compare the motions of objects. In this view, *time* does not refer to any kind of entity that "flows," that objects "move through," or that is a "container" for events. Spatial measurements are used to quantify the extent of and distances between objects, and temporal measurements are used to quantify the durations of and between events. Time was designated by Kant as the purest possible schema of a pure concept or category.

Henri Bergson believed that time was neither a real homogeneous medium nor a mental construct, but possesses what he referred to as *Duration*. Duration, in Bergson's view, was creativity and memory as an essential component of reality.[69]

According to Martin Heidegger we do not exist inside time, we *are* time. Hence, the relationship to the past is a present awareness of *having been*, which allows the past to exist in the present. The relationship to the future is the state of anticipating a potential possibility, task, or engagement. It is related to the human propensity for caring and being concerned, which causes "being ahead of oneself" when thinking of a pending occurrence. Therefore, this concern for a potential occurrence also allows the future to exist in the present. The present becomes an experience, which is qualitative instead of quantitative. Heidegger seems to think this is the way that a linear relationship with time, or temporal existence, is broken or transcended.[70] We are not stuck in sequential time. We are able to remember the past and project into the future—we have a kind of random access to our representation of temporal existence; we can, in our thoughts, step out of (ecstasis) sequential time.[71]

### 11.4.1 Time as "unreal"

In 5th century BC Greece, Antiphon the Sophist, in a fragment preserved from his chief work *On Truth*, held that:

*"Time is not a reality (hypostasis), but a concept (noêma) or a measure (metron)."* Parmenides went further, maintaining that time, motion, and change were illusions, leading to the paradoxes of his follower Zeno.[72] Time as an illusion is also a common theme in Buddhist thought.[73][74]

J. M. E. McTaggart's 1908 *The Unreality of Time* argues that, since every event has the characteristic of being both present and not present (i.e., future or past), that time is a self-contradictory idea (see also The flow of time).

These arguments often center around what it means for something to be *unreal*. Modern physicists generally believe that time is as *real* as space—though others, such as Julian Barbour in his book *The End of Time*, argue that quantum equations of the universe take their true form when expressed in the timeless realm containing every possible *now* or momentary configuration of the universe, called 'platonia' by Barbour.[75] (See also: Eternalism (philosophy of time))

## 11.5 Physical definition

Main article: Time in physics

Until Einstein's profound reinterpretation of the physical concepts associated with time and space, time was considered to be the same everywhere in the universe, with all observers measuring the same time interval for any event.[76] Non-relativistic classical mechanics is based on this Newtonian idea of time.

Einstein, in his special theory of relativity,[77] postulated the constancy and finiteness of the speed of light for all observers. He showed that this postulate, together with a reasonable definition for what it means for two events to be simultaneous, requires that distances appear compressed and time intervals appear lengthened for events associated with objects in motion relative to an inertial observer.

The theory of special relativity finds a convenient formulation in Minkowski spacetime, a mathematical structure that combines three dimensions of space with a single dimension of time. In this formalism, distances in space can be measured by how long light takes to travel that distance, e.g., a light-year is a measure of distance, and a meter is now defined in terms of how far light travels in a certain amount of time. Two events in Minkowski spacetime are separated by an *invariant interval*, which can be either space-like, light-like, or time-like. Events that have a time-like separation cannot be simultaneous in any frame of reference, there must be a temporal component (and possibly a spatial one) to their separation. Events that have a space-like separation will be simultaneous in some frame of reference, and there is no frame of reference in which they do not have a spatial separation. Different observers may calculate different distances and different time intervals between two events, but the *invariant interval* between the events is independent of the observer (and his velocity).

### 11.5.1 Classical mechanics

In non-relativistic classical mechanics, Newton's concept of "relative, apparent, and common time" can be used in the formulation of a prescription for the synchronization of clocks. Events seen by two different observers in motion relative to each other produce a mathematical concept of time that works sufficiently well for describing the everyday phenomena of most people's experience. In the late nineteenth century, physicists encountered problems with the classical understanding of time, in connection with the behavior of electricity and magnetism. Einstein resolved these problems by invoking a method of synchronizing clocks using the constant, finite speed of light as the maximum signal velocity. This led directly to the result that observers in motion relative to one another measure different elapsed times for the same event.

### 11.5.2 Spacetime

Main article: Spacetime

Time has historically been closely related with space, the two together merging into spacetime in Einstein's special relativity and general relativity. According to these theories, the concept of time depends on the spatial reference frame of the observer, and the human perception as well as the measurement by instruments such as clocks are different for observers in relative motion. For example, if a spaceship carrying a clock flies through space at (very nearly) the speed of light, its crew does not notice a change in the speed of time on board their vessel because everything traveling at the same speed slows down at the same rate (including the clock, the crew's thought processes, and the functions of their bodies). However, to a stationary observer watching the spaceship fly by, the spaceship appears flattened in the direction it is traveling and the clock on board the spaceship appears to move very slowly.

On the other hand, the crew on board the spaceship also perceives the observer as slowed down and flattened along the spaceship's direction of travel, because both are moving at very nearly the speed of light relative to each other. Because the outside universe appears flattened to the spaceship, the crew perceives themselves as quickly traveling between regions of space that (to the stationary observer) are many light years apart. This is reconciled by the fact that

the crew's perception of time is different from the stationary observer's; what seems like seconds to the crew might be hundreds of years to the stationary observer. In either case, however, causality remains unchanged: the past is the set of events that can send light signals to an entity and the future is the set of events to which an entity can send light signals.[78][79][80]

### 11.5.3 Time dilation

Main article: Time dilation

Einstein showed in his thought experiments that people travelling at different speeds, while agreeing on cause and effect, measure different time separations between events, and can even observe different chronological orderings between non-causally related events. Though these effects are typically minute in the human experience, the effect becomes much more pronounced for objects moving at speeds approaching the speed of light. Many subatomic particles exist for only a fixed fraction of a second in a lab relatively at rest, but some that travel close to the speed of light can be measured to travel farther and survive much longer than expected (a muon is one example). According to the special theory of relativity, in the high-speed particle's frame of reference, it exists, on the average, for a standard amount of time known as its mean lifetime, and the distance it travels in that time is zero, because its velocity is zero. Relative to a frame of reference at rest, time seems to "slow down" for the particle. Relative to the high-speed particle, distances seem to shorten. Einstein showed how both temporal and spatial dimensions can be altered (or "warped") by high-speed motion.

Einstein (*The Meaning of Relativity*): "Two events taking place at the points A and B of a system K are simultaneous if they appear at the same instant when observed from the middle point, M, of the interval AB. Time is then defined as the ensemble of the indications of similar clocks, at rest relatively to K, which register the same simultaneously."

Einstein wrote in his book, *Relativity*, that simultaneity is also relative, i.e., two events that appear simultaneous to an observer in a particular inertial reference frame need not be judged as simultaneous by a second observer in a different inertial frame of reference.

### 11.5.4 Relativistic time versus Newtonian time

The animations visualise the different treatments of time in the Newtonian and the relativistic descriptions. At the heart of these differences are the Galilean and Lorentz transformations applicable in the Newtonian and relativistic theories, respectively.

In the figures, the vertical direction indicates time. The horizontal direction indicates distance (only one spatial dimension is taken into account), and the thick dashed curve is the spacetime trajectory ("world line") of the observer. The small dots indicate specific (past and future) events in spacetime.

The slope of the world line (deviation from being vertical) gives the relative velocity to the observer. Note how in both pictures the view of spacetime changes when the observer accelerates.

In the Newtonian description these changes are such that *time* is absolute:[81] the movements of the observer do not influence whether an event occurs in the 'now' (i.e., whether an event passes the horizontal line through the observer).

However, in the relativistic description the *observability of events* is absolute: the movements of the observer do not influence whether an event passes the "light cone" of the observer. Notice that with the change from a Newtonian to a relativistic description, the concept of *absolute time* is no longer applicable: events move up-and-down in the figure depending on the acceleration of the observer.

### 11.5.5 Arrow of time

Main article: Arrow of time

Time appears to have a direction—the past lies behind, fixed and immutable, while the future lies ahead and is not necessarily fixed. Yet for the most part the laws of physics do not specify an arrow of time, and allow any process to proceed both forward and in reverse. This is generally a consequence of time being modeled by a parameter in the system being analyzed, where there is no "proper time": the direction of the arrow of time is sometimes arbitrary. Examples of this include the Second law of thermodynamics, which states that entropy must increase over time (see Entropy); the cosmological arrow of time, which points away from the Big Bang, CPT symmetry, and the radiative arrow of time, caused by light only traveling forwards in time (see light cone). In particle physics, the violation of CP symmetry implies that there should be a small counterbalancing time asymmetry to preserve CPT symmetry as stated above. The standard description of measurement in quantum mechanics is also time asymmetric (see Measurement in quantum mechanics).

### 11.5.6 Quantized time

See also: Chronon

Time quantization is a hypothetical concept. In the modern established physical theories (the Standard Model of Particles and Interactions and General Relativity) time is not quantized.

Planck time ($\sim 5.4 \times 10^{-44}$ seconds) is the unit of time in the system of natural units known as Planck units. Current established physical theories are believed to fail at this time scale, and many physicists expect that the Planck time might be the smallest unit of time that could ever be measured, even in principle. Tentative physical theories that describe this time scale exist; see for instance loop quantum gravity.

## 11.6 Time and the Big Bang theory

Stephen Hawking in particular has addressed a connection between time and the Big Bang. In *A Brief History of Time* and elsewhere, Hawking says that even if time did not begin with the Big Bang and there were another time frame before the Big Bang, no information from events then would be accessible to us, and nothing that happened then would have any effect upon the present time-frame.[82] Upon occasion, Hawking has stated that time actually began with the Big Bang, and that questions about what happened *before* the Big Bang are *meaningless*.[83][84][85] This less-nuanced, but commonly repeated formulation has received criticisms from philosophers such as Aristotelian philosopher Mortimer J. Adler.[86][87]

Scientists have come to some agreement on descriptions of events that happened $10^{-35}$ seconds after the Big Bang, but generally agree that descriptions about what happened before one Planck time ($5 \times 10^{-44}$ seconds) after the Big Bang are likely to remain pure speculation.

### 11.6.1 Speculative physics beyond the Big Bang

While the Big Bang model is well established in cosmology, it is likely to be refined in the future. Little is known about the earliest moments of the universe's history. The Penrose–Hawking singularity theorems require the existence of a singularity at the beginning of cosmic time. However, these theorems assume that general relativity is correct, but general relativity must break down before the universe reaches the Planck temperature, and a correct treatment of quantum gravity may avoid the singularity.[88]

If inflation has indeed occurred, it is likely that there are parts of the universe so distant that they cannot be observed in principle, as exponential expansion would push large regions of space beyond our observable horizon.

Some proposals, each of which entails untested hypotheses, are:

- Models including the Hartle–Hawking boundary condition in which the whole of space-time is finite; the Big Bang does represent the limit of time, but without the need for a singularity.[89]

- Brane cosmology models[90] in which inflation is due to the movement of branes in string theory; the pre-big bang model; the ekpyrotic model, in which the Big Bang is the result of a collision between branes; and the cyclic model, a variant of the ekpyrotic model in which collisions occur periodically.[91][92][93]

- Chaotic inflation, in which inflation events start here and there in a random quantum-gravity foam, each leading to a *bubble universe* expanding from its own big bang.[94]

Proposals in the last two categories see the Big Bang as an event in a much larger and older universe, or multiverse, and not the literal beginning.

## 11.7  Time travel

Main article: Time travel
See also: Time travel in fiction, Wormhole, and Twin paradox

Time travel is the concept of moving backwards or forwards to different points in time, in a manner analogous to moving through space, and different from the normal "flow" of time to an earthbound observer. In this view, all points in time (including future times) "persist" in some way. Time travel has been a plot device in fiction since the 19th century. Traveling backwards in time has never been verified, presents many theoretic problems, and may be an impossibility.[95] Any technological device, whether fictional or hypothetical, that is used to achieve time travel is known as a time machine.

A central problem with time travel to the past is the violation of causality; should an effect precede its cause, it would give rise to the possibility of a temporal paradox. Some interpretations of time travel resolve this by accepting the possibility of travel between branch points, parallel realities, or universes.

Another solution to the problem of causality-based temporal paradoxes is that such paradoxes cannot arise simply because they have not arisen. As illustrated in numerous works of fiction, free will either ceases to exist in the past or the outcomes of such decisions are predetermined. As such, it would not be possible to enact the grandfather paradox because it is a historical fact that your grandfather was not killed before his child (your parent) was conceived. This view doesn't simply hold that history is an unchangeable constant, but that any change made by a hypothetical future time traveler would already have happened in his or her past, resulting in the reality that the traveler moves from. More elaboration on this view can be found in the Novikov self-consistency principle.

## 11.8  Time perception

Main article: Time perception

The specious present refers to the time duration wherein one's perceptions are considered to be in the present. The experienced present is said to be 'specious' in that, unlike the objective present, it is an interval and not a durationless instant. The term *specious present* was first introduced by the psychologist E.R. Clay, and later developed by William James.[96]

### 11.8.1  Biopsychology

The brain's judgment of time is known to be a highly distributed system, including at least the cerebral cortex, cerebellum and basal ganglia as its components. One particular component, the suprachiasmatic nuclei, is responsible for the circadian (or daily) rhythm, while other cell clusters appear capable of shorter-range (ultradian) timekeeping.

Psychoactive drugs can impair the judgment of time. Stimulants can lead both humans and rats to overestimate time intervals,[97][98] while depressants can have the opposite effect.[99] The level of activity in the brain of neurotransmitters such as dopamine and norepinephrine may be the reason for this.[100] Such chemicals will either excite or inhibit the firing of neurons in the brain, with a greater firing rate allowing the brain to register the occurrence of more events within a given interval (speed up time) and a decreased firing rate reducing the brain's capacity to distinguish events occurring within a given interval (slow down time).[101]

Mental chronometry is the use of response time in perceptual-motor tasks to infer the content, duration, and temporal sequencing of cognitive operations.

### 11.8.2 Development of awareness and understanding of time in children

Children's expanding cognitive abilities allow them to understand time more clearly. Two- and three-year-olds' understanding of time is mainly limited to "now and not now." Five- and six-year-olds can grasp the ideas of past, present, and future. Seven- to ten-year-olds can use clocks and calendars.[102]

### 11.8.3 Alterations

In addition to psychoactive drugs, judgments of time can be altered by temporal illusions (like the kappa effect),[103] age,[104] and hypnosis.[105] The sense of time is impaired in some people with neurological diseases such as Parkinson's disease and attention deficit disorder.

Psychologists assert that time seems to go faster with age, but the literature on this age-related perception of time remains controversial.[106] Those who support this notion argue that young people, having more excitatory neurotransmitters, are able to cope with faster external events.[101]

## 11.9 Use of time

See also: Time management and Time discipline

In sociology and anthropology, time discipline is the general name given to social and economic rules, conventions, customs, and expectations governing the measurement of time, the social currency and awareness of time measurements, and people's expectations concerning the observance of these customs by others. Arlie Russell Hochschild[107][108] and Norbert Elias[109] have written on the use of time from a sociological perspective.

The use of time is an important issue in understanding human behavior, education, and travel behavior. Time-use research is a developing field of study. The question concerns how time is allocated across a number of activities (such as time spent at home, at work, shopping, etc.). Time use changes with technology, as the television or the Internet created new opportunities to use time in different ways. However, some aspects of time use are relatively stable over long periods of time, such as the amount of time spent traveling to work, which despite major changes in transport, has been observed to be about 20–30 minutes one-way for a large number of cities over a long period.

Time management is the organization of tasks or events by first estimating how much time a task requires and when it must be completed, and adjusting events that would interfere with its completion so it is done in the appropriate amount of time. Calendars and day planners are common examples of time management tools.

A sequence of events, or series of events, is a sequence of items, facts, events, actions, changes, or procedural steps, arranged in time order (chronological order), often with causality relationships among the items.[110][111][112] Because of causality, cause precedes effect, or cause and effect may appear together in a single item, but effect never precedes cause. A sequence of events can be presented in text, tables, charts, or timelines. The description of the items or events may include a timestamp. A sequence of events that includes the time along with place or location information to describe a sequential path may be referred to as a world line.

Uses of a sequence of events include stories,[113] historical events (chronology), directions and steps in procedures,[114] and timetables for scheduling activities. A sequence of events may also be used to help describe processes in science, technology, and medicine. A sequence of events may be focused on past events (e.g., stories, history, chronology), on future events that must be in a predetermined order (e.g., plans, schedules, procedures, timetables), or focused on the observation of past events with the expectation that the events will occur in the future (e.g., processes). The use of a sequence of events occurs in fields as diverse as machines (cam timer), documentaries (*Seconds From Disaster*), law (choice of law), computer simulation (discrete event simulation), and electric power transmission[115] (sequence of events recorder). A specific example of a sequence of events is the timeline of the Fukushima Daiichi nuclear disaster.

## 11.10 See also

- Era
- Horology
- International System of Quantities
- Kairos
- Term (time)

### 11.10.1 Books

- *A Brief History of Time* by Stephen Hawking
- *About Time: Einstein's Unfinished Revolution* by Paul Davies
- *From Eternity to Here: The Quest for the Ultimate Theory of Time* by Sean M. Carroll

- *The Physical Basis of The Direction of Time* by Heinz-Dieter Zeh

- *An Experiment with Time* by John William Dunne

- *Einstein's Dreams* by Alan Lightman

- *Being and Time* by Martin Heidegger

### 11.10.2  Organizations

*Leading scholarly organizations for researchers on the history and technology of time and timekeeping*

- Antiquarian Horological Society—AHS (United Kingdom)

- Chronometrophilia (Switzerland)

- Deutsche Gesellschaft für Chronometrie—DGC (Germany)

- National Association of Watch and Clock Collectors—NAWCC (United States)

## 11.11  References

[1] "Oxford Dictionaries:Time". Oxford University Press. 2011. Retrieved 18 December 2011. the indefinite continued progress of existence and events in the past, present, and future regarded as a whole

[2]  - "Webster's New World College Dictionary". 2010. Retrieved 9 April 2011. 1.indefinite, unlimited duration in which things are considered as happening in the past, present, or future; every moment there has ever been or ever will be... a system of measuring duration 2.the period between two events or during which something exists, happens, or acts; measured or measurable interval

   - "The American Heritage Stedman's Medical Dictionary @dictionary.com". 2002. Retrieved 9 April 2011. A duration or relation of events expressed in terms of past, present, and future, and measured in units such as minutes, hours, days, months, or years.

   - "Collins Language.com". HarperCollins. 2011. Retrieved 18 December 2011. 1. The continuous passage of existence in which events pass from a state of potentiality in the future, through the present, to a state of finality in the past. 2. *physics* a quantity measuring duration, usually with reference to a periodic process such as the rotation of the earth or the frequency of electromagnetic radiation emitted from certain atoms. In classical mechanics, time is absolute in the sense that the time of an event is independent of the observer. According to the theory of relativity it depends on the observer's frame of reference. Time is considered as a fourth coordinate required, along with three spatial coordinates, to specify an event.

   - "The American Heritage Science Dictionary @dictionary.com". 2002. Retrieved 9 April 2011. 1. A continuous, measurable quantity in which events occur in a sequence proceeding from the past through the present to the future. 2a. An interval separating two points of this quantity; a duration. 2b. A system or reference frame in which such intervals are measured or such quantities are calculated.

   - "Eric Weisstein's World of Science". 2007. Retrieved 9 April 2011. A quantity used to specify the order in which events occurred and measure the amount by which one event preceded or followed another. In special relativity, ct (where c is the speed of light and t is time), plays the role of a fourth dimension.

[3] "Time". *The American Heritage Dictionary of the English Language* (Fourth ed.) (Houghton Mifflin Company). 2011. A nonspatial continuum in which events occur in apparently irreversible succession from the past through the present to the future.

[4] Merriam-Webster Dictionary the measured or measurable period during which an action, process, or condition exists or continues : duration; a nonspatial continuum which is measured in terms of events that succeed one another from past through present to future

[5] Compact Oxford English Dictionary A limited stretch or space of continued existence, as the interval between two successive events or acts, or the period through which an action, condition, or state continues. (1971)

[6]  - "Internet Encyclopedia of Philosophy". 2010. Retrieved 9 April 2011. Time is what clocks measure. We use time to place events in sequence one after the other, and we use time to compare how long events last... Among philosophers of physics, the most popular short answer to the question "What is physical time?" is that it is not a substance or object but rather a special system of relations among instantaneous events. This working definition is offered by Adolf Grünbaum who applies the contemporary mathematical theory of continuity to physical processes, and he says time is a linear continuum of instants and is a distinguished one-dimensional subspace of four-dimensional spacetime.

   - "Dictionary.com Unabridged, based on Random House Dictionary". 2010. Retrieved 9 April 2011. 1. the system of those sequential relations that any event has to any other, as past, present, or future; indefinite and continuous duration regarded as that in which events succeed one another.... 3. (sometimes initial capital letter) a system or method of measuring or reckoning the passage of time: mean time; apparent time; Greenwich Time. 4. a limited period or interval, as between two successive events: a long time....

14. a particular or definite point in time, as indicated by a clock: What time is it? ... 18. an indefinite, frequently prolonged period or duration in the future: Time will tell if what we have done here today was right.

- Ivey, Donald G.; Hume, J.N.P. (1974). *Physics* 1. Ronald Press. p. 65. Our operational definition of time is that time is what clocks measure.

[7] Le Poidevin, Robin (Winter 2004). "The Experience and Perception of Time". In Edward N. Zalta. *The Stanford Encyclopedia of Philosophy*. Retrieved 9 April 2011.

[8] "Newton did for time what the Greek geometers did for space, idealized it into an exactly measurable dimension." *About Time: Einstein's Unfinished Revolution*, Paul Davies, p. 31, Simon & Schuster, 1996, ISBN 978-0684818221

[9] Sean M Carroll (2009). *From Eternity to Here: The Quest for the Ultimate Theory of Time*. Dutton. ISBN 978-0-525-95133-9.

[10] Adam Frank, *Cosmology and Culture at the Twilight of the Big Bang*, "the time we imagined from the cosmos and the time we imagined into the human experience turn out to be woven so tightly together that we have lost the ability to see each of them for what it is." p. xv, Free Press, 2011, ISBN 978-1439169599

[11] St. Augustine, *Confessions*, Simon & Brown, 2012, ISBN 978-1613823262

[12] Official Baseball Rules, 2011 Edition (2011). "Rules 8.03 and 8.04" (Free PDF download). Major League Baseball. Retrieved 7 July 2012. Rule 8.03 Such preparatory pitches shall not consume more than one minute of time...Rule 8.04 When the bases are unoccupied, the pitcher shall deliver the ball to the batter within 12 seconds...The 12-second timing starts when the pitcher is in possession of the ball and the batter is in the box, alert to the pitcher. The timing stops when the pitcher releases the ball

[13] "Guinness Book of Baseball World Records". Guinness World Records, Ltd. Retrieved 7 July 2012. The record for the fastest time for circling the bases is 13.3 seconds, set by Evar Swanson at Columbus, Ohio in 1932...The greatest reliably recorded speed at which a baseball has been pitched is 100.9 mph by Lynn Nolan Ryan (California Angels) at Anaheim Stadium in California on 20 August 1974.

[14] Zeigler, Kenneth (2008). *Getting organized at work : 24 lessons to set goals, establish priorities, and manage your time*. McGraw-Hill. ISBN 9780071591386. 108 pages

[15] Burnham, Douglas : Staffordshire University (2006). "Gottfried Wilhelm Leibniz (1646–1716) Metaphysics – 7. Space, Time, and Indiscernibles". *The Internet Encyclopedia of Philosophy*. Retrieved 9 April 2011. First of all, Leibniz finds the idea that space and time might be substances or substance-like absurd (see, for example, "Correspondence with Clarke," Leibniz's Fourth Paper, §8ff). In short, an empty space would be a substance with no properties; it will be a substance that even God cannot modify or destroy.... That is, space and time are internal or intrinsic features of the complete concepts of things, not extrinsic.... Leibniz's view has two major implications. First, there is no absolute location in either space or time; location is always the situation of an object or event relative to other objects and events. Second, space and time are not in themselves real (that is, not substances). Space and time are, rather, ideal. Space and time are just metaphysically illegitimate ways of perceiving certain virtual relations between substances. They are phenomena or, strictly speaking, illusions (although they are illusions that are well-founded upon the internal properties of substances).... It is sometimes convenient to think of space and time as something "out there," over and above the entities and their relations to each other, but this convenience must not be confused with reality. Space is nothing but the order of co-existent objects; time nothing but the order of successive events. This is usually called a relational theory of space and time.

[16] Cummings, Raymond King (1922). *The Girl in the Golden Atom*. U of Nebraska Press. p. 46. ISBN 978-0-8032-6457-1. Retrieved 9 April 2011. Chapter 5. Cummings repeated this sentence in several of his novellas. Sources, such as this one, attribute it to his earlier work, *The Time Professor*, in 1921. Before taking book form, several of Cummings's stories appeared serialized in magazines. The first eight chapters of his *The Girl in the Golden Atom* appeared in *All-Story Magazine* on 15 March 1919. In the novel version the quote about time appears in Chapter V.

[17] International, Rotary (Aug 1973). "The Rotarian". Published by Rotary International: 47. ISSN 0035-838X. Retrieved 9 April 2011., What does a man possess? page 47

[18] Daintith, John (2008). *Biographical Encyclopedia of Scientists* (third ed.). CRC Press. p. 796. ISBN 1-4200-7271-4. Retrieved 9 April 2011., Page 796, quoting Wheeler from the American Journal of Physics, 1978

[19] Davies, Davies (1995). *About time: Einstein's unfinished revolution*. Simon & Schuster. p. 236. ISBN 0-671-79964-9. Retrieved 9 April 2011.

[20] Rynasiewicz, Robert : Johns Hopkins University (12 August 2004). "Newton's Views on Space, Time, and Motion". *Stanford Encyclopedia of Philosophy*. Stanford University. Retrieved 5 February 2012. Newton did not regard space and time as genuine substances (as are, paradigmatically, bodies and minds), but rather as real entities with their own manner of existence as necessitated by God's existence... To paraphrase: Absolute, true, and mathematical time, from its own nature, passes equably without relation to anything external, and thus without reference to any change or way of measuring of time (e.g., the hour, day, month, or year).

[21] Markosian, Ned. "Time". In Edward N. Zalta. *The Stanford Encyclopedia of Philosophy (Winter 2002 Edition)*. Retrieved 23 September 2011. The opposing view, normally referred to either as "Platonism with Respect to Time" or

as "Absolutism with Respect to Time," has been defended by Plato, Newton, and others. On this view, time is like an empty container into which events may be placed; but it is a container that exists independently of whether or not anything is placed in it.

[22] Mattey, G. J. : UC Davis (22 January 1997). "Critique of Pure Reason, Lecture notes: Philosophy 175 UC Davis". Retrieved 9 April 2011. What is correct in the Leibnizian view was its anti-metaphysical stance. Space and time do not exist in and of themselves, but in some sense are the product of the way we represent things. The[y] are ideal, though not in the sense in which Leibniz thought they are ideal (figments of the imagination). The ideality of space is its mind-dependence: it is only a condition of sensibility.... Kant concluded "absolute space is not an object of outer sensation; it is rather a fundamental concept which first of all makes possible all such outer sensation."...Much of the argumentation pertaining to space is applicable, mutatis mutandis, to time, so I will not rehearse the arguments. As space is the form of outer intuition, so time is the form of inner intuition.... Kant claimed that time is real, it is "the real form of inner intuition."

[23] McCormick, Matt : California State University, Sacramento (2006). "Immanuel Kant (1724–1804) Metaphysics: 4. Kant's Transcendental Idealism". *The Internet Encyclopedia of Philosophy*. Retrieved 9 April 2011. Time, Kant argues, is also necessary as a form or condition of our intuitions of objects. The idea of time itself cannot be gathered from experience because succession and simultaneity of objects, the phenomena that would indicate the passage of time, would be impossible to represent if we did not already possess the capacity to represent objects in time.... Another way to put the point is to say that the fact that the mind of the knower makes the *a priori* contribution does not mean that space and time or the categories are mere figments of the imagination. Kant is an empirical realist about the world we experience; we can know objects as they appear to us. He gives a robust defense of science and the study of the natural world from his argument about the mind's role in making nature. All discursive, rational beings must conceive of the physical world as spatially and temporally unified, he argues.

[24] Duff, Okun, Veneziano, *ibid*. p. 3. "There is no well established terminology for the fundamental constants of Nature. ... The absence of accurately defined terms or the uses (i.e., actually misuses) of ill-defined terms lead to confusion and proliferation of wrong statements."

[25] Carrol, Sean, Chapter One, Section Two, Plume, 2010. *From Eternity to Here*. ISBN 978-0452296541. As human beings we 'feel' the passage of time.

[26] Lehar, Steve. (2000). The Function of Conscious Experience: An Analogical Paradigm of Perception and Behavior, *Consciousness and Cognition*.

[27] Richards, E. G. (1998). *Mapping Time: The Calendar and its History*. Oxford University Press. pp. 3–5.

[28] Rudgley, Richard (1999). *The Lost Civilizations of the Stone Age*. New York: Simon & Schuster. pp. 86–105.

[29] Van Stone, Mark. "The Maya Long Count Calendar: An Introduction." Archaeoastronomy 24.(2011): 8-11. Academic Search Complete. Web. 20 Feb. 2016.

[30] "French Republican Calendar | Chronology." Encyclopedia Britannica Online. Encyclopedia Britannica, n.d. Web. 21 Feb. 2016.

[31] Barnett, Jo Ellen *Time's Pendulum: The Quest to Capture Time—from Sundials to Atomic Clocks* Plenum, 1998 ISBN 0-306-45787-3 p.28

[32] Lombardi, Michael A. "Why Is a Minute Divided into 60 Seconds, an Hour into 60 Minutes, Yet There Are Only 24 Hours in a Day?" Scientific American. Springer Nature, 5 Mar. 2007. Web. 21 Feb. 2016.

[33] Barnett, *ibid*, p.37

[34] Laurence Bergreen, *Over the Edge of the World: Magellan's Terrifying Circumnavigation of the Globe*, HarperCollins Publishers, 2003, hardcover 480 pages, ISBN 0-06-621173-5

[35] North, J. (2004) *God's Clockmaker: Richard of Wallingford and the Invention of Time*. Oxbow Books. ISBN 1-85285-451-0

[36] Watson, E (1979) "The St Albans Clock of Richard of Wallingford". *Antiquarian Horology* 372–384.

[37] "History of Clocks." About.com Inventors. About.com, n.d. Web. 21 Feb. 2016.

[38] "NIST Unveils Chip-Scale Atomic Clock". 27 August 2004. Retrieved 9 June 2011.

[39] "New atomic clock can keep time for 200 million years: Super-precise instruments vital to deep space navigation". *Vancouver Sun*. 16 February 2008. Retrieved 9 April 2011.

[40] "NIST-F1 Cesium Fountain Clock". Retrieved 24 July 2015.

[41] "Byrhtferth of Ramsey". (2008). In Encyclopædia Britannica. Retrieved 2008-09-15, from Encyclopædia Britannica Online: http://search.eb.com/eb/article-9438957

[42] "atom", Oxford English Dictionary, Draft Revision September 2008 (contains relevant citations from Byrhtferth's *Enchiridion*)

[43] "12 attoseconds is the world record for shortest controllable time". 12 May 2010. Retrieved 19 April 2012.

[44] Bacon, Roger. *Opera quaedam hactenus inedita*. Harvard University. Retrieved 5 July 2014.

[45] McCarthy, Dennis D.; Seidelmann, P. Kenneth (2009). *Time: from Earth rotation to atomic physics*. Wiley-VCH. p. 18. ISBN 3-527-40780-4. Extract of page 18

[46] Jones, Floyd Nolen (2005). *The Chronology Of The Old Testament* (15th ed.). New Leaf Publishing Group. p. 287. ISBN 0-89051-416-X. Extract of page 287

[47] Cohen, K.M.; Finney, S.; Gibbard, P.L. (2013), *International Chronostratigraphic Chart* (PDF), International Commission on Stratigraphy, retrieved 23 September 2013

[48] http://starchild.gsfc.nasa.gov/docs/StarChild/questions/ question18.html NASA - StarChild Question of the Month for February 2000

[49] "Aeon - Definition and More from the Free Merriam-Webster Dictionary". Merriam-webster.com. 31 August 2012. Retrieved 24 September 2013.

[50] Organisation Intergouvernementale de la Convention du Métre (1998). *The International System of Units (SI), 7th Edition* (PDF). Retrieved 9 April 2011.

[51] "Base unit definitions: Second". NIST. Retrieved 9 April 2011.

[52] IEC 60050-113:2011, item 113-01-08

[53] IEC 60050-113:2011, item 113-01-010; ISO 80000-3:2006, item 3-7

[54] IEC 60050-113:2011, item 113-01-012: "mark attributed to an instant by means of a specified time scale

[55] IEC 60050-113:2011, item 113-01-013: "range of a time interval (113-01-10)"

[56] ISO 80000-3:2006, item 3-7

[57] Rust, Eric Charles (1981). *Religion, Revelation and Reason.* Mercer University Press. p. 60. ISBN 9780865540583. Retrieved 2015-08-20. Profane time, as Eliade points out, is linear. As man dwelt increasingly in the profane and a sense of history developed, the desire to escape into the sacred began to drop in the background. The myths, tied up with cyclic time, were not so easily operative. [...] So secular man became content with his linear time. He could not return to cyclic time and re-enter sacred space though its myths. [...] Just here, as Eliade sees it, a new religious structure became available. In the Judaeo-Christian religions - Judaism, Christianity, Islam - history is taken seriously, and linear time is accepted. The cyclic time of the primordial mythical consciousness has been transformed into the time of profane man, but the mythical consciousness remains. It has been historicized. The Christian mythos and its accompanying ritual are bound up, for example, with history and center in authentic history, especially the Christ-event. Sacred space, the Transcendent Presence, is thus opened up to secular man because it meets him where he is, in the linear flow of secular time. The Christian myth gives such time a beginning in creation, a center in the Christ-event, and an end in the final consummation.

[58] Betz, Hans Dieter, ed. (2008). *Religion Past & Present: Encyclopedia of Theology and Religion.* 4: Dev-Ezr (4 ed.). Brill. p. 101. ISBN 9789004146884. Retrieved 2015-08-20. [...] God produces a creation with a directional time structure [...].

[59] Lundin, Roger; Thiselton, Anthony C.; Walhout, Clarence (1999). *The Promise of Hermeneutics.* Wm. B. Eerdmans Publishing. p. 121. ISBN 9780802846358. Retrieved 2015-08-20. We need to note the close ties between teleology, eschatology, and utopia. In Christian theology, the understanding of the teleology of particular actions is ultimately related to the teleology of history in general, which is the concern of eschatology.

[60] Hus, Bo'az; Pasi, Marco; Von Stuckrad, Kocku (2011). *Kabbalah and Modernity: Interpretations, Transformations, Adaptations.* BRILL. ISBN 9004182845.

[61] Wolfson, Elliot R. (2006). *Alef, Mem, Tau: Kabbalistic Musings on Time, Truth, and Death.* University of California Press. p. 111. ISBN 0-520-93231-5. Extract of page 111

[62] Navratil, Gerhard (2009). *Research Trends in Geographic Information Science.* Springer Japan. p. 217. ISBN 3-540-88243-X. Retrieved 9 April 2011.

[63] Layton, Robert (1994). *Who needs the past?: indigenous values and archaeology* (2nd ed.). Routledge. p. 7. ISBN 0-415-09558-1. Retrieved 9 April 2011., Introduction, p. 7

[64] Dagobert Runes, *Dictionary of Philosophy,* p. 318

[65] Hardie, R. P.; Gaye, R. K. "Physics by Aristotle". MIT. Retrieved 4 May 2014."*Time then is a kind of number. (Number, we must note, is used in two senses-both of what is counted or the countable and also of that with which we count. Time obviously is what is counted, not that with which we count: there are different kinds of thing.) [...] It is clear, then, that time is 'number of movement in respect of the before and after', and is continuous since it is an attribute of what is continuous.* "

[66] Augustine of Hippo. *Confessions.* Retrieved 9 April 2011. Book 11, Chapter 14.

[67] Gottfried Martin, *Kant's Metaphysics and Theory of Science*

[68] Kant, Immanuel (1787). *The Critique of Pure Reason, 2nd edition.* Retrieved 9 April 2011. translated by J. M. D. Meiklejohn, eBooks@Adelaide, 2004

[69] Bergson, Henri (1907) *Creative Evolution.* trans. by Arthur Mitchell. Mineola: Dover, 1998.

[70] Balslev, Anindita N.; Jitendranath Mohanty (November 1992). *Religion and Time.* Studies in the History of Religions, 54. The Netherlands: Brill Academic Publishers. pp. 53, 54, 55, 56, 57, 58, and 59. ISBN 978-90-04-09583-0.

[71] Martin Heidegger (1962). "V". *Being and Time.* p. 425. ISBN 978-0-631-19770-6.

[72] Harry Foundalis. "You are about to disappear". Retrieved 9 April 2011.

[73] Huston, Tom. "Buddhism and the illusion of time". Retrieved 9 April 2011.

[74] Garfield, Jay L. (1995). *The fundamental wisdom of the middle way: Nāgārjuna's Mūlamadhyamakakārikā*. New York: Oxford University Press. ISBN 978-0-19-509336-0.

[75] "Time is an illusion?". Retrieved 9 April 2011.

[76] Herman M. Schwartz, *Introduction to Special Relativity*, McGraw-Hill Book Company, 1968, hardcover 442 pages, see ISBN 0-88275-478-5 (1977 edition), pp. 10–13

[77] A. Einstein, H. A. Lorentz, H. Weyl, H. Minkowski, *The Principle of Relativity*, Dover Publications, Inc, 2000, softcover 216 pages, ISBN 0-486-60081-5, See pp. 37–65 for an English translation of Einstein's original 1905 paper.

[78] "Albert Einstein's Theory of Relativity". YouTube. 30 November 2011. Retrieved 24 September 2013.

[79] "Time Travel: Einstein's big idea (Theory of Relativity)". YouTube. 9 January 2007. Retrieved 24 September 2013.

[80] Hours, After (11 February 2012). "7 Theories on Time That Would Make Doc Brown's Head Explode". Cracked.com. Retrieved 24 September 2013.

[81] Knudsen, Jens M.; Hjorth, Poul (2012). *Elements of Newtonian Mechanics* (illustrated ed.). Springer Science & Business Media. p. 30. ISBN 978-3-642-97599-8. Extract of page 30

[82] Hawking, Stephen (1996). "The Beginning of Time". University of Cambridge. Retrieved 8 July 2012. Since events before the Big Bang have no observational consequences, one may as well cut them out of the theory, and say that time began at the Big Bang. Events before the Big Bang, are simply not defined, because there's no way one could measure what happened at them. This kind of beginning to the universe, and of time itself, is very different to the beginnings that had been considered earlier.

[83] Hawking, Stephen (1996). "The Beginning of Time". University of Cambridge. Retrieved 8 July 2012. The conclusion of this lecture is that the universe has not existed forever. Rather, the universe, and time itself, had a beginning in the Big Bang, about 15 billion years ago.

[84] Hawking, Stephen (27 February 2006). "Professor Stephen Hawking lectures on the origin of the universe". University of Oxford. Retrieved 5 December 2012. Suppose the beginning of the universe was like the South Pole of the earth, with degrees of latitude playing the role of time. The universe would start as a point at the South Pole. As one moves north, the circles of constant latitude, representing the size of the universe, would expand. To ask what happened before the beginning of the universe would become a meaningless question because there is nothing south of the South Pole.

[85] Ghandchi, Sam : Editor/Publisher (16 January 2004). "Space and New Thinking". Retrieved 9 April 2011. and as Stephen Hawking puts it, asking what was before Big Bang is like asking what is North of North Pole, a meaningless question.

[86] Adler, Mortimer J., PhD. "Natural Theology, Chance, and God". Retrieved 9 April 2011. Hawking could have avoided the error of supposing that time had a beginning with the Big Bang if he had distinguished time as it is measured by physicists from time that is not measurable by physicists.... an error shared by many other great physicists in the twentieth century, the error of saying that what cannot be measured by physicists does not exist in reality. "The Great Ideas Today". *Encyclopædia Britannica*. 1992.

[87] Adler, Mortimer J., PhD. "Natural Theology, Chance, and God". Retrieved 9 April 2011. Where Einstein had said that what is not measurable by physicists is of no interest to them, Hawking flatly asserts that what is not measurable by physicists does not exist—has no reality whatsoever.
With respect to time, that amounts to the denial of psychological time which is not measurable by physicists, and also to everlasting time—time before the Big Bang—which physics cannot measure. Hawking does not know that both Aquinas and Kant had shown that we cannot rationally establish that time is either finite or infinite. "The Great Ideas Today". *Encyclopædia Britannica*. 1992.

[88] Hawking, Stephen; and Ellis, G. F. R. (1973). *The Large Scale Structure of Space-Time*. Cambridge: Cambridge University Press. ISBN 0-521-09906-4.

[89] J. Hartle and S. W. Hawking (1983). "Wave function of the universe". *Phys. Rev. D* **28** (12): 2960. Bibcode:1983PhRvD..28.2960H. doi:10.1103/PhysRevD.28.2960.

[90] Langlois, David (2002). "Brane cosmology: an introduction". *Progress of Theoretical Physics Supplement* **148**: 181. arXiv:hep-th/0209261. Bibcode:2002PThPS.148..181L. doi:10.1143/PTPS.148.181.

[91] Linde, Andre (2002). "Inflationary Theory versus Ekpyrotic/Cyclic Scenario". *In: the future of theoretical physics and cosmology. Edited by G. W. Gibbons*: 801. arXiv:hep-th/0205259. Bibcode:2003ftpc.book..801L.

[92] "Recycled Universe: Theory Could Solve Cosmic Mystery". Space.com. 8 May 2006. Retrieved 9 April 2011.

[93] "What Happened Before the Big Bang?". Archived from the original on 4 July 2007. Retrieved 9 April 2011.

[94] A. Linde (1986). "Eternal chaotic inflation". *Mod. Phys. Lett.* **A1** (2): 81. Bibcode:1986MPLA....1...81L. doi:10.1142/S0217732386000129. A. Linde (1986). "Eternally existing self-reproducing chaotic inflationary universe". *Phys. Lett.* **B175** (4): 395–400. Bibcode:1986PhLB..175..395L. doi:10.1016/0370-2693(86)90611-8.

[95] G. Quznetsov, Prespacetime Journal, March 2010, Vol.1, Issue 2, Page 274-275

[96] Andersen, Holly; Rick Grush (2009). "A brief history of time-consciousness: historical precursors to James and Husserl" (PDF) **47** (2). Journal of the History of Philosophy: 277–307. Retrieved 9 April 2011.

[97] Wittmann, M.; Leland DS; Churan J; Paulus MP. (8 October 2007). "Impaired time perception and motor timing in stimulant-dependent subjects" (online abstract). *Drug Alcohol Depend.* **90** (2–3): 183–92. doi:10.1016/j.drugalcdep.2007.03.005. PMC 1997301. PMID 17434690.

[98] Cheng, Ruey-Kuang; Macdonald, Christopher J.; Meck, Warren H. (2006). "Differential effects of cocaine and ketamine on time estimation: Implications for neurobiological models of interval timing" (online abstract). *Pharmacology, biochemistry and behavior* **85** (1): 114–122. doi:10.1016/j.pbb.2006.07.019. PMID 16920182. Retrieved 9 April 2011.

[99] Tinklenberg, Jared R.; Walton T. Roth1; Bert S. Kopell (January 1976). "Marijuana and ethanol: Differential effects on time perception, heart rate, and subjective response". *Psychopharmacology* **49** (3): 275–279. doi:10.1007/BF00426830. PMID 826945. Retrieved 9 April 2011.

[100] Arzy, Shahar; Istvan Molnar-Szakacs; Olaf Blanke (18 June 2008). "Self in Time: Imagined Self-Location Influences Neural Activity Related to Mental Time Travel" (Abstract). *The Journal of Neuroscience* **28** (25): 6502–6507. doi:10.1523/JNEUROSCI.5712-07.2008. PMID 18562621. Retrieved 9 April 2011.

[101] Carter, Rita (2009). *The Human Brain Book*. Dorling Kindersley Publishing. pp. 186–187. ISBN 978-0-7566-5441-2.

[102] Kennedy-Moore, Eileen (28 March 2014). "Time Management for Kids". Psychology Today. Retrieved 26 April 2014.

[103] Wada Y, Masuda T, Noguchi K, 2005, "Temporal illusion called 'kappa effect' in event perception" Perception 34 ECVP Abstract Supplement

[104] Robert, Adler. "Look how time flies..". Retrieved 9 April 2011.

[105] Bowers, Kenneth; Brenneman, HA (January 1979). "Hypnosis and the perception of time". *International Journal of Clinical and Experimental Hypnosis* (International Journal of Clinical and Experimental Hypnosis) 27 (1): 29–41. doi:10.1080/00207147908407540. PMID 541126.

[106] Gruber, Ronald P.; Wagner, Lawrence F.; Block, Richard A. (2000). "Subjective Time Versus Proper (Clock) Time". In Buccheri, R.; Di Gesù, V.; Saniga, Metod. *Studies on the structure of time: from physics to psycho(patho)logy.* Springer. p. 54. ISBN 0-306-46439-X. Retrieved 9 April 2011. Extract of page 54

[107] Russell Hochschild, Arlie (1997). *The time bind: when work becomes home and home becomes work.* New York: Metropolitan Books. ISBN 9780805044713

[108] Russell Hochschild, Arlie (20 April 1997). "There's no place like work". *New York Times Magazine* (The New York Times).

[109] Elias, Norbert (1992). *Time: an essay.* Oxford, UK Cambridge, USA: Blackwell. ISBN 9780631157984.

[110] "Sequence - Order of Important Events" (PDF). Austin Independent School District. 2009.

[111] "Sequence of Events Worksheets". Reference.com.

[112] Compiled by David Luckham and Roy Schulte. "Event Processing Glossary—Version 2.0". Complex Event Processing.

[113] Richard Nordquist. "narrative". About.com.

[114] David J. Piasecki. "Inventory Accuracy Glossary". AccuracyBook.com (OPS Publishing).

[115] "Utility Communications Architecture (UCA) glossary". NettedAutomation.

## 11.12 Further reading

- Barbour, Julian (1999). *The End of Time: The Next Revolution in Our Understanding of the Universe.* Oxford University Press. ISBN 0-19-514592-5.

- Landes, David (2000). *Revolution in Time.* Harvard University Press. ISBN 0-674-00282-2.

- Das, Tushar Kanti (1990). *The Time Dimension: An Interdisciplinary Guide.* New York: Praeger. ISBN 0-275-92681-8.- Research bibliography

- Davies, Paul (1996). *About Time: Einstein's Unfinished Revolution.* New York: Simon & Schuster Paperbacks. ISBN 0-684-81822-1.

- Feynman, Richard (1994) [1965]. *The Character of Physical Law.* Cambridge (Mass): The MIT Press. pp. 108–126. ISBN 0-262-56003-8.

- Galison, Peter (1992). *Einstein's Clocks and Poincaré's Maps: Empires of Time.* New York: W. W. Norton. ISBN 0-393-02001-0.

- Highfield, Roger (1992). *Arrow of Time: A Voyage through Science to Solve Time's Greatest Mystery.* Random House. ISBN 0-449-90723-6.

- Mermin, N. David (2005). *It's About Time: Understanding Einstein's Relativity*. Princeton University Press. ISBN 0-691-12201-6.

- Penrose, Roger (1999) [1989]. *The Emperor's New Mind: Concerning Computers, Minds, and the Laws of Physics*. New York: Oxford University Press. pp. 391–417. ISBN 0-19-286198-0. Retrieved 9 April 2011.

- Price, Huw (1996). *Time's Arrow and Archimedes' Point*. Oxford University Press. ISBN 0-19-511798-0. Retrieved 9 April 2011.

- Reichenbach, Hans (1999) [1956]. *The Direction of Time*. New York: Dover. ISBN 0-486-40926-0.

- Stiegler, Bernard, *Technics and Time, 1: The Fault of Epimetheus*

- Quznetsov, Gunn A. (2006). *Logical Foundation of Theoretical Physics*. Nova Sci. Publ. ISBN 1-59454-948-6.

- Whitrow, Gerald J. (1973). *The Nature of Time*. Holt, Rinehart and Wilson (New York).

- Whitrow, Gerald J. (1980). *The Natural Philosophy of Time*. Clarendon Press (Oxford).

- Whitrow, Gerald J. (1988). *Time in History. The evolution of our general awareness of time and temporal perspective*. Oxford University Press. ISBN 0-19-285211-6.

- Rovelli, Carlo (2006). *What is time? What is space?*. Rome: Di Renzo Editore. ISBN 88-8323-146-5.

- Charlie Gere, (2005) *Art, Time and Technology: Histories of the Disappearing Body*, Berg

- Craig Callendar, *Introducing Time*, Icon Books, 2010, ISBN 978-1848311206

- Benjamin Gal-Or, *Cosmology, Physics and Philosophy*, Springer Verlag, 1981, 1983, 1987, ISBN 0-387-90581-2, ISBN 0-387-96526-2.

- Roberto Mangabeira Unger and Lee Smolin, *The Singular Universe and the Reality of Time*, Cambridge University Press, 2014, ISBN 978-1-107-07406-4.

- Different systems of measuring time

- 

- Time on *In Our Time* at the BBC. (listen now)

- Time in the *Internet Encyclopedia of Philosophy*, by Bradley Dowden.

- Le Poidevin, Robin (Winter 2004). "The Experience and Perception of Time". In Edward N. Zalta. *The Stanford Encyclopedia of Philosophy*. Retrieved 9 April 2011.

- Time at Open Directory

## 11.13 External links

- Accurate time vs. PC Clock Difference

- Exploring Time from Planck Time to the lifespan of the universe

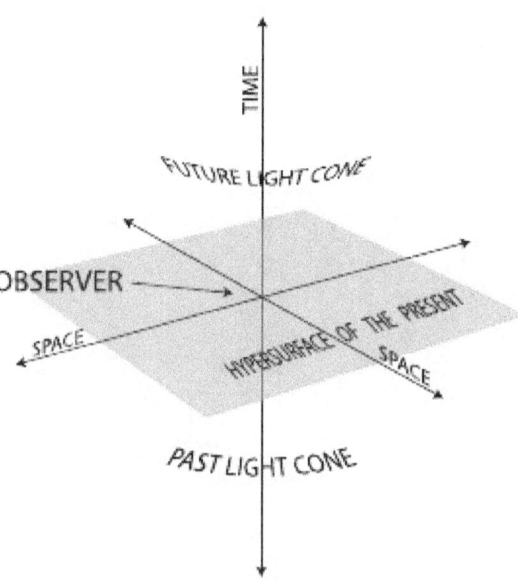

*Two-dimensional space depicted in three-dimensional spacetime. The past and future light cones are absolute, the "present" is a relative concept different for observers in relative motion.*

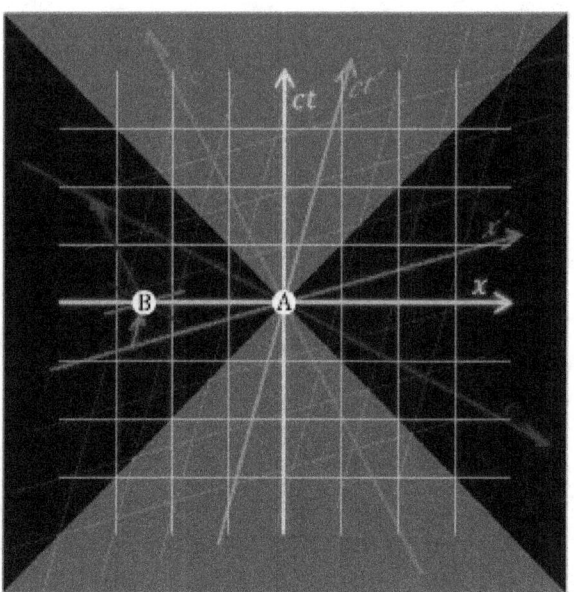

*Relativity of simultaneity: Event B is simultaneous with A in the green reference frame, but it occurred before in the blue frame, and occurs later in the red frame.*

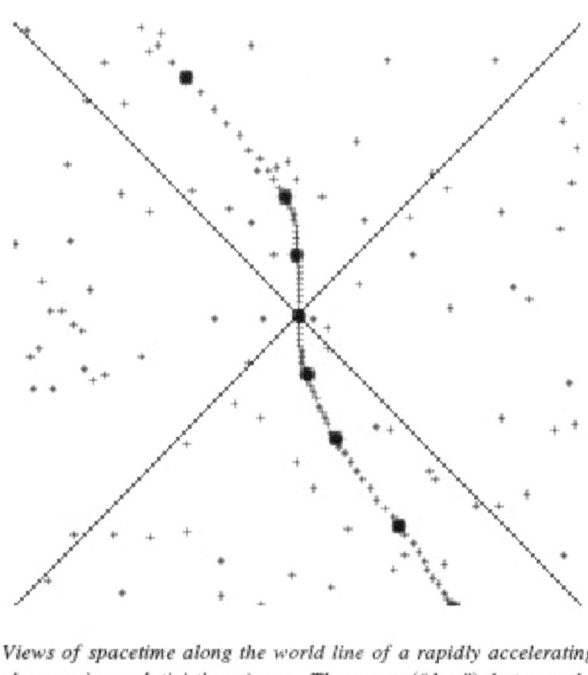

*Views of spacetime along the world line of a rapidly accelerating observer in a relativistic universe. The events ("dots") that pass the two diagonal lines in the bottom half of the image (the past light cone of the observer in the origin) are the events visible to the observer.*

*Philosopher and psychologist William James*

*A graphical representation of the expansion of the universe with the inflationary epoch represented as the dramatic expansion of the metric seen on the left*

*Time's mortal aspect is personified in this bronze statue by Charles van der Stappen*

# Chapter 12

# Manifold

For other uses, see Manifold (disambiguation).

In mathematics, a **manifold** is a topological space that

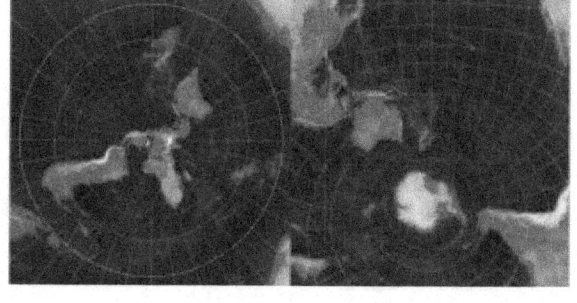

*The real projective plane is a two-dimensional manifold that cannot be realized in three dimensions without self-intersection, shown here as Boy's surface.*

*The surface of the Earth requires (at least) two charts to include every point. Here the globe is decomposed into charts around the North and South Poles.*

locally resembles Euclidean space near each point. More

precisely, each point of an $n$-dimensional manifold has a neighbourhood that is homeomorphic to the Euclidean space of dimension $n$.

One-dimensional manifolds include lines and circles, but not figure eights (because they have *crossing points* which are not locally homeomorphic to Euclidean 1-space). Two-dimensional manifolds are also called surfaces. Examples include the plane, the sphere, and the torus, which can all be embedded (formed without self-intersections) in three dimensional real space, but also the Klein bottle and real projective plane which will always self-intersect when immersed in real space.

Although a manifold locally resembles Euclidean space, globally it may not. For example, the surface of the sphere is not a Euclidean space, but in a region it can be charted by means of map projections of the region into the Euclidean plane (in the context of manifolds they are called *charts*). When a region appears in two neighbouring charts, the two representations do not coincide exactly and a transformation is needed to pass from one to the other, called a *transition map*.

The concept of a manifold is central to many parts of geometry and modern mathematical physics because it allows more complicated structures to be described and understood in terms of the relatively well-understood properties of Euclidean space. Manifolds naturally arise as solution sets of systems of equations and as graphs of functions. Manifolds may have additional features. One important class of manifolds is the class of differentiable manifolds. This differentiable structure allows calculus to be done on manifolds. A Riemannian metric on a manifold allows distances and angles to be measured. Symplectic manifolds serve as the phase spaces in the Hamiltonian formalism of classical mechanics, while four-dimensional Lorentzian manifolds model spacetime in general relativity.

## 12.1 Motivational examples

## 12.1.1 Circle

Main article: Circle

After a line, the circle is the simplest example of a topologi-

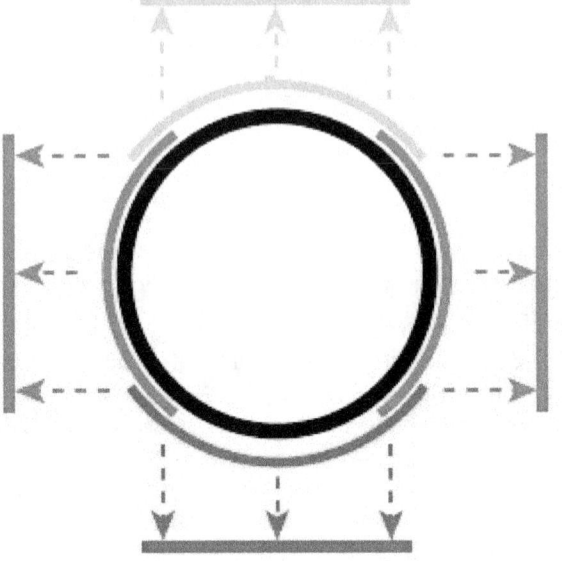

*Figure 1: The four charts each map part of the circle to an open interval, and together cover the whole circle.*

cal manifold. Topology ignores bending, so a small piece of a circle is treated exactly the same as a small piece of a line. Consider, for instance, the top part of the unit circle, $x^2 + y^2 = 1$, where the $y$-coordinate is positive (indicated by the yellow circular arc in *Figure 1*). Any point of this arc can be uniquely described by its $x$-coordinate. So, projection onto the first coordinate is a continuous, and invertible, mapping from the upper arc to the open interval $(-1,1)$:

$$\chi_{top}(x, y) = x.$$

Such functions along with the open regions they map are called *charts*. Similarly, there are charts for the bottom (red), left (blue), and right (green) parts of the circle:

$$\chi_{bottom}(x, y) = x$$

$$\chi_{left}(x, y) = y$$

$$\chi_{right}(x, y) = y.$$

Together, these parts cover the whole circle and the four charts form an atlas for the circle.

The top and right charts overlap: their intersection lies in the quarter of the circle where both the $x$- and the $y$-coordinates are positive. The two charts $\chi_{top}$ and $\chi_{right}$ each map this part

into the interval $(0, 1)$. Thus a function $T$ from $(0, 1)$ to itself can be constructed, which first uses the inverse of the top chart to reach the circle and then follows the right chart back to the interval. Let $a$ be any number in $(0, 1)$, then:

$$T(a) = \chi_{right}\left(\chi_{top}^{-1}[a]\right)$$
$$= \chi_{right}\left(a, \sqrt{1 - a^2}\right)$$
$$= \sqrt{1 - a^2}$$

Such a function is called a *transition map*.

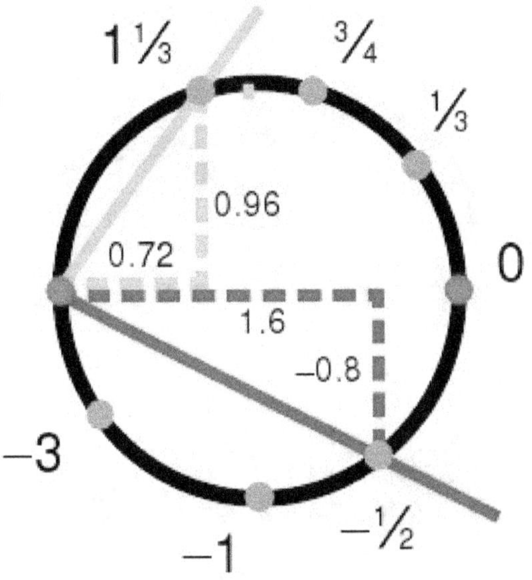

*Figure 2: A circle manifold chart based on slope, covering all but one point of the circle.*

The top, bottom, left, and right charts show that the circle is a manifold, but they do not form the only possible atlas. Charts need not be geometric projections, and the number of charts is a matter of some choice. Consider the charts

$$\chi_{minus}(x, y) = s = \frac{y}{1 + x}$$

and

$$\chi_{plus}(x, y) = t = \frac{y}{1 - x}$$

Here $s$ is the slope of the line through the point at coordinates $(x,y)$ and the fixed pivot point $(-1, 0)$; $t$ follows similarly, but with pivot point $(+1, 0)$. The inverse mapping from $s$ to $(x, y)$ is given by

$$x = \frac{1 - s^2}{1 + s^2}$$
$$y = \frac{2s}{1 + s^2}$$

It can easily be confirmed that $x^2 + y^2 = 1$ for all values of the slope $s$. These two charts provide a second atlas for the circle, with

$$t = \frac{1}{s}$$

Each chart omits a single point, either $(-1, 0)$ for $s$ or $(+1, 0)$ for $t$, so neither chart alone is sufficient to cover the whole circle. It can be proved that it is not possible to cover the full circle with a single chart. For example, although it is possible to construct a circle from a single line interval by overlapping and "gluing" the ends, this does not produce a chart; a portion of the circle will be mapped to both ends at once, losing invertibility.

## 12.1.2   Enriched circle

Viewed using calculus, the circle transition function $T$ is simply a function between open intervals, which gives a meaning to the statement that $T$ is differentiable. The transition map $T$, and all the others, are differentiable on $(0, 1)$; therefore, with this atlas the circle is a *differentiable manifold*. It is also *smooth* and *analytic* because the transition functions have these properties as well.

Other circle properties allow it to meet the requirements of more specialized types of manifold. For example, the circle has a notion of distance between two points, the arc-length between the points; hence it is a *Riemannian manifold*.

## 12.1.3   Sphere

The sphere is an example of a manifold of dimension 2. The unit sphere of implicit equation

$$x^2 + y^2 + z^2 - 1 = 0$$

may be covered by an atlas of six charts: the plane $z = 0$ divides the sphere into two half spheres ($z > 0$ and $z < 0$), which may both be mapped on the disc $x^2 + y^2 < 1$ by the projection on the $xy$ plane of coordinates. This provides two charts; the four other charts are provided by a similar construction with the two other coordinate planes.

As for the circle, one may define one chart that covers the whole sphere excluding one point. Thus two charts are sufficient, but the sphere cannot be covered by a single chart.

This example is historically significant, as it has motivated the terminology; it became apparent that the whole surface of the Earth cannot have a plane representation consisting of a single map (also called "chart", see nautical chart), and therefore one needs atlases for covering the whole Earth surface.

## 12.1.4   Other curves

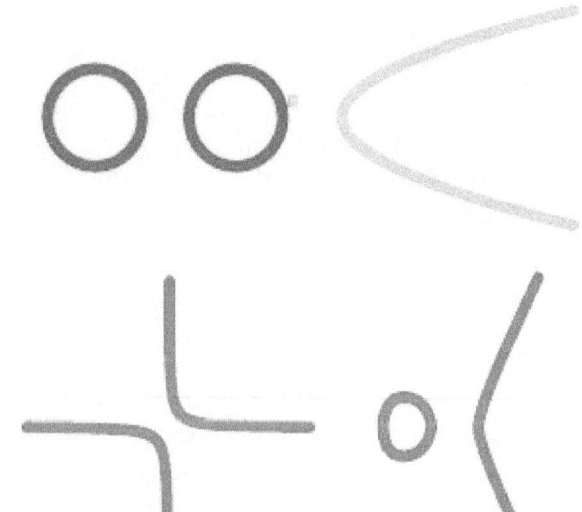

*Four manifolds from algebraic curves:*
*⬚ circles, ⬚ parabola, ⬚ hyperbola, ⬚ cubic.*

Manifolds need not be connected (all in "one piece"); an example is a pair of separate circles.

Manifolds need not be closed; thus a line segment without its end points is a manifold. And they are never countable, unless the dimension of the manifold is 0. Putting these freedoms together, other examples of manifolds are a parabola, a hyperbola (two open, infinite pieces), and the locus of points on a cubic curve $y^2 = x^3 - x$ (a closed loop piece and an open, infinite piece).

However, excluded are examples like two touching circles that share a point to form a figure-8; at the shared point a satisfactory chart cannot be created. Even with the bending allowed by topology, the vicinity of the shared point looks like a "+", not a line. A "+" is not homeomorphic to a closed interval (line segment), since deleting the center point from the "+" gives a space with four components (i.e. pieces), whereas deleting a point from a closed interval gives a space with at most two pieces; topological operations always preserve the number of pieces.

## 12.2 History

For more details on this topic, see History of manifolds and varieties.

The study of manifolds combines many important areas of mathematics: it generalizes concepts such as curves and surfaces as well as ideas from linear algebra and topology.

### 12.2.1 Early development

Before the modern concept of a manifold there were several important results.

Non-Euclidean geometry considers spaces where Euclid's parallel postulate fails. Saccheri first studied such geometries in 1733 but sought only to disprove them. Gauss, Bolyai and Lobachevsky independently discovered them 100 years later. Their research uncovered two types of spaces whose geometric structures differ from that of classical Euclidean space; these gave rise to hyperbolic geometry and elliptic geometry. In the modern theory of manifolds, these notions correspond to Riemannian manifolds with constant negative and positive curvature, respectively.

Carl Friedrich Gauss may have been the first to consider abstract spaces as mathematical objects in their own right. His theorema egregium gives a method for computing the curvature of a surface without considering the ambient space in which the surface lies. Such a surface would, in modern terminology, be called a manifold; and in modern terms, the theorem proved that the curvature of the surface is an intrinsic property. Manifold theory has come to focus exclusively on these intrinsic properties (or invariants), while largely ignoring the extrinsic properties of the ambient space.

Another, more topological example of an intrinsic property of a manifold is its Euler characteristic. Leonhard Euler showed that for a convex polytope in the three-dimensional Euclidean space with $V$ vertices (or corners), $E$ edges, and $F$ faces,

$$V - E + F = 2.$$

The same formula will hold if we project the vertices and edges of the polytope onto a sphere, creating a topological map with $V$ vertices, $E$ edges, and $F$ faces, and in fact, will remain true for any spherical map, even if it does not arise from any convex polytope.[1] Thus 2 is a topological invariant of the sphere, called its **Euler characteristic**. On the other hand, a torus can be sliced open by its 'parallel' and 'meridian' circles, creating a map with $V = 1$ vertex, $E = 2$ edges, and $F = 1$ face. Thus the Euler characteristic of the torus is $1 - 2 + 1 = 0$. The Euler characteristic of other surfaces is a useful topological invariant, which can be extended to higher dimensions using Betti numbers. In the mid nineteenth century, the Gauss–Bonnet theorem linked the Euler characteristic to the Gaussian curvature.

### 12.2.2 Synthesis

Investigations of Niels Henrik Abel and Carl Gustav Jacobi on inversion of elliptic integrals in the first half of 19th century led them to consider special types of complex manifolds, now known as Jacobians. Bernhard Riemann further contributed to their theory, clarifying the geometric meaning of the process of analytic continuation of functions of complex variables.

Another important source of manifolds in 19th century mathematics was analytical mechanics, as developed by Siméon Poisson, Jacobi, and William Rowan Hamilton. The possible states of a mechanical system are thought to be points of an abstract space, phase space in Lagrangian and Hamiltonian formalisms of classical mechanics. This space is, in fact, a high-dimensional manifold, whose dimension corresponds to the degrees of freedom of the system and where the points are specified by their generalized coordinates. For an unconstrained movement of free particles the manifold is equivalent to the Euclidean space, but various conservation laws constrain it to more complicated formations, e.g. Liouville tori. The theory of a rotating solid body, developed in the 18th century by Leonhard Euler and Joseph-Louis Lagrange, gives another example where the manifold is nontrivial. Geometrical and topological aspects of classical mechanics were emphasized by Henri Poincaré, one of the founders of topology.

Riemann was the first one to do extensive work generalizing the idea of a surface to higher dimensions. The name *manifold* comes from Riemann's original German term, *Mannigfaltigkeit*, which William Kingdon Clifford translated as "manifoldness". In his Göttingen inaugural lecture, Riemann described the set of all possible values of a variable with certain constraints as a *Mannigfaltigkeit*, because the variable can have *many* values. He distinguishes between *stetige Mannigfaltigkeit* and *diskrete Mannigfaltigkeit* (*continuous manifoldness* and *discontinuous manifoldness*), depending on whether the value changes continuously or not. As continuous examples, Riemann refers to not only colors and the locations of objects in space, but also the possible shapes of a spatial figure. Using induction, Riemann constructs an *n-fach ausgedehnte Mannigfaltigkeit* (*n times extended manifoldness* or *n-dimensional manifoldness*) as a continuous stack of (n−1) dimensional manifoldnesses. Riemann's intuitive notion of a *Mannigfaltigkeit* evolved into what is today formalized as a manifold. Riemannian manifolds and Riemann surfaces are named after Riemann.

### 12.2.3   Poincaré's definition

In his very influential paper, Analysis Situs,[2] Henri Poincaré gave a definition of a (differentiable) manifold (*variété*) which served as a precursor to the modern concept of a manifold.[3]

In the first section of Analysis Situs, Poincaré defines a manifold as the level set of a continuously differentiable function between Euclidean spaces that satisfies the non-degeneracy hypothesis of the implicit function theorem. In the third section, he begins by remarking that the graph of a continuously differentiable function is a manifold in the latter sense. He then proposes a new, more general, definition of manifold based on a 'chain of manifolds' (*une chaîne des variétés*).

Poincaré's notion of a 'chain of manifolds' is a precursor to the modern notion of atlas. In particular, he considers two manifolds defined respectively as graphs of functions $\theta(y)$ and $\theta'(y')$ . If these manifolds overlap (*a une partie commune*), then he requires that the coordinates $y$ depend continuously differentiably on the coordinates $y'$ and vice versa ('...*les  y sont fonctions analytiques des  y' et inversement*'). In this way he introduces a precursor to the notion of a chart and of a transition map. Note that it is implicit in Analysis Situs that a manifold obtained as a 'chain' is a subset of Euclidean space.

For example, the unit circle in the plane can be thought of as the graph of the function $y = \sqrt{1 - x^2}$ or else the function $y = -\sqrt{1 - x^2}$ in a neighborhood of every point except the points $(1,0)$ and $(-1,0)$; and in a neighborhood of those points, it can be thought of as the graph of, respectively, $x = \sqrt{1 - y^2}$ and $x = -\sqrt{1 - y^2}$ . The reason the circle can be represented by a graph in the neighborhood of every point is because the left hand side of its defining equation $x^2 + y^2 - 1 = 0$ has nonzero gradient at every point of the circle. By the implicit function theorem, every submanifold of Euclidean space is locally the graph of a function.

Hermann Weyl gave an intrinsic definition for differentiable manifolds in his lecture course on Riemann surfaces in 1911–1912, opening the road to the general concept of a topological space that followed shortly. During the 1930s Hassler Whitney and others clarified the foundational aspects of the subject, and thus intuitions dating back to the latter half of the 19th century became precise, and developed through differential geometry and Lie group theory. Notably, the Whitney embedding theorem[4] showed that the intrinsic definition in terms of charts was equivalent to Poincaré's definition in terms of subsets of Euclidean space.

### 12.2.4   Topology of manifolds: highlights

Two-dimensional manifolds, also known as a 2D *surfaces* embedded in our common 3D space, were considered by Riemann under the guise of Riemann surfaces, and rigorously classified in the beginning of the 20th century by Poul Heegaard and Max Dehn. Henri Poincaré pioneered the study of three-dimensional manifolds and raised a fundamental question about them, today known as the Poincaré conjecture. After nearly a century of effort by many mathematicians, starting with Poincaré himself, a consensus among experts (as of 2006) is that Grigori Perelman has proved the Poincaré conjecture (see the Solution of the Poincaré conjecture). William Thurston's geometrization program, formulated in the 1970s, provided a far-reaching extension of the Poincaré conjecture to the general three-dimensional manifolds. Four-dimensional manifolds were brought to the forefront of mathematical research in the 1980s by Michael Freedman and in a different setting, by Simon Donaldson, who was motivated by the then recent progress in theoretical physics (Yang–Mills theory), where they serve as a substitute for ordinary 'flat' spacetime. Andrey Markov Jr. showed in 1960 that no algorithm exists for classifying four-dimensional manifolds. Important work on higher-dimensional manifolds, including analogues of the Poincaré conjecture, had been done earlier by René Thom, John Milnor, Stephen Smale and Sergei Novikov. One of the most pervasive and flexible techniques underlying much work on the topology of manifolds is Morse theory.

## 12.3   Mathematical definition

For more details on this topic, see Categories of manifolds.

Informally, a manifold is a space that is "modeled on" Euclidean space.

There are many different kinds of manifolds and generalizations. In geometry and topology, all manifolds are topological manifolds, possibly with additional structure, most often a differentiable structure. In terms of constructing manifolds via patching, a manifold has an additional structure if the transition maps between different patches satisfy axioms beyond just continuity. For instance, differentiable manifolds have homeomorphisms on overlapping neighborhoods diffeomorphic with each other, so that the manifold has a well-defined set of functions which are differentiable in each neighborhood, and so differentiable on the manifold as a whole.

Formally, a **topological manifold**[5] is a second countable Hausdorff space that is locally homeomorphic to Euclidean space.

*Second countable* and *Hausdorff* are point-set conditions; *second countable* excludes spaces which are in some sense 'too large' such as the long line, while *Hausdorff* excludes spaces such as "the line with two origins" (these generalizations of manifolds are discussed in non-Hausdorff manifolds).

*Locally homeomorphic* to Euclidean space means[6] that every point has a neighborhood homeomorphic to an open Euclidean *n*-ball,

$$\mathbf{B}^n = \{(x_1, x_2, \ldots, x_n) \in \mathbb{R}^n \mid x_1^2 + x_2^2 + \cdots + x_n^2 < 1\}.$$

Generally manifolds are taken to have a fixed dimension (the space must be locally homeomorphic to a fixed *n*-ball), and such a space is called an **n-manifold**; however, some authors admit manifolds where different points can have different dimensions.[7] If a manifold has a fixed dimension, it is called a **pure manifold**. For example, the sphere has a constant dimension of 2 and is therefore a pure manifold whereas the disjoint union of a sphere and a line in three-dimensional space is *not* a pure manifold. Since dimension is a local invariant (i.e. the map sending each point to the dimension of its neighbourhood over which a chart is defined, is locally constant), each connected component has a fixed dimension.

Scheme-theoretically, a manifold is a locally ringed space, whose structure sheaf is locally isomorphic to the sheaf of continuous (or differentiable, or complex-analytic, etc.) functions on Euclidean space. This definition is mostly used when discussing analytic manifolds in algebraic geometry.

### 12.3.1   Broad definition

Main article: Banach manifold

The broadest common definition of manifold is a topological space locally homeomorphic to a topological vector space over the reals. This omits the point-set axioms, allowing higher cardinalities and non-Hausdorff manifolds; and it omits finite dimension, allowing structures such as Hilbert manifolds to be modeled on Hilbert spaces, Banach manifolds to be modeled on Banach spaces, and Fréchet manifolds to be modeled on Fréchet spaces. Usually one relaxes one or the other condition: manifolds with the point-set axioms are studied in general topology, while infinite-dimensional manifolds are studied in functional analysis.

## 12.4   Charts, atlases, and transition maps

Main article: Atlas (topology)
See also: Differentiable manifold

The spherical Earth is navigated using flat maps or charts, collected in an atlas. Similarly, a differentiable manifold can be described using mathematical maps, called *coordinate charts*, collected in a mathematical *atlas*. It is not generally possible to describe a manifold with just one chart, because the global structure of the manifold is different from the simple structure of the charts. For example, no single flat map can represent the entire Earth without separation of adjacent features across the map's boundaries or duplication of coverage. When a manifold is constructed from multiple overlapping charts, the regions where they overlap carry information essential to understanding the global structure.

### 12.4.1   Charts

Main article: Coordinate chart

A **coordinate map**, a **coordinate chart**, or simply a **chart**, of a manifold is an invertible map between a subset of the manifold and a simple space such that both the map and its inverse preserve the desired structure.[8] For a topological manifold, the simple space is some Euclidean space $\mathbf{R}^n$ and interest focuses on the topological structure. This structure is preserved by homeomorphisms, invertible maps that are continuous in both directions.

In the case of a differentiable manifold, a set of **charts** called an **atlas** allows us to do calculus on manifolds. Polar coordinates, for example, form a chart for the plane $\mathbf{R}^2$ minus the positive *x*-axis and the origin. Another example of a chart is the map $\chi_{\text{top}}$ mentioned in the section above, a chart for the circle.

### 12.4.2   Atlases

Main article: Atlas (topology)

The description of most manifolds requires more than one chart (a single chart is adequate for only the simplest manifolds). A specific collection of charts which covers a manifold is called an atlas. An atlas is not unique as all manifolds can be covered multiple ways using different combinations of charts. Two atlases are said to be $C^k$-equivalent if their union is also a $C^k$ atlas.

The atlas containing all possible charts consistent with a given atlas is called the **maximal atlas** (i.e. an equivalence class containing that given atlas (under the already defined equivalence relation given in the previous paragraph)). Unlike an ordinary atlas, the maximal atlas of a given manifold is unique. Though it is useful for definitions, it is an abstract object and not used directly (e.g. in calculations).

### 12.4.3 Transition maps

Charts in an atlas may overlap and a single point of a manifold may be represented in several charts. If two charts overlap, parts of them represent the same region of the manifold, just as a map of Europe and a map of Asia may both contain Moscow. Given two overlapping charts, a **transition function** can be defined which goes from an open ball in $\mathbf{R}^n$ to the manifold and then back to another (or perhaps the same) open ball in $\mathbf{R}^n$. The resultant map, like the map $T$ in the circle example above, is called a **change of coordinates**, a **coordinate transformation**, a **transition function**, or a **transition map**.

### 12.4.4 Additional structure

An atlas can also be used to define additional structure on the manifold. The structure is first defined on each chart separately. If all the transition maps are compatible with this structure, the structure transfers to the manifold.

This is the standard way differentiable manifolds are defined. If the transition functions of an atlas for a topological manifold preserve the natural differential structure of $\mathbf{R}^n$ (that is, if they are diffeomorphisms), the differential structure transfers to the manifold and turns it into a differentiable manifold. Complex manifolds are introduced in an analogous way by requiring that the transition functions of an atlas are holomorphic functions. For symplectic manifolds, the transition functions must be symplectomorphisms.

The structure on the manifold depends on the atlas, but sometimes different atlases can be said to give rise to the same structure. Such atlases are called **compatible**.

These notions are made precise in general through the use of pseudogroups.

## 12.5 Manifold with boundary

See also: Topological manifold § Manifolds with boundary

A **manifold with boundary** is a manifold with an edge.

For example, a sheet of paper is a 2-manifold with a 1-dimensional boundary. The boundary of an $n$-manifold with boundary is an $(n - 1)$-manifold. A disk (circle plus interior) is a 2-manifold with boundary. Its boundary is a circle, a 1-manifold. A square with interior is also a 2-manifold with boundary. A ball (sphere plus interior) is a 3-manifold with boundary. Its boundary is a sphere, a 2-manifold. (See also Boundary (topology)).

In technical language, a manifold with boundary is a space containing both interior points and boundary points. Every interior point has a neighborhood homeomorphic to the open $n$-ball $\{(x_1, x_2, ..., xn) \mid \Sigma\, xi^2 < 1\}$. Every boundary point has a neighborhood homeomorphic to the "half" $n$-ball $\{(x_1, x_2, ..., xn) \mid \Sigma\, xi^2 < 1 \text{ and } x_1 \geq 0\}$. The homeomorphism must send each boundary point to a point with $x_1 = 0$.

### 12.5.1 Boundary and interior

Let $M$ be a manifold with boundary. The **interior** of $M$, denoted Int $M$, is the set of points in $M$ which have neighborhoods homeomorphic to an open subset of $\mathbf{R}^n$. The **boundary** of $M$, denoted $\partial M$, is the complement of Int $M$ in $M$. The boundary points can be characterized as those points which land on the boundary hyperplane ($xn = 0$) of $\mathbf{R}^n{}_+$ under some coordinate chart.

If $M$ is a manifold with boundary of dimension $n$, then Int $M$ is a manifold (without boundary) of dimension $n$ and $\partial M$ is a manifold (without boundary) of dimension $n - 1$.

## 12.6 Construction

A single manifold can be constructed in different ways, each stressing a different aspect of the manifold, thereby leading to a slightly different viewpoint.

### 12.6.1 Charts

Perhaps the simplest way to construct a manifold is the one used in the example above of the circle. First, a subset of $\mathbf{R}^2$ is identified, and then an atlas covering this subset is constructed. The concept of *manifold* grew historically from constructions like this. Here is another example, applying this method to the construction of a sphere:

**Sphere with charts**

A sphere can be treated in almost the same way as the circle. In mathematics a sphere is just the surface (not the solid interior), which can be defined as a subset of $\mathbf{R}^3$:

*The chart maps the part of the sphere with positive z coordinate to a disc.*

## 12.6.2 Patchwork

For more details on this topic, see Surgery theory.

A manifold can be constructed by gluing together pieces in a consistent manner, making them into overlapping charts. This construction is possible for any manifold and hence it is often used as a characterisation, especially for differentiable and Riemannian manifolds. It focuses on an atlas, as the patches naturally provide charts, and since there is no exterior space involved it leads to an intrinsic view of the manifold.

The manifold is constructed by specifying an atlas, which is itself defined by transition maps. A point of the manifold is therefore an equivalence class of points which are mapped to each other by transition maps. Charts map equivalence classes to points of a single patch. There are usually strong demands on the consistency of the transition maps. For topological manifolds they are required to be homeomorphisms; if they are also diffeomorphisms, the resulting manifold is a differentiable manifold.

This can be illustrated with the transition map $t = {}^1/s$ from the second half of the circle example. Start with two copies of the line. Use the coordinate $s$ for the first copy, and $t$ for the second copy. Now, glue both copies together by identifying the point $t$ on the second copy with the point $s = {}^1/t$ on the first copy (the points $t = 0$ and $s = 0$ are not identified with any point on the first and second copy, respectively). This gives a circle.

$$S = \{(x, y, z) \in \mathbf{R}^3 \mid x^2 + y^2 + z^2 = 1\}.$$

The sphere is two-dimensional, so each chart will map part of the sphere to an open subset of $\mathbf{R}^2$. Consider the northern hemisphere, which is the part with positive $z$ coordinate (coloured red in the picture on the right). The function $\chi$ defined by

$$\chi(x, y, z) = (x, y),$$

maps the northern hemisphere to the open unit disc by projecting it on the $(x, y)$ plane. A similar chart exists for the southern hemisphere. Together with two charts projecting on the $(x, z)$ plane and two charts projecting on the $(y, z)$ plane, an atlas of six charts is obtained which covers the entire sphere.

This can be easily generalized to higher-dimensional spheres.

### Intrinsic and extrinsic view

The first construction and this construction are very similar, but they represent rather different points of view. In the first construction, the manifold is seen as embedded in some Euclidean space. This is the *extrinsic view*. When a manifold is viewed in this way, it is easy to use intuition from Euclidean spaces to define additional structure. For example, in a Euclidean space it is always clear whether a vector at some point is tangential or normal to some surface through that point.

The patchwork construction does not use any embedding, but simply views the manifold as a topological space by itself. This abstract point of view is called the *intrinsic view*. It can make it harder to imagine what a tangent vector might be, and there is no intrinsic notion of a normal bundle, but instead there is an intrinsic stable normal bundle.

### *n*-Sphere as a patchwork

The *n*-sphere $S^n$ is a generalisation of the idea of a circle (1-sphere) and sphere (2-sphere) to higher dimensions. An *n*-sphere $S^n$ can be constructed by gluing together two copies of $\mathbf{R}^n$. The transition map between them is defined as

$$\mathbf{R}^n \setminus \{0\} \to \mathbf{R}^n \setminus \{0\} : x \mapsto x/\|x\|^2.$$

This function is its own inverse and thus can be used in both directions. As the transition map is a smooth function, this atlas defines a smooth manifold. In the case $n = 1$, the example simplifies to the circle example given earlier.

### 12.6.3 Identifying points of a manifold

Main articles: Orbifold and Group action

It is possible to define different points of a manifold to be same. This can be visualized as gluing these points together in a single point, forming a quotient space. There is, however, no reason to expect such quotient spaces to be manifolds. Among the possible quotient spaces that are not necessarily manifolds, orbifolds and CW complexes are considered to be relatively well-behaved. An example of a quotient space of a manifold that is also a manifold is the real projective space identified as a quotient space of the corresponding sphere.

One method of identifying points (gluing them together) is through a right (or left) action of a group, which acts on the manifold. Two points are identified if one is moved onto the other by some group element. If $M$ is the manifold and $G$ is the group, the resulting quotient space is denoted by $M / G$ (or $G \setminus M$).

Manifolds which can be constructed by identifying points include tori and real projective spaces (starting with a plane and a sphere, respectively).

### 12.6.4 Gluing along boundaries

Main article: Quotient space (topology)

Two manifolds with boundaries can be glued together along a boundary. If this is done the right way, the result is also a manifold. Similarly, two boundaries of a single manifold can be glued together.

Formally, the gluing is defined by a bijection between the two boundaries. Two points are identified when they are mapped onto each other. For a topological manifold this bijection should be a homeomorphism, otherwise the result

will not be a topological manifold. Similarly for a differentiable manifold it has to be a diffeomorphism. For other manifolds other structures should be preserved.

A finite cylinder may be constructed as a manifold by starting with a strip $[0, 1] \times [0, 1]$ and gluing a pair of opposite edges on the boundary by a suitable diffeomorphism. A projective plane may be obtained by gluing a sphere with a hole in it to a Möbius strip along their respective circular boundaries.

### 12.6.5 Cartesian products

The Cartesian product of manifolds is also a manifold.

The dimension of the product manifold is the sum of the dimensions of its factors. Its topology is the product topology, and a Cartesian product of charts is a chart for the product manifold. Thus, an atlas for the product manifold can be constructed using atlases for its factors. If these atlases define a differential structure on the factors, the corresponding atlas defines a differential structure on the product manifold. The same is true for any other structure defined on the factors. If one of the factors has a boundary, the product manifold also has a boundary. Cartesian products may be used to construct tori and finite cylinders, for example, as $S^1 \times S^1$ and $S^1 \times [0, 1]$, respectively.

*A finite cylinder is a manifold with boundary.*

## 12.7 Manifolds with additional structure

Main article: Categories of manifolds

### 12.7.1 Topological manifolds

Main article: topological manifold

The simplest kind of manifold to define is the topological manifold, which looks locally like some "ordinary" Euclidean space $\mathbf{R}^n$. Formally, a topological manifold is a topological space locally homeomorphic to a Euclidean space. This means that every point has a neighbourhood for which there exists a homeomorphism (a bijective continuous function whose inverse is also continuous) mapping that neighbourhood to $\mathbf{R}^n$. These homeomorphisms are the charts of the manifold.

It is to be noted that a *topological* manifold looks locally like a Euclidean space in a rather weak manner: while for each individual chart it is possible to distinguish differentiable functions or measure distances and angles, merely by virtue of being a topological manifold a space does not have any *particular* and *consistent* choice of such concepts. In order to discuss such properties for a manifold, one needs to specify further structure and consider differentiable manifolds and Riemannian manifolds discussed below. In particular, the same underlying topological manifold can have several mutually incompatible classes of differentiable functions and an infinite number of ways to specify distances and angles.

Usually additional technical assumptions on the topological space are made to exclude pathological cases. It is customary to require that the space be Hausdorff and second countable.

The *dimension* of the manifold at a certain point is the dimension of the Euclidean space that the charts at that point map to (number $n$ in the definition). All points in a connected manifold have the same dimension. Some authors require that all charts of a topological manifold map to Euclidean spaces of same dimension. In that case every topological manifold has a topological invariant, its dimension. Other authors allow disjoint unions of topological manifolds with differing dimensions to be called manifolds.

### 12.7.2 Differentiable manifolds

Main article: Differentiable manifold

For most applications a special kind of topological manifold, namely a **differentiable manifold**, is used. If the local charts on a manifold are compatible in a certain sense, one can define directions, tangent spaces, and differentiable functions on that manifold. In particular it is possible to use calculus on a differentiable manifold. Each point of an $n$-dimensional differentiable manifold has a tangent space.

This is an $n$-dimensional Euclidean space consisting of the tangent vectors of the curves through the point.

Two important classes of differentiable manifolds are **smooth** and **analytic manifolds**. For smooth manifolds the transition maps are smooth, that is infinitely differentiable. Analytic manifolds are smooth manifolds with the additional condition that the transition maps are analytic (they can be expressed as power series). The sphere can be given analytic structure, as can most familiar curves and surfaces.

There are also topological manifolds, i.e., locally Euclidean spaces, which possess no differentiable structures at all.[9]

A rectifiable set generalizes the idea of a piecewise smooth or rectifiable curve to higher dimensions; however, rectifiable sets are not in general manifolds.

### 12.7.3 Riemannian manifolds

Main article: Riemannian manifold

To measure distances and angles on manifolds, the manifold must be Riemannian. A 'Riemannian manifold' is a differentiable manifold in which each tangent space is equipped with an inner product $\langle \cdot, \cdot \rangle$ in a manner which varies smoothly from point to point. Given two tangent vectors $\mathbf{u}$ and $\mathbf{v}$, the inner product $\langle \mathbf{u}, \mathbf{v} \rangle$ gives a real number. The dot (or scalar) product is a typical example of an inner product. This allows one to define various notions such as length, angles, areas (or volumes), curvature and divergence of vector fields.

All differentiable manifolds (of constant dimension) can be given the structure of a Riemannian manifold. The Euclidean space itself carries a natural structure of Riemannian manifold (the tangent spaces are naturally identified with the Euclidean space itself and carry the standard scalar product of the space). Many familiar curves and surfaces, including for example all $n$-spheres, are specified as subspaces of a Euclidean space and inherit a metric from their embedding in it.

### 12.7.4 Finsler manifolds

Main article: Finsler manifold

A **Finsler manifold** allows the definition of distance but does not require the concept of angle; it is an analytic manifold in which each tangent space is equipped with a norm, $\|\cdot\|$, in a manner which varies smoothly from point to point. This norm can be extended to a metric, defining the length of a curve; but it cannot in general be used to define an inner product.

Any Riemannian manifold is a Finsler manifold.

### 12.7.5 Lie groups

Main article: Lie group

**Lie groups**, named after Sophus Lie, are differentiable manifolds that carry also the structure of a group which is such that the group operations are defined by smooth maps.

A Euclidean vector space with the group operation of vector addition is an example of a non-compact Lie group. A simple example of a compact Lie group is the circle: the group operation is simply rotation. This group, known as U(1), can be also characterised as the group of complex numbers of modulus 1 with multiplication as the group operation. Other examples of Lie groups include special groups of matrices, which are all subgroups of the general linear group, the group of $n$ by $n$ matrices with non-zero determinant. If the matrix entries are real numbers, this will be an $n^2$-dimensional disconnected manifold. The orthogonal groups, the symmetry groups of the sphere and hyperspheres, are $n(n-1)/2$ dimensional manifolds, where $n-1$ is the dimension of the sphere. Further examples can be found in the table of Lie groups.

### 12.7.6 Other types of manifolds

Main articles: Complex manifold and Symplectic manifold

- A 'complex manifold' is a manifold modeled on $\mathbb{C}^n$ with holomorphic transition functions on chart overlaps. These manifolds are the basic objects of study in complex geometry. A one-complex-dimensional manifold is called a Riemann surface. Note that an $n$-dimensional complex manifold has dimension $2n$ as a real differentiable manifold.

- A 'CR manifold' is a manifold modeled on boundaries of domains in $\mathbb{C}^n$ .

- 'Infinite dimensional manifolds': to allow for infinite dimensions, one may consider Banach manifolds which are locally homeomorphic to Banach spaces. Similarly, Fréchet manifolds are locally homeomorphic to Fréchet spaces.

- A 'symplectic manifold' is a kind of manifold which is used to represent the phase spaces in classical mechanics. They are endowed with a 2-form that defines the Poisson bracket. A closely related type of manifold is a contact manifold.

- A 'combinatorial manifold' is a kind of manifold which is discretization of a manifold. It usually means a piecewise linear manifold made by simplicial complexes.

- A 'digital manifold' is a special kind of combinatorial manifold which is defined in digital space. See digital topology

## 12.8 Classification and invariants

For more details on this topic, see Classification of manifolds.

Different notions of manifolds have different notions of classification and invariant; in this section we focus on smooth closed manifolds.

The classification of smooth closed manifolds is well-understood *in principle*, except in dimension 4: in low dimensions (2 and 3) it is geometric, via the uniformization theorem and the solution of the Poincaré conjecture, and in high dimension (5 and above) it is algebraic, via surgery theory. This is a classification in principle: the general question of whether two smooth manifolds are diffeomorphic is not computable in general. Further, specific computations remain difficult, and there are many open questions.

Orientable surfaces can be visualized, and their diffeomorphism classes enumerated, by genus. Given two orientable surfaces, one can determine if they are diffeomorphic by computing their respective genera and comparing: they are diffeomorphic if and only if the genera are equal, so the genus forms a complete set of invariants.

This is much harder in higher dimensions: higher-dimensional manifolds cannot be directly visualized (though visual intuition is useful in understanding them), nor can their diffeomorphism classes be enumerated, nor can one in general determine if two different descriptions of a higher-dimensional manifold refer to the same object.

However, one can determine if two manifolds are *different* if there is some intrinsic characteristic that differentiates them. Such criteria are commonly referred to as invariants, because, while they may be defined in terms of some presentation (such as the genus in terms of a triangulation), they are the same relative to all possible descriptions of a particular manifold: they are *invariant* under different descriptions.

Naively, one could hope to develop an arsenal of invariant criteria that would definitively classify all manifolds up to isomorphism. Unfortunately, it is known that for manifolds of dimension 4 and higher, no program exists that can decide whether two manifolds are diffeomorphic.

Smooth manifolds have a rich set of invariants, coming from point-set topology, classic algebraic topology, and geometric topology. The most familiar invariants, which are visible for surfaces, are orientability (a normal invariant, also detected by homology) and genus (a homological invariant).

Smooth closed manifolds have no local invariants (other than dimension), though geometric manifolds have local invariants, notably the curvature of a Riemannian manifold and the torsion of a manifold equipped with an affine connection. This distinction between local invariants and no local invariants is a common way to distinguish between geometry and topology. All invariants of a smooth closed manifold are thus global.

Algebraic topology is a source of a number of important global invariant properties. Some key criteria include the *simply connected* property and orientability (see below). Indeed, several branches of mathematics, such as homology and homotopy theory, and the theory of characteristic classes were founded in order to study invariant properties of manifolds.

## 12.9   Examples of surfaces

### 12.9.1   Orientability

Main article: Orientable manifold

In dimensions two and higher, a simple but important invariant criterion is the question of whether a manifold admits a meaningful orientation. Consider a topological manifold with charts mapping to $\mathbf{R}^n$. Given an ordered basis for $\mathbf{R}^n$, a chart causes its piece of the manifold to itself acquire a sense of ordering, which in 3-dimensions can be viewed as either right-handed or left-handed. Overlapping charts are not required to agree in their sense of ordering, which gives manifolds an important freedom. For some manifolds, like the sphere, charts can be chosen so that overlapping regions agree on their "handedness"; these are *orientable* manifolds. For others, this is impossible. The latter possibility is easy to overlook, because any closed surface embedded (without self-intersection) in three-dimensional space is orientable.

Some illustrative examples of non-orientable manifolds include: (1) the Möbius strip, which is a manifold with boundary, (2) the Klein bottle, which must intersect itself in its 3-space representation, and (3) the real projective plane, which arises naturally in geometry.

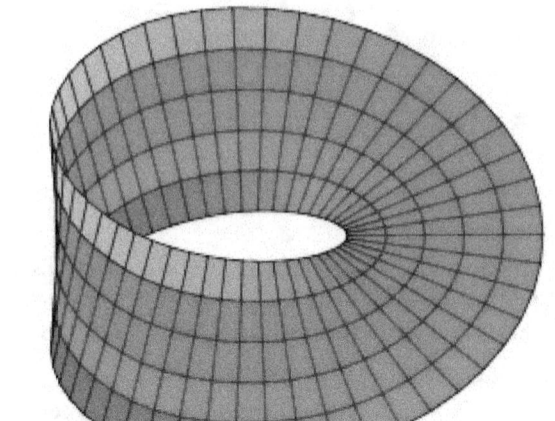

*Möbius strip*

## Möbius strip

Main article: Möbius strip

Begin with an infinite circular cylinder standing vertically, a manifold without boundary. Slice across it high and low to produce two circular boundaries, and the cylindrical strip between them. This is an orientable manifold with boundary, upon which "surgery" will be performed. Slice the strip open, so that it could unroll to become a rectangle, but keep a grasp on the cut ends. Twist one end 180°, making the inner surface face out, and glue the ends back together seamlessly. This results in a strip with a permanent half-twist: the Möbius strip. Its boundary is no longer a pair of circles, but (topologically) a single circle; and what was once its "inside" has merged with its "outside", so that it now has only a *single* side.

## Klein bottle

Main article: Klein bottle

Take two Möbius strips; each has a single loop as a boundary. Straighten out those loops into circles, and let the strips distort into cross-caps. Gluing the circles together will produce a new, closed manifold without boundary, the Klein bottle. Closing the surface does nothing to improve the lack of orientability, it merely removes the boundary. Thus, the Klein bottle is a closed surface with no distinction between inside and outside. Note that in three-dimensional space, a Klein bottle's surface must pass through itself. Building a Klein bottle which is not self-intersecting requires four or more dimensions of space.

## 12.9.2 Genus and the Euler characteristic

For two dimensional manifolds a key invariant property is the genus, or the "number of handles" present in a surface. A torus is a sphere with one handle, a double torus is a sphere with two handles, and so on. Indeed, it is possible to fully characterize compact, two-dimensional manifolds on the basis of genus and orientability. In higher-dimensional manifolds genus is replaced by the notion of Euler characteristic, and more generally Betti numbers and homology and cohomology.

## 12.10 Maps of manifolds

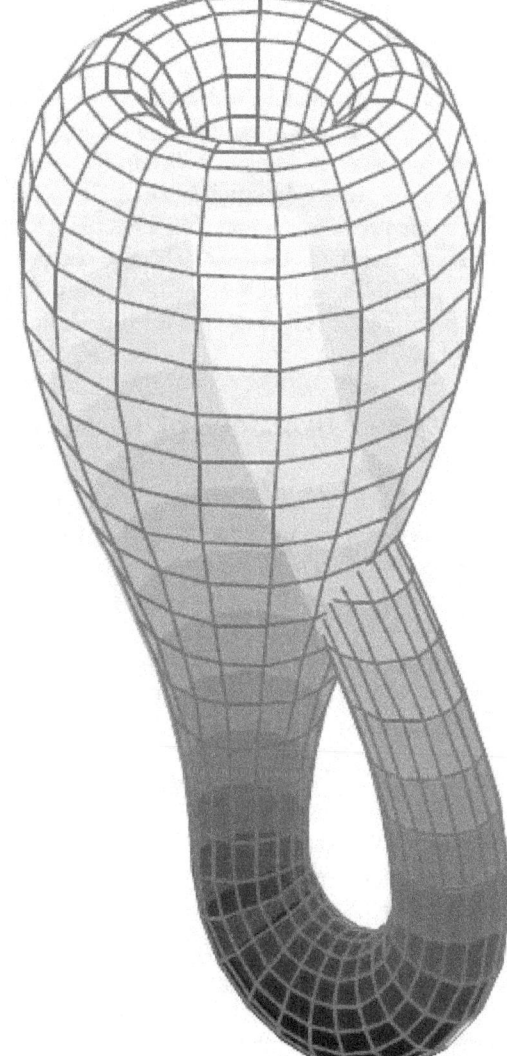

*The Klein bottle immersed in three-dimensional space*

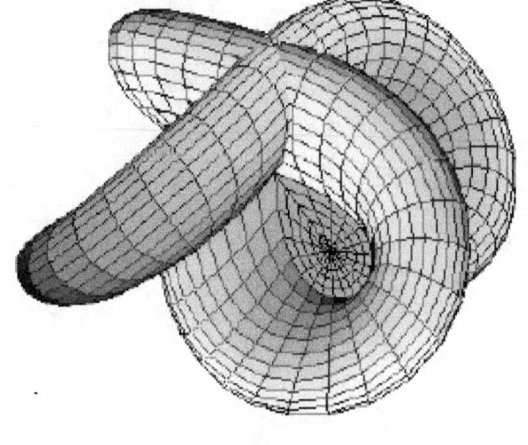

*A Morin surface, an immersion used in sphere eversion*

Main article: Maps of manifolds

Just as there are various types of manifolds, there are various types of maps of manifolds. In addition to continuous functions and smooth functions generally, there are maps with special properties. In geometric topology a basic type are embeddings, of which knot theory is a central example, and generalizations such as immersions, submersions, covering spaces, and ramified covering spaces. Basic results include the Whitney embedding theorem and Whitney immersion theorem.

In Riemannian geometry, one may ask for maps to preserve the Riemannian metric, leading to notions of isometric embeddings, isometric immersions, and Riemannian submersions; a basic result is the Nash embedding theorem.

**Real projective plane**

Main article: Real projective space

Begin with a sphere centered on the origin. Every line through the origin pierces the sphere in two opposite points called *antipodes*. Although there is no way to do so physically, it is possible (by considering a quotient space) to mathematically merge each antipode pair into a single point. The closed surface so produced is the real projective plane, yet another non-orientable surface. It has a number of equivalent descriptions and constructions, but this route explains its name: all the points on any given line through the origin project to the same "point" on this "plane".

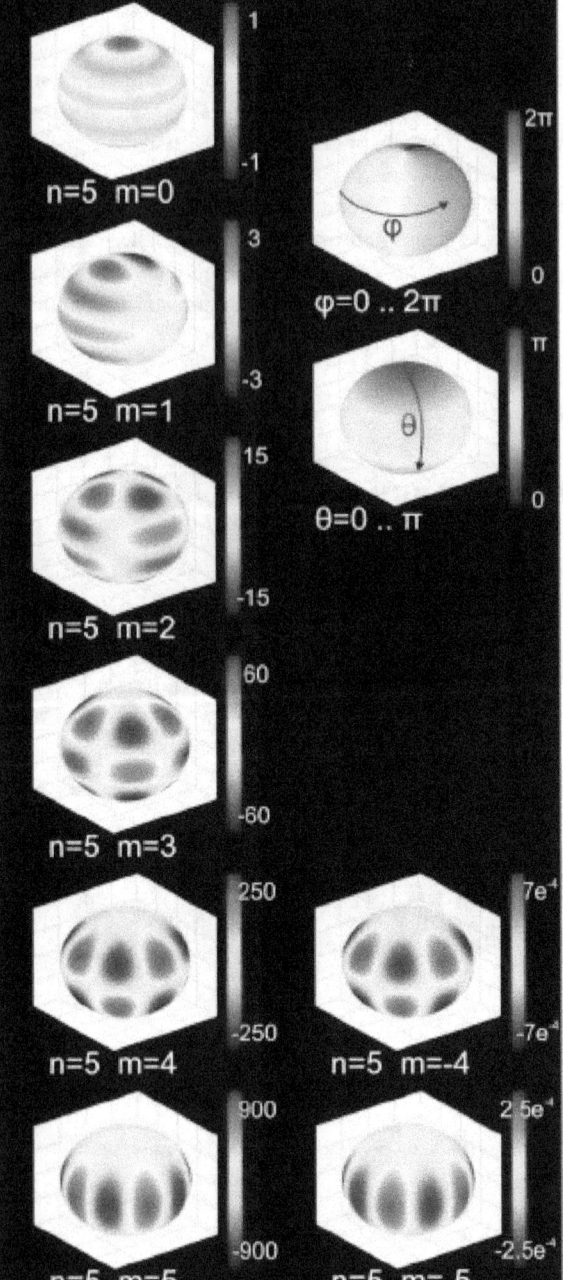

*3D color plot of the spherical harmonics of degree* $n = 5$

### 12.10.1   Scalar-valued functions

A basic example of maps between manifolds are scalar-valued functions on a manifold,

$$f : M \to \mathbf{R} \text{ or } f : M \to \mathbf{C},$$

sometimes called regular functions or functionals, by analogy with algebraic geometry or linear algebra. These are of interest both in their own right, and to study the underlying manifold.

In geometric topology, most commonly studied are Morse functions, which yield handlebody decompositions, while in mathematical analysis, one often studies solution to partial differential equations, an important example of which is harmonic analysis, where one studies harmonic functions: the kernel of the Laplace operator. This leads to such functions as the spherical harmonics, and to heat kernel methods of studying manifolds, such as hearing the shape of a drum and some proofs of the Atiyah–Singer index theorem.

## 12.11   Generalizations of manifolds

- **Orbifolds**: An orbifold is a generalization of manifold allowing for certain kinds of "singularities" in the topology. Roughly speaking, it is a space which locally looks like the quotients of some simple space (*e.g.* Euclidean space) by the actions of various finite groups. The singularities correspond to fixed points of the group actions, and the actions must be compatible in a certain sense.

- **Algebraic varieties and schemes**: Non-singular algebraic varieties over the real or complex numbers are manifolds. One generalizes this first by allowing singularities, secondly by allowing different fields, and thirdly by emulating the patching construction of manifolds: just as a manifold is glued together from open subsets of Euclidean space, an algebraic variety is glued together from affine algebraic varieties, which are zero sets of polynomials over algebraically closed fields. Schemes are likewise glued together from affine schemes, which are a generalization of algebraic varieties. Both are related to manifolds, but are constructed algebraically using sheaves instead of atlases.

  Because of singular points, a variety is in general not a manifold, though linguistically the French *variété*, German *Mannigfaltigkeit* and English *manifold* are largely synonymous. In French an algebraic variety is called *une variété algébrique* (an *algebraic variety*), while a smooth manifold is called *une variété différentielle* (a *differential variety*).

- **Stratified space**: A "stratified space" is a space that can be divided into pieces ("strata"), with each stratum a manifold, with the strata fitting together in prescribed ways (formally, a filtration by closed subsets). There are various technical definitions, notably a **Whitney stratified space** (see Whitney conditions) for smooth manifolds and a topologically stratified space for topological manifolds. Basic examples include manifold with boundary (top dimensional manifold and codimension 1 boundary) and manifold with corners (top dimensional manifold, codimension 1 boundary, codimension 2 corners). Whitney stratified spaces are a broad class of spaces, including algebraic varieties, analytic varieties, semialgebraic sets, and subanalytic sets.

- **CW-complexes**: A CW complex is a topological space formed by gluing disks of different dimensionality together. In general the resulting space is singular, and hence not a manifold. However, they are of central interest in algebraic topology, especially in homotopy theory, as they are easy to compute with and singularities are not a concern.

- **Homology manifolds**: A homology manifold is a space that behaves like a manifold from the point of view of homology theory. These are not all manifolds, but (in high dimension) can be analyzed by surgery theory similarly to manifolds, and failure to be a manifold is a local obstruction, as in surgery theory.[10]

- **Differential spaces**: Let $M$ be a nonempty set. Suppose that some family of real functions on $M$ was chosen. Denote it by $C \subseteq \mathbb{R}^M$. It is an algebra with respect to the pointwise addition and multiplication. Let $M$ be equipped with the topology induced by $C$. Suppose also that the following conditions hold. First: for every $H \in C^\infty(\mathbb{R}^i)$, where $i \in \mathbb{N}$, and arbitrary $f_1, \ldots, f_n \in C$, the composition $H \circ (f_1, \ldots, f_n) \in C$. Second: every function, which in every point of $M$ locally coincides with some function from $C$, also belongs to $C$. A pair $(M, C)$ for which the above conditions hold, is called a Sikorski differential space.[11][12]

## 12.12 See also

- Affine geodesic: paths on manifolds

- Directional statistics: statistics on manifolds

- List of manifolds

- Mathematics of general relativity

- Submanifold

### 12.12.1 By dimension

- Curve (1-manifold)

- Surface (2-manifold)

- 3-manifold

- 4-manifold

- 5-manifold

- Banach manifold

- Fréchet manifold

- Manifolds of mappings

## 12.13 Notes

[1] The notion of a map can formalized as a cell decomposition.

[2] Poincaré, H.: Analysis Situs. (French) *Journal de l'Ecole Polytechnique*, Serié 11 Gauthier-Villars (1895).

[3] Arnol'd, V. I.: On the teaching of mathematics.(Russian) Uspekhi Mat. Nauk 53 (1998), no. 1(319), 229–234; translation in Russian Math. Surveys 53 (1998), no. 1, 229–236

[4] Whitney H., *Differentiable manifolds*, Ann. of Math. (2), *37* (1936), 645–680.

[5] In the narrow sense of requiring point-set axioms and finite dimension.

[6] Formally, locally homeomorphic means that each point $m$ in the manifold $M$ has a neighborhood homeomorphic to a *neighborhood* in Euclidean space, not to the unit ball specifically. However, given such a homeomorphism, the preimage of an $\epsilon$-ball gives a homeomorphism between the unit ball and a smaller neighborhood of $m$, so this is no loss of generality. For topological or differentiable manifolds, one can also ask that every point have a neighborhood homeomorphic to all of Euclidean space (as this is diffeomorphic to the unit ball), but this cannot be done for complex manifolds, as the complex unit ball is not holomorphic to complex space.

[7] E.g. see Riaza, Ricardo (2008), *Differential-Algebraic Systems: Analytical Aspects and Circuit Applications*, World Scientific, p. 110, ISBN 9789812791818; Gunning, R. C. (1990), *Introduction to Holomorphic Functions of Several Variables, Volume 2*, CRC Press, p. 73, ISBN 9780534133092.

[8] Shigeyuki Morita, Teruko Nagase, Katsumi Nomizu (2001). *Geometry of Differential Forms*. American Mathematical Society Bookstore. p. 12. ISBN 0-8218-1045-6.

[9] Kervaire M., *A Manifold which does not admit any differentiable structure*, Comment. Math. Helv., *35* (1961), 1–14.

[10] J. Bryant, S. Ferry, W. Mio, and S. Weinberger, *Topology of homology manifolds*, Annals of Maths. 143, 435–467 (1996)

[11] R. Sikorski, *Abstract covariant derivative*, Coll. Math. 18, 251-272 (1967)

[12] K. Drachal, *Introduction to d–spaces theory*, Math. Aeterna 3, 753-770 (2013)

## 12.14 References

- Freedman, Michael H., and Quinn, Frank (1990) *Topology of 4-Manifolds*. Princeton University Press. ISBN 0-691-08577-3.

- Guillemin, Victor and Pollack, Alan (1974) *Differential Topology*. Prentice-Hall. ISBN 0-13-212605-2. Inspired by Milnor and commonly used in undergraduate courses.

- Hempel, John (1976) *3-Manifolds*. Princeton University Press. ISBN 0-8218-3695-1.

- Hirsch, Morris, (1997) *Differential Topology*. Springer Verlag. ISBN 0-387-90148-5. The most complete account, with historical insights and excellent, but difficult, problems. The standard reference for those wishing to have a deep understanding of the subject.

- Kirby, Robion C. and Siebenmann, Laurence C. (1977) *Foundational Essays on Topological Manifolds. Smoothings, and Triangulations*. Princeton University Press. ISBN 0-691-08190-5. A detailed study of the category of topological manifolds.

- Lee, John M. (2000) *Introduction to Topological Manifolds*. Springer-Verlag. ISBN 0-387-98759-2.

- Lee, John M. (2002), *Introduction to Smooth Manifolds*, Springer, ISBN 978-0-387-95448-6

- Lee, John M. (2003) *Introduction to Smooth Manifolds*. Springer-Verlag. ISBN 0-387-95495-3.

- Massey, William S. (1977) *Algebraic Topology: An Introduction*. Springer-Verlag. ISBN 0-387-90271-6.

- Milnor, John (1997) *Topology from the Differentiable Viewpoint*. Princeton University Press. ISBN 0-691-04833-9.

- Munkres, James R. (2000) *Topology*. Prentice Hall. ISBN 0-13-181629-2.

- Neuwirth, L. P., ed. (1975) *Knots, Groups, and 3-Manifolds. Papers Dedicated to the Memory of R. H. Fox*. Princeton University Press. ISBN 978-0-691-08170-0.

- Riemann, Bernhard, *Gesammelte mathematische Werke und wissenschaftlicher Nachlass*, Sändig Reprint. ISBN 3-253-03059-8.

  - *Grundlagen für eine allgemeine Theorie der Functionen einer veränderlichen complexen Grösse*. The 1851 doctoral thesis in which "manifold" (*Mannigfaltigkeit*) first appears.

  - *Ueber die Hypothesen, welche der Geometrie zu Grunde liegen*. The 1854 Göttingen inaugural lecture (*Habilitationsschrift*).

- Spivak, Michael (1965) *Calculus on Manifolds: A Modern Approach to Classical Theorems of Advanced Calculus*. HarperCollins Publishers. ISBN 0-8053-9021-9. The standard graduate text.

## 12.15 External links

- Hazewinkel, Michiel, ed. (2001), "Manifold", *Encyclopedia of Mathematics*, Springer, ISBN 978-1-55608-010-4

- Dimensions-math.org (A film explaining and visualizing manifolds up to fourth dimension.)

- The manifold atlas project of the Max Planck Institute for Mathematics in Bonn

# Chapter 13

# Inertial frame of reference

For other uses, see Framing.

In physics, an **inertial frame of reference** (also **inertial reference frame** or **inertial frame**, **Galilean reference frame** or **inertial space**) is a frame of reference that describes time and space homogeneously, isotropically, and in a time-independent manner.[1]

All inertial frames are in a state of constant, rectilinear motion with respect to one another; an accelerometer moving with any of them would detect zero acceleration. Measurements in one inertial frame can be converted to measurements in another by a simple transformation (the Galilean transformation in Newtonian physics and the Lorentz transformation in special relativity). In general relativity, in any region small enough for the curvature of spacetime and tidal forces[2] to be negligible, one can find a set of inertial frames that approximately describe that region.[3][4]

Physical laws take the same form in all inertial frames.[5] By contrast, in a non-inertial reference frame the laws of physics vary depending on the acceleration of that frame with respect to an inertial frame, and the usual physical forces must be supplemented by fictitious forces.[6][7] For example, a ball dropped towards the ground does not go exactly straight down because the Earth is rotating. Someone rotating with the Earth must account for the Coriolis effect—in this case thought of as a force—to predict the horizontal motion. Another example of such a fictitious force associated with rotating reference frames is the centrifugal effect, or centrifugal force.

## 13.1  Introduction

The motion of a body can only be described relative to something else - other bodies, observers, or a set of spacetime coordinates. These are called frames of reference. If the coordinates are chosen badly, the laws of motion may be more complex than necessary. For example, suppose a free body (one having no external forces on it) is at rest at some instant. In many coordinate systems, it would begin to move at the next instant, even though there are no forces on it. However, a frame of reference can always be chosen in which it remains stationary. Similarly, if space is not described uniformly or time independently, a coordinate system could describe the simple flight of a free body in space as a complicated zig-zag in its coordinate system. Indeed, an intuitive summary of inertial frames can be given as: In an inertial reference frame, the laws of mechanics take their simplest form.[1]

In an inertial frame, Newton's first law (the *law of inertia*) is satisfied: Any free motion has a constant magnitude and direction.[1] Newton's second law for a particle takes the form:

$$\mathbf{F} = m\mathbf{a} ,$$

with $\mathbf{F}$ the net force (a vector), $m$ the mass of a particle and $\mathbf{a}$ the acceleration of the particle (also a vector) which would be measured by an observer at rest in the frame. The force $\mathbf{F}$ is the vector sum of all "real" forces on the particle, such as electromagnetic, gravitational, nuclear and so forth. In contrast, Newton's second law in a rotating frame of reference, rotating at angular rate $\Omega$ about an axis, takes the form:

$$\mathbf{F'} = m\mathbf{a} ,$$

which looks the same as in an inertial frame, but now the force $\mathbf{F'}$ is the resultant of not only $\mathbf{F}$, but also additional terms (the paragraph following this equation presents the main points without detailed mathematics):

$$\mathbf{F'} = \mathbf{F} - 2m\ \times \mathbf{v}_B - m\ \times (\ \times \mathbf{x}_B) - m\frac{d}{dt} \times \mathbf{x}_B ,$$

where the angular rotation of the frame is expressed by the vector $\Omega$ pointing in the direction of the axis of rotation, and with magnitude equal to the angular rate of rotation

$\Omega$, symbol × denotes the vector cross product, vector $\mathbf{x}B$ locates the body and vector $\mathbf{v}B$ is the velocity of the body according to a rotating observer (different from the velocity seen by the inertial observer).

The extra terms in the force $\mathbf{F'}$ are the "fictitious" forces for this frame. (The first extra term is the Coriolis force, the second the centrifugal force, and the third the Euler force.) These terms all have these properties: they vanish when $\Omega$ = 0; that is, they are zero for an inertial frame (which, of course, does not rotate); they take on a different magnitude and direction in every rotating frame, depending upon its particular value of $\Omega$; they are ubiquitous in the rotating frame (affect every particle, regardless of circumstance); and they have no apparent source in identifiable physical sources, in particular, matter. Also, fictitious forces do not drop off with distance (unlike, for example, nuclear forces or electrical forces). For example, the centrifugal force that appears to emanate from the axis of rotation in a rotating frame increases with distance from the axis.

All observers agree on the real forces, $\mathbf{F}$; only non-inertial observers need fictitious forces. The laws of physics in the inertial frame are simpler because unnecessary forces are not present.

In Newton's time the fixed stars were invoked as a reference frame, supposedly at rest relative to absolute space. In reference frames that were either at rest with respect to the fixed stars or in uniform translation relative to these stars, Newton's laws of motion were supposed to hold. In contrast, in frames accelerating with respect to the fixed stars, an important case being frames rotating relative to the fixed stars, the laws of motion did not hold in their simplest form, but had to be supplemented by the addition of fictitious forces, for example, the Coriolis force and the centrifugal force. Two interesting experiments were devised by Newton to demonstrate how these forces could be discovered, thereby revealing to an observer that they were not in an inertial frame: the example of the tension in the cord linking two spheres rotating about their center of gravity, and the example of the curvature of the surface of water in a rotating bucket. In both cases, application of Newton's second law would not work for the rotating observer without invoking centrifugal and Coriolis forces to account for their observations (tension in the case of the spheres; parabolic water surface in the case of the rotating bucket).

As we now know, the fixed stars are not fixed. Those that reside in the Milky Way turn with the galaxy, exhibiting proper motions. Those that are outside our galaxy (such as nebulae once mistaken to be stars) participate in their own motion as well, partly due to expansion of the universe, and partly due to peculiar velocities.[8] (The Andromeda galaxy is on collision course with the Milky Way at a speed of 117 km/s.[9]) The concept of inertial frames of reference is no longer tied to either the fixed stars or to absolute space. Rather, the identification of an inertial frame is based upon the simplicity of the laws of physics in the frame. In particular, the absence of fictitious forces is their identifying property.[10]

In practice, although not a requirement, using a frame of reference based upon the fixed stars as though it were an inertial frame of reference introduces very little discrepancy. For example, the centrifugal acceleration of the Earth because of its rotation about the Sun is about thirty million times greater than that of the Sun about the galactic center.[11]

To illustrate further, consider the question: "Does our Universe rotate?" To answer, we might attempt to explain the shape of the Milky Way galaxy using the laws of physics.[12] (Other observations might be more definitive (that is, provide larger discrepancies or less measurement uncertainty), like the anisotropy of the microwave background radiation or Big Bang nucleosynthesis.[13][14]) Just how flat the disc of the Milky Way is depends on its rate of rotation in an inertial frame of reference. If we attribute its apparent rate of rotation entirely to rotation in an inertial frame, a different "flatness" is predicted than if we suppose part of this rotation actually is due to rotation of the Universe and should not be included in the rotation of the galaxy itself. Based upon the laws of physics, a model is set up in which one parameter is the rate of rotation of the Universe. If the laws of physics agree more accurately with observations in a model with rotation than without it, we are inclined to select the best-fit value for rotation, subject to all other pertinent experimental observations. If no value of the rotation parameter is successful and theory is not within observational error, a modification of physical law is considered. (For example, dark matter is invoked to explain the galactic rotation curve.) So far, observations show any rotation of the Universe is very slow (no faster than once every $60 \cdot 10^{12}$ years ($10^{-13}$ rad/yr)[15]), and debate persists over whether there is *any* rotation. However, if rotation were found, interpretation of observations in a frame tied to the Universe would have to be corrected for the fictitious forces inherent in such rotation. Evidently, such an approach adopts the view that "an inertial frame of reference is one where our laws of physics apply" (or need the least modification).

When quantum effects are important, there are additional conceptual complications that arise in quantum reference frames.

## 13.2    Background

A brief comparison of inertial frames in special relativity and in Newtonian mechanics, and the role of absolute space is next.

## 13.2.1  A set of frames where the laws of physics are simple

According to the first postulate of special relativity, all physical laws take their simplest form in an inertial frame, and there exist multiple inertial frames interrelated by uniform translation: [16]

> Special principle of relativity: If a system of coordinates K is chosen so that, in relation to it, physical laws hold good in their simplest form, the same laws hold good in relation to any other system of coordinates K' moving in uniform translation relatively to K.
> — Albert Einstein: *The foundation of the general theory of relativity*, Section A, §1

The principle of simplicity can be used within Newtonian physics as well as in special relativity; see Nagel[17] and also Blagojević.[18]

> The laws of Newtonian mechanics do not always hold in their simplest form...If, for instance, an observer is placed on a disc rotating relative to the earth, he/she will sense a 'force' pushing him/her toward the periphery of the disc, which is not caused by any interaction with other bodies. Here, the acceleration is not the consequence of the usual force, but of the so-called inertial force. Newton's laws hold in their simplest form only in a family of reference frames, called inertial frames. This fact represents the essence of the Galilean principle of relativity:
> The laws of mechanics have the same form in all inertial frames.
> — Milutin Blagojević: *Gravitation and Gauge Symmetries*, p. 4

In practical terms, the equivalence of inertial reference frames means that scientists within a box moving uniformly cannot determine their absolute velocity by any experiment (otherwise the differences would set up an absolute standard reference frame).[19][20] According to this definition, supplemented with the constancy of the speed of light, inertial frames of reference transform among themselves according to the Poincaré group of symmetry transformations, of which the Lorentz transformations are a subgroup.[21] In

Newtonian mechanics, which can be viewed as a limiting case of special relativity in which the speed of light is infinite, inertial frames of reference are related by the Galilean group of symmetries.

## 13.2.2  Absolute space

Main article: Absolute space and time

Newton posited an absolute space considered well approximated by a frame of reference stationary relative to the fixed stars. An inertial frame was then one in uniform translation relative to absolute space. However, some scientists (called "relativists" by Mach[22]), even at the time of Newton, felt that absolute space was a defect of the formulation, and should be replaced.

Indeed, the expression *inertial frame of reference* (German: *Inertialsystem*) was coined by Ludwig Lange in 1885, to replace Newton's definitions of "absolute space and time" by a more operational definition.[23][24] As translated by Iro, Lange proposed the following definition:[25]

> A reference frame in which a mass point thrown from the same point in three different (non co-planar) directions follows rectilinear paths each time it is thrown, is called an inertial frame.

A discussion of Lange's proposal can be found in Mach.[22]

The inadequacy of the notion of "absolute space" in Newtonian mechanics is spelled out by Blagojević:[26]

> • The existence of absolute space contradicts the internal logic of classical mechanics since, according to Galilean principle of relativity, none of the inertial frames can be singled out.
>
> • Absolute space does not explain inertial forces since they are related to acceleration with respect to any one of the inertial frames.
>
> • Absolute space acts on physical objects by inducing their resistance to acceleration but it cannot be acted upon.
> — Milutin Blagojević: *Gravitation and Gauge Symmetries*, p. 5

The utility of operational definitions was carried much further in the special theory of relativity.[27] Some historical background including Lange's definition is provided by DiSalle, who says in summary:[28]

The original question, "relative to what frame of reference do the laws of motion hold?" is revealed to be wrongly posed. For the laws of motion essentially determine a class of reference frames, and (in principle) a procedure for constructing them.

— Robert DiSalle *Space and Time: Inertial Frames*

## 13.3   Newton's inertial frame of reference

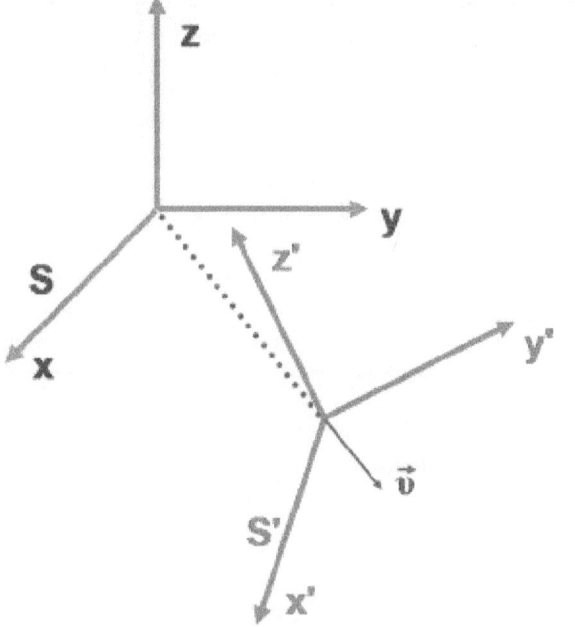

*Figure 1: Two frames of reference moving with relative velocity $\vec{v}$ . Frame S' has an arbitrary but fixed rotation with respect to frame S. They are both* inertial frames *provided a body not subject to forces appears to move in a straight line. If that motion is seen in one frame, it will also appear that way in the other.*

Within the realm of Newtonian mechanics, an inertial frame of reference, or inertial reference frame, is one in which Newton's first law of motion is valid.[29] However, the principle of special relativity generalizes the notion of inertial frame to include all physical laws, not simply Newton's first law.

Newton viewed the first law as valid in any reference frame that is in uniform motion relative to the fixed stars;[30] that is, neither rotating nor accelerating relative to the stars.[31] Today the notion of "absolute space" is abandoned, and an inertial frame in the field of classical mechanics is defined as:[32][33]

An inertial frame of reference is one in which the motion of a particle not subject to forces is in a straight line at constant speed.

Hence, with respect to an inertial frame, an object or body accelerates only when a physical force is applied, and (following Newton's first law of motion), in the absence of a net force, a body at rest will remain at rest and a body in motion will continue to move uniformly—that is, in a straight line and at constant speed. Newtonian inertial frames transform among each other according to the Galilean group of symmetries.

If this rule is interpreted as saying that straight-line motion is an indication of zero net force, the rule does not identify inertial reference frames because straight-line motion can be observed in a variety of frames. If the rule is interpreted as defining an inertial frame, then we have to be able to determine when zero net force is applied. The problem was summarized by Einstein:[34]

The weakness of the principle of inertia lies in this, that it involves an argument in a circle: a mass moves without acceleration if it is sufficiently far from other bodies; we know that it is sufficiently far from other bodies only by the fact that it moves without acceleration.

— Albert Einstein: *The Meaning of Relativity*, p. 58

There are several approaches to this issue. One approach is to argue that all real forces drop off with distance from their sources in a known manner, so we have only to be sure that a body is far enough away from all sources to ensure that no force is present.[35] A possible issue with this approach is the historically long-lived view that the distant universe might affect matters (Mach's principle). Another approach is to identify all real sources for real forces and account for them. A possible issue with this approach is that we might miss something, or account inappropriately for their influence, perhaps, again, due to Mach's principle and an incomplete understanding of the universe. A third approach is to look at the way the forces transform when we shift reference frames. Fictitious forces, those that arise due to the acceleration of a frame, disappear in inertial frames, and have complicated rules of transformation in general cases. On the basis of universality of physical law and the request for frames where the laws are most simply expressed, inertial frames are distinguished by the absence of such fictitious forces.

Newton enunciated a principle of relativity himself in one of his corollaries to the laws of motion:[36][37]

> The motions of bodies included in a given space are the same among themselves, whether that space is at rest or moves uniformly forward in a straight line.
> — Isaac Newton: *Principia*, Corollary V, p. 88 in Andrew Motte translation

This principle differs from the special principle in two ways: first, it is restricted to mechanics, and second, it makes no mention of simplicity. It shares with the special principle the invariance of the form of the description among mutually translating reference frames.[38] The role of fictitious forces in classifying reference frames is pursued further below.

## 13.4 Separating non-inertial from inertial reference frames

### 13.4.1 Theory

Main article: Fictitious force
See also: Non-inertial frame, Rotating spheres, and Bucket argument

Inertial and non-inertial reference frames can be distinguished by the absence or presence of fictitious forces, as explained shortly.[6][7]

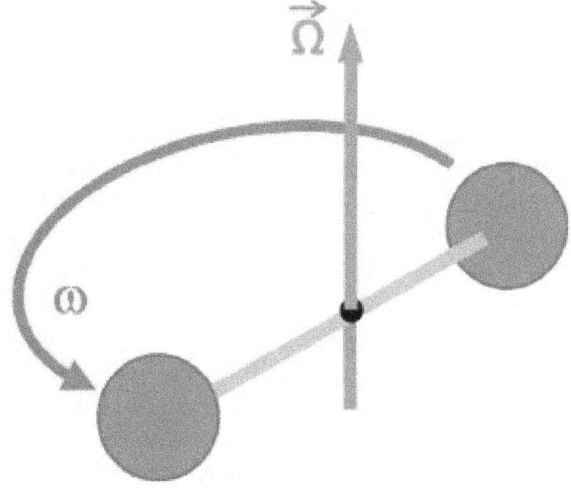

*Figure 2: Two spheres tied with a string and rotating at an angular rate ω. Because of the rotation, the string tying the spheres together is under tension.*

> The effect of this being in the noninertial frame is to require the observer to introduce a

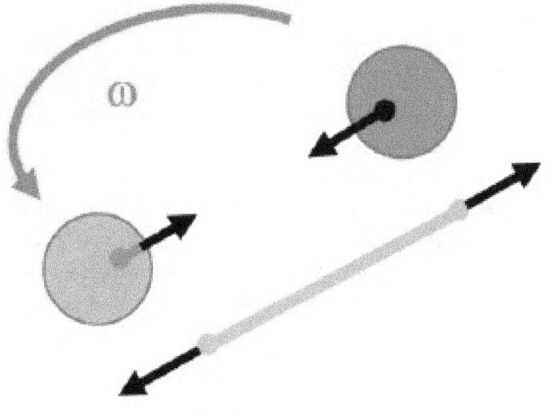

*Figure 3: Exploded view of rotating spheres in an inertial frame of reference showing the centripetal forces on the spheres provided by the tension in the tying string.*

> fictitious force into his calculations....
> — Sidney Borowitz and Lawrence A Bornstein in *A Contemporary View of Elementary Physics*, p. 138

The presence of fictitious forces indicates the physical laws are not the simplest laws available so, in terms of the special principle of relativity, a frame where fictitious forces are present is not an inertial frame:[39]

> The equations of motion in a non-inertial system differ from the equations in an inertial system by additional terms called inertial forces. This allows us to detect experimentally the non-inertial nature of a system.
> — V. I. Arnol'd: *Mathematical Methods of Classical Mechanics* Second Edition, p. 129

Bodies in non-inertial reference frames are subject to so-called *fictitious* forces (pseudo-forces); that is, forces that result from the acceleration of the reference frame itself and not from any physical force acting on the body. Examples of fictitious forces are the centrifugal force and the Coriolis force in rotating reference frames.

How then, are "fictitious" forces to be separated from "real" forces? It is hard to apply the Newtonian definition of an inertial frame without this separation. For example, consider a stationary object in an inertial frame. Being at rest, no net force is applied. But in a frame rotating about a fixed axis, the object appears to move in a circle, and is subject to centripetal force (which is made up of the Coriolis force and the centrifugal force). How can we decide that the rotating frame is a non-inertial frame? There are two approaches to this resolution: one approach is to look for the origin of the fictitious forces (the Coriolis force and the centrifugal force). We will find there are no sources for these forces, no associated force carriers, no originating bodies.[40] A second approach is to look at a variety of frames of reference. For any inertial frame, the Coriolis force and the centrifugal force disappear, so application of the principle of special relativity would identify these frames where the forces disappear as sharing the same and the simplest physical laws, and hence rule that the rotating frame is not an inertial frame.

Newton examined this problem himself using rotating spheres, as shown in Figure 2 and Figure 3. He pointed out that if the spheres are not rotating, the tension in the tying string is measured as zero in every frame of reference.[41] If the spheres only appear to rotate (that is, we are watching stationary spheres from a rotating frame), the zero tension in the string is accounted for by observing that the centripetal force is supplied by the centrifugal and Coriolis forces in combination, so no tension is needed. If the spheres really are rotating, the tension observed is exactly the centripetal force required by the circular motion. Thus, measurement of the tension in the string identifies the inertial frame: it is the one where the tension in the string provides exactly the centripetal force demanded by the motion as it is observed in that frame, and not a different value. That is, the inertial frame is the one where the fictitious forces vanish.

So much for fictitious forces due to rotation. However, for linear acceleration, Newton expressed the idea of undetectability of straight-line accelerations held in common:[37]

> If bodies, any how moved among themselves, are urged in the direction of parallel lines by equal accelerative forces, they will continue to move among themselves, after the same manner as if they had been urged by no such forces.
> — Isaac Newton: *Principia* Corollary VI, p. 89, in Andrew Motte translation

This principle generalizes the notion of an inertial frame. For example, an observer confined in a free-falling lift will assert that he himself is a valid inertial frame, even if he is accelerating under gravity, so long as he has no knowledge about anything outside the lift. So, strictly speaking, inertial frame is a relative concept. With this in mind, we can define inertial frames collectively as a set of frames which are stationary or moving at constant velocity with respect to each other, so that a single inertial frame is defined as an element of this set.

For these ideas to apply, everything observed in the frame has to be subject to a base-line, common acceleration shared by the frame itself. That situation would apply, for example, to the elevator example, where all objects are subject to the same gravitational acceleration, and the elevator itself accelerates at the same rate.

In 1899 the astronomer Karl Schwarzschild pointed out an observation about double stars. The motion of two stars orbiting each other is planar, the two orbits of the stars of the system lie in a plane. In the case of sufficiently near double star systems, it can be seen from Earth whether the perihelion of the orbits of the two stars remains pointing in the same direction with respect to the solar system. Schwarzschild pointed out that that was invariably seen: the direction of the angular momentum of all observed double star systems remains fixed with respect to the direction of the angular momentum of the Solar system. The logical inference is that just like gyroscopes, the angular momentum of all celestial bodies is angular momentum with respect to a universal inertial space.[42]

## 13.4.2  Applications

Inertial navigation systems used a cluster of gyroscopes and accelerometers to determine accelerations relative to inertial space. After a gyroscope is spun up in a particular orientation in inertial space, the law of conservation of angular momentum requires that it retain that orientation as long as no external forces are applied to it.[43]:59 Three orthogonal gyroscopes establish an inertial reference frame, and the accelerators measure acceleration relative to that frame. The accelerations, along with a clock, can then be used to calculate the change in position. Thus, inertial navigation is a form of dead reckoning that requires no external input, and therefore cannot be jammed by any external or internal signal source.[44]

A gyrocompass, employed for navigation of seagoing vessels, finds the geometric north. It does so, not by sensing the Earth's magnetic field, but by using inertial space as its reference. The outer casing of the gyrocompass device is held in such a way that it remains aligned with the local plumb line. When the gyroscope wheel inside the gyrocompass device is spun up, the way the gyroscope wheel is suspended causes the gyroscope wheel to gradually align its spinning axis with the Earth's axis. Alignment with the Earth's axis is the only direction for which the gyroscope's spinning axis can be stationary with respect to the Earth and not be required to change direction with respect to inertial space. After being spun up, a gyrocompass can reach the direction of alignment with the Earth's axis in as little as a quarter of an hour.[45]

## 13.5 Newtonian mechanics

Main article: Newton's laws of motion

Classical mechanics, which includes relativity, assumes the equivalence of all inertial reference frames. Newtonian mechanics makes the additional assumptions of absolute space and absolute time. Given these two assumptions, the coordinates of the same event (a point in space and time) described in two inertial reference frames are related by a Galilean transformation.

$$\mathbf{r}' = \mathbf{r} - \mathbf{r}_0 - \mathbf{v}t$$

$$t' = t - t_0$$

where $\mathbf{r}_0$ and $t_0$ represent shifts in the origin of space and time, and $\mathbf{v}$ is the relative velocity of the two inertial reference frames. Under Galilean transformations, the time $t_2 - t_1$ between two events is the same for all inertial reference frames and the distance between two simultaneous events (or, equivalently, the length of any object, $|\mathbf{r}_2 - \mathbf{r}_1|$) is also the same.

## 13.6 Special relativity

Main articles: Special relativity and Introduction to special relativity

Einstein's theory of special relativity, like Newtonian mechanics, assumes the equivalence of all inertial reference frames, but makes an additional assumption, foreign to Newtonian mechanics, namely, that in free space light always is propagated with the speed of light $c_0$, a defined value independent of its direction of propagation and its frequency, and also independent of the state of motion of the emitting body. This second assumption has been verified experimentally and leads to counter-intuitive deductions including:

- time dilation (moving clocks tick more slowly)

- length contraction (moving objects are shortened in the direction of motion)

- relativity of simultaneity (simultaneous events in one reference frame are not simultaneous in almost all frames moving relative to the first).

These deductions are logical consequences of the stated assumptions, and are general properties of space-time, typ-

ically without regard to a consideration of properties pertaining to the structure of individual objects like atoms or stars, nor to the mechanisms of clocks.

These effects are expressed mathematically by the Lorentz transformation

$$x' = \gamma\,(x - vt)$$

$$y' = y$$

$$z' = z$$

$$t' = \gamma\left(t - \frac{vx}{c_0^2}\right)$$

where shifts in origin have been ignored, the relative velocity is assumed to be in the $x$ -direction and the Lorentz factor $\gamma$ is defined by:

$$\gamma \overset{\text{def}}{=} \frac{1}{\sqrt{1 - (v/c_0)^2}} \geq 1.$$

The Lorentz transformation is equivalent to the Galilean transformation in the limit $c_0 \to \infty$ (a hypothetical case) or $v \to 0$ (low speeds).

Under Lorentz transformations, the time and distance between events may differ among inertial reference frames; however, the Lorentz scalar distance $s$ between two events is the same in all inertial reference frames

$$s^2 = (x_2 - x_1)^2 + (y_2 - y_1)^2 + (z_2 - z_1)^2 - c_0^2\,(t_2 - t_1)^2$$

From this perspective, the speed of light is only accidentally a property of light, and is rather a property of spacetime, a conversion factor between conventional time units (such as seconds) and length units (such as meters).

Incidentally, because of the limitations on speeds faster than the speed of light, notice that in a rotating frame of reference (which is a non-inertial frame, of course) stationarity is not possible at arbitrary distances because at large radius the object would move faster than the speed of light.[46]

## 13.7 General relativity

Main articles: General relativity and Introduction to general relativity
See also: Equivalence principle and Eötvös experiment

General relativity is based upon the principle of equivalence:[47][48]

There is no experiment observers can perform to distinguish whether an acceleration arises because of a gravitational force or because their reference frame is accelerating.
— Douglas C. Giancoli, *Physics for Scientists and Engineers with Modern Physics*, p. 155.

This idea was introduced in Einstein's 1907 article "Principle of Relativity and Gravitation" and later developed in 1911.[49] Support for this principle is found in the Eötvös experiment, which determines whether the ratio of inertial to gravitational mass is the same for all bodies, regardless of size or composition. To date no difference has been found to a few parts in $10^{11}$.[50] For some discussion of the subtleties of the Eötvös experiment, such as the local mass distribution around the experimental site (including a quip about the mass of Eötvös himself), see Franklin.[51]

Einstein's general theory modifies the distinction between nominally "inertial" and "noninertial" effects by replacing special relativity's "flat" Minkowski Space with a metric that produces non-zero curvature. In general relativity, the principle of inertia is replaced with the principle of geodesic motion, whereby objects move in a way dictated by the curvature of spacetime. As a consequence of this curvature, it is not a given in general relativity that inertial objects moving at a particular rate with respect to each other will continue to do so. This phenomenon of geodesic deviation means that inertial frames of reference do not exist globally as they do in Newtonian mechanics and special relativity.

However, the general theory reduces to the special theory over sufficiently small regions of spacetime, where curvature effects become less important and the earlier inertial frame arguments can come back into play.[52][53] Consequently, modern special relativity is now sometimes described as only a "local theory".[54]

## 13.8    See also

## 13.9    References

[1] Landau, L. D.; Lifshitz, E. M. (1960). *Mechanics*. Pergamon Press. pp. 4–6.

[2] Cheng, Ta-Pei (2013). *Einstein's Physics: Atoms, Quanta, and Relativity - Derived, Explained, and Appraised* (illustrated ed.). OUP Oxford. p. 219. ISBN 978-0-19-966991-2. Extract of page 219

[3] Albert Einstein (2001) [Reprint of edition of 1920 translated by RQ Lawson]. *Relativity: The Special and General Theory* (3rd ed.). Courier Dover Publications. p. 71. ISBN 0-486-41714-X.

[4] Domenico Giulini (2005). *Special Relativity*. Cambridge University Press. p. 19. ISBN 0-19-856746-4.

[5] Assuming the coordinate systems have the same handedness.

[6] Milton A. Rothman (1989). *Discovering the Natural Laws: The Experimental Basis of Physics*. Courier Dover Publications. p. 23. ISBN 0-486-26178-6.

[7] Sidney Borowitz & Lawrence A. Bornstein (1968). *A Contemporary View of Elementary Physics*. McGraw-Hill. p. 138. ASIN B000GQB02A.

[8] Amedeo Balbi (2008). *The Music of the Big Bang*. Springer. p. 59. ISBN 3-540-78726-7.

[9] Abraham Loeb, Mark J. Reid, Andreas Brunthaler, Heino Falcke (2005). "Constraints on the proper motion of the Andromeda galaxy based on the survival of its satellite M33" (PDF). *The Astrophysical Journal* **633** (2): 894–898. arXiv:astro-ph/0506609. Bibcode:2005ApJ...633..894L. doi:10.1086/491644.

[10] John J. Stachel (2002). *Einstein from "B" to "Z"*. Springer. pp. 235–236. ISBN 0-8176-4143-2.

[11] Peter Graneau & Neal Graneau (2006). *In the Grip of the Distant Universe*. World Scientific. p. 147. ISBN 981-256-754-2.

[12] Henning Genz (2001). *Nothingness*. Da Capo Press. p. 275. ISBN 0-7382-0610-5.

[13] J Garcío-Bellido (2005). "The Paradigm of Inflation". In J. M. T. Thompson. *Advances in Astronomy*. Imperial College Press. p. 32, §9. ISBN 1-86094-577-5.

[14] Wlodzimierz Godlowski and Marek Szydlowski (2003). "Dark energy and global rotation of the Universe". *General Relativity and Gravitation* **35** (12): 2171. arXiv:astro-ph/0303248. Bibcode:2003GReGr..35.2171G. doi:10.1023/A:1027301723533.

[15] P Birch *Is the Universe rotating?* Nature 298, 451 - 454 (29 July 1982)

[16] Einstein, A., Lorentz, H. A., Minkowski, H., & Weyl, H. (1952). *The Principle of Relativity: a collection of original memoirs on the special and general theory of relativity*. Courier Dover Publications. p. 111. ISBN 0-486-60081-5.

[17] Ernest Nagel (1979). *The Structure of Science*. Hackett Publishing. p. 212. ISBN 0-915144-71-9.

[18] Milutin Blagojević (2002). *Gravitation and Gauge Symmetries*. CRC Press. p. 4. ISBN 0-7503-0767-6.

[19] Albert Einstein (1920). *Relativity: The Special and General Theory*. H. Holt and Company. p. 17.

[20] Richard Phillips Feynman (1998). *Six not-so-easy pieces: Einstein's relativity, symmetry, and space-time*. Basic Books. p. 73. ISBN 0-201-32842-9.

[21] Armin Wachter & Henning Hoeber (2006). *Compendium of Theoretical Physics*. Birkhäuser. p. 98. ISBN 0-387-25799-3.

[22] Ernst Mach (1915). *The Science of Mechanics*. The Open Court Publishing Co. p. 38.

[23] Lange, Ludwig (1885). "Über die wissenschaftliche Fassung des Galileischen Beharrungsgesetzes". *Philosophische Studien 2*.

[24] Julian B. Barbour (2001). *The Discovery of Dynamics* (Reprint of 1989 *Absolute or Relative Motion?* ed.). Oxford University Press. pp. 645–646. ISBN 0-19-513202-5.

[25] L. Lange (1885) as quoted by Max von Laue in his book (1921) *Die Relativitätstheorie*, p. 34, and translated by Harald Iro (2002). *A Modern Approach to Classical Mechanics*. World Scientific. p. 169. ISBN 981-238-213-5.

[26] Milutin Blagojević (2002). *Gravitation and Gauge Symmetries*. CRC Press. p. 5. ISBN 0-7503-0767-6.

[27] NMJ Woodhouse (2003). *Special relativity*. London: Springer. p. 58. ISBN 1-85233-426-6.

[28] Robert DiSalle (Summer 2002). "Space and Time: Inertial Frames". In Edward N. Zalta. *The Stanford Encyclopedia of Philosophy*.

[29] C Møller (1976). *The Theory of Relativity* (Second ed.). Oxford UK: Oxford University Press. p. 1. ISBN 0-19-560539-X.

[30] The question of "moving uniformly relative to what?" was answered by Newton as "relative to absolute space". As a practical matter, "absolute space" was considered to be the fixed stars. For a discussion of the role of fixed stars, see Henning Genz (2001). *Nothingness: The Science of Empty Space*. Da Capo Press. p. 150. ISBN 0-7382-0610-5.

[31] Robert Resnick, David Halliday, Kenneth S. Krane (2001). *Physics* (5th ed.). Wiley. Volume 1, Chapter 3. ISBN 0-471-32057-9.

[32] RG Takwale (1980). *Introduction to classical mechanics*. New Delhi: Tata McGraw-Hill. p. 70. ISBN 0-07-096617-6.

[33] NMJ Woodhouse (2003). *Special relativity*. London/Berlin: Springer. p. 6. ISBN 1-85233-426-6.

[34] A Einstein (1950). *The Meaning of Relativity*. Princeton University Press. p. 58.

[35] William Geraint Vaughan Rosser (1991). *Introductory Special Relativity*. CRC Press. p. 3. ISBN 0-85066-838-7.

[36] Richard Phillips Feynman (1998). *Six not-so-easy pieces: Einstein's relativity, symmetry, and space-time*. Basic Books. p. 50. ISBN 0-201-32842-9.

[37] See the *Principia* on line at Andrew Motte Translation

[38] However, in the Newtonian system the Galilean transformation connects these frames and in the special theory of relativity the Lorentz transformation connects them. The two transformations agree for speeds of translation much less than the speed of light.

[39] V. I. Arnol'd (1989). *Mathematical Methods of Classical Mechanics*. Springer. p. 129. ISBN 978-0-387-96890-2.

[40] For example, there is no body providing a gravitational or electrical attraction.

[41] That is, the universality of the laws of physics requires the same tension to be seen by everybody. For example, it cannot happen that the string breaks under extreme tension in one frame of reference and remains intact in another frame of reference, just because we choose to look at the string from a different frame.

[42] In the Shadow of the Relativity Revolution Section 3: The Work of Karl Schwarzschild (2.2 MB PDF-file)

[43] Chatfield, Averil B. (1997). *Fundamentals of High Accuracy Inertial Navigation, Volume 174*. AIAA. ISBN 9781600864278.

[44] Kennie, edited by T.J.M.; Petrie, G. (1993). *Engineering Surveying Technology* (Pbk. ed.). Hoboken: Taylor & Francis. p. 95. ISBN 9780203860748.

[45] "The gyroscope pilots ships & planes". *Life*: 80–83. Mar 15, 1943.

[46] LD Landau & LM Lifshitz (1975). *The Classical Theory of Fields* (4th Revised English ed.). Pergamon Press. pp. 273–274. ISBN 978-0-7506-2768-9.

[47] David Morin (2008). *Introduction to Classical Mechanics*. Cambridge University Press. p. 649. ISBN 0-521-87622-2.

[48] Douglas C. Giancoli (2007). *Physics for Scientists and Engineers with Modern Physics*. Pearson Prentice Hall. p. 155. ISBN 0-13-149508-9.

[49] A. Einstein, "On the influence of gravitation on the propagation of light", *Annalen der Physik*, vol. 35, (1911) : 898-908

[50] National Research Council (US) (1986). *Physics Through the Nineteen Nineties: Overview*. National Academies Press. p. 15. ISBN 0-309-03579-1.

[51] Allan Franklin (2007). *No Easy Answers: Science and the Pursuit of Knowledge*. University of Pittsburgh Press. p. 66. ISBN 0-8229-5968-2.

[52] Green, Herbert S. (2000). *Information Theory and Quantum Physics: Physical Foundations for Understanding the Conscious Process*. Springer. p. 154. ISBN 354066517X. Extract of page 154

[53] Bandyopadhyay, Nikhilendu (2000). *Theory of Special Relativity*. Academic Publishers. p. 116. ISBN 8186358528. Extract of page 116

[54] Liddle, Andrew R.; Lyth, David H. (2000). *Cosmological Inflation and Large-Scale Structure*. Cambridge University Press. p. 329. ISBN 0-521-57598-2. Extract of page 329

## 13.10   Further reading

- Edwin F. Taylor and John Archibald Wheeler, *Spacetime Physics*, 2nd ed. (Freeman, NY, 1992)

- Albert Einstein, *Relativity, the special and the general theories*, 15th ed. (1954)

- Poincaré, Henri (1900). "La théorie de Lorentz et le Principe de Réaction". *Archives Neerlandaises* V: 253–78.

- Albert Einstein, *On the Electrodynamics of Moving Bodies*, included in *The Principle of Relativity*, page 38. Dover 1923

### Rotation of the Universe

- Julian B. Barbour, Herbert Pfister (1998). *Mach's Principle: From Newton's Bucket to Quantum Gravity*. Birkhäuser. p. 445. ISBN 0-8176-3823-7.

- PJ Nahin (1999). *Time Machines*. Springer. p. 369; Footnote 12. ISBN 0-387-98571-9.

- B Ciobanu, I Radinchi *Modeling the electric and magnetic fields in a rotating universe* Rom. Journ. Phys., Vol. 53, Nos. 1–2, P. 405–415, Bucharest, 2008

- Yuri N. Obukhov, Thoralf Chrobok, Mike Scherfner *Shear-free rotating inflation* Phys. Rev. D 66, 043518 (2002) [5 pages]

- Yuri N. Obukhov *On physical foundations and observational effects of cosmic rotation* (2000)

- Li-Xin Li *Effect of the Global Rotation of the Universe on the Formation of Galaxies* General Relativity and Gravitation, **30** (1998) doi:10.1023/A:1018867011142

- P Birch *Is the Universe rotating?* Nature 298, 451 - 454 (29 July 1982)

- Kurt Gödel *An example of a new type of cosmological solutions of Einstein's field equations of gravitation* Rev. Mod. Phys., Vol. 21, p. 447, 1949.

## 13.11   External links

- Stanford Encyclopedia of Philosophy entry

- Animation clip on YouTube showing scenes as viewed from both an inertial frame and a rotating frame of reference, visualizing the Coriolis and centrifugal forces.

# Chapter 14

# Pseudo-Riemannian manifold

In differential geometry, a **pseudo-Riemannian manifold**[1][2] (also called a **semi-Riemannian manifold**) is a generalization of a Riemannian manifold in which the metric tensor need not be positive-definite. Instead a weaker condition of nondegeneracy is imposed on the metric tensor.

Every tangent space of a pseudo-Riemannian manifold is a pseudo-Euclidean space described by an isotropic quadratic form.

A special case of great importance to general relativity is a **Lorentzian manifold**, in which one dimension has a sign opposite to that of the rest. This allows tangent vectors to be classified into timelike, null, and spacelike. Spacetime can be modeled as a 4-dimensional Lorentzian manifold.

## 14.1 Introduction

### 14.1.1 Manifolds

Main articles: Manifold and Differentiable manifold

In differential geometry, a differentiable manifold is a space which is locally similar to a Euclidean space. In an n-dimensional Euclidean space any point can be specified by n real numbers. These are called the coordinates of the point.

An n-dimensional differentiable manifold is a generalisation of n-dimensional Euclidean space. In a manifold it may only be possible to define coordinates *locally*. This is achieved by defining coordinate patches: subsets of the manifold which can be mapped into n-dimensional Euclidean space.

See Manifold, differentiable manifold, coordinate patch for more details.

### 14.1.2 Tangent spaces and metric tensors

Main articles: Tangent space and metric tensor

Associated with each point $p$ in an $n$ -dimensional differentiable manifold $M$ is a tangent space (denoted $T_pM$ ). This is an $n$ -dimensional vector space whose elements can be thought of as equivalence classes of curves passing through the point $p$ .

A metric tensor is a non-degenerate, smooth, symmetric, bilinear map which assigns a real number to pairs of tangent vectors at each tangent space of the manifold. Denoting the metric tensor by $g$ we can express this as

$$g : T_pM \times T_pM \to \mathbb{R}.$$

The map is symmetric and bilinear so if $X,Y,Z \in T_pM$ are tangent vectors at a point $p$ to the manifold $M$ then we have

- $g(X,Y) = g(Y,X)$

- $g(aX + Y, Z) = ag(X,Z) + g(Y,Z)$

for any real number $a \in \mathbb{R}$ .

That $g$ is non-degenerate means there are no non-zero $X \in T_pM$ such that $g(X,Y) = 0$ for all $Y \in T_pM$ .

### 14.1.3 Metric signatures

Main article: Metric signature

Given a metric tensor $g$ on an $n$-dimensional real manifold, the quadratic form $q(x) = g(x, x)$ associated with the metric tensor applied to each vector of any orthogonal basis produces $n$ real values. By Sylvester's law of inertia, the number of each positive, negative and zero values produced in this manner are invariants of the metric tensor, independent of the choice of orthogonal basis. The signature $(p, q, r)$ of the metric tensor gives these numbers, shown in the same order. A non-degenerate metric tensor has $r = 0$ and the signature may be denoted $(p, q)$, where $p + q = n$.

## 14.2   Definition

A **pseudo-Riemannian manifold** $(M, g)$ is a differentiable manifold $M$ equipped with a non-degenerate, smooth, symmetric metric tensor $g$ .

Such a metric is called a **pseudo-Riemannian metric** and its values can be positive, negative or zero.

The signature of a pseudo-Riemannian metric is $(p, q)$, where both $p$ and $q$ are non-negative.

## 14.3   Lorentzian manifold

A **Lorentzian manifold** is an important special case of a pseudo-Riemannian manifold in which the signature of the metric is (1, n−1) (or sometimes (n−1, 1), see sign convention). Such metrics are called **Lorentzian metrics**. They are named after the physicist Hendrik Lorentz.

### 14.3.1   Applications in physics

After Riemannian manifolds, Lorentzian manifolds form the most important subclass of pseudo-Riemannian manifolds. They are important in applications of general relativity.

A principal basis of general relativity is that spacetime can be modeled as a 4-dimensional Lorentzian manifold of signature (3, 1) or, equivalently, (1, 3). Unlike Riemannian manifolds with positive-definite metrics, a signature of (p, 1) or (1, q) allows tangent vectors to be classified into *timelike*, *null* or *spacelike* (see Causal structure).

## 14.4   Properties of pseudo-Riemannian manifolds

Just as Euclidean space $\mathbb{R}^n$ can be thought of as the model Riemannian manifold, Minkowski space $\mathbb{R}^{n-1,1}$ with the flat Minkowski metric is the model Lorentzian manifold. Likewise, the model space for a pseudo-Riemannian manifold of signature (p, q) is $\mathbb{R}^{p,q}$ with the metric

$$g = dx_1^2 + \cdots + dx_p^2 - dx_{p+1}^2 - \cdots - dx_{p+q}^2$$

Some basic theorems of Riemannian geometry can be generalized to the pseudo-Riemannian case. In particular, the fundamental theorem of Riemannian geometry is true of pseudo-Riemannian manifolds as well. This allows one to speak of the Levi-Civita connection on a pseudo-Riemannian manifold along with the associated curvature

tensor. On the other hand, there are many theorems in Riemannian geometry which do not hold in the generalized case. For example, it is *not* true that every smooth manifold admits a pseudo-Riemannian metric of a given signature; there are certain topological obstructions. Furthermore, a submanifold does not always inherit the structure of a pseudo-Riemannian manifold; for example, the metric tensor becomes zero on any light-like curve. The Clifton–Pohl torus provides an example of a pseudo-Riemannian manifold that is compact but not complete, a combination of properties that the Hopf–Rinow theorem disallows for Riemannian manifolds.[3]

## 14.5   See also

- Spacetime

- Orientable manifold

- Hyperbolic partial differential equation

- Causality conditions

- Globally hyperbolic manifold

## 14.6   Notes

[1]  Benn & Tucker (1987), p. 172.

[2]  Bishop & Goldberg (1968), p. 208

[3]  O'Neill (1983), p. 193.

## 14.7   References

- Benn, I.M.; Tucker, R.W. (1987), *An introduction to Spinors and Geometry with Applications in Physics* (First published 1987 ed.), Adam Hilger, ISBN 0-85274-169-3

- Bishop, Richard L.; Goldberg, Samuel I. (1968), *Tensor Analysis on Manifolds* (First Dover 1980 ed.), The Macmillan Company, ISBN 0-486-64039-6

- Chen, Bang-Yen (2011), *Pseudo-Riemannian Geometry, [delta]-invariants and Applications*, World Scientific Publisher, ISBN 978-981-4329-63-7

- O'Neill, Barrett (1983), *Semi-Riemannian Geometry With Applications to Relativity*, Pure and Applied Mathematics **103**, Academic Press, ISBN 9780080570570

- Vrănceanu, G.; Roşca, R. (1976), *Introduction to Relativity and Pseudo-Riemannian Geometry*, Bucarest: Editura Academiei Republicii Socialiste România.

# Chapter 15

# Tensor

This article is about tensors on a single vector space. For tensor fields, see Tensor field. For other uses, see Tensor (disambiguation).

**Tensors** are geometric objects that describe linear re-

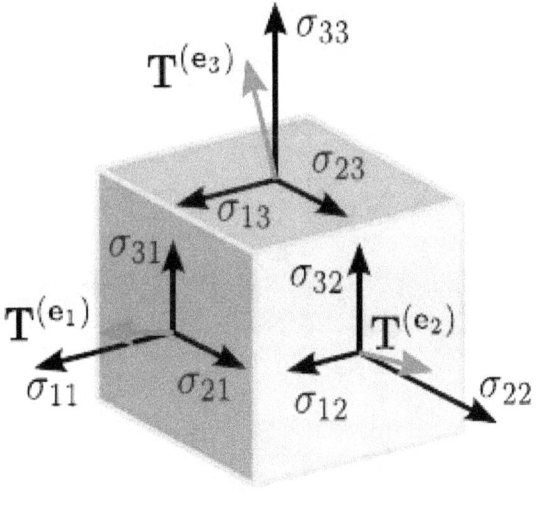

*Cauchy stress tensor, a second-order tensor. The tensor's components, in a three-dimensional Cartesian coordinate system, form the matrix*

$$\sigma = \left[ \mathbf{T}^{(e_1)} \mathbf{T}^{(e_2)} \mathbf{T}^{(e_3)} \right]$$

$$= \begin{bmatrix} \sigma_{11} & \sigma_{12} & \sigma_{13} \\ \sigma_{21} & \sigma_{22} & \sigma_{23} \\ \sigma_{31} & \sigma_{32} & \sigma_{33} \end{bmatrix}$$

*whose columns are the stresses (forces per unit area) acting on the $e_1$, $e_2$, and $e_3$ faces of the cube.*

lations between geometric vectors, scalars, and other tensors. Elementary examples of such relations include the dot product, the cross product, and linear maps. Euclidean vectors, often used in physics and engineering applications, and scalars themselves are also tensors.[1] A more sophisticated example is the Cauchy stress tensor $\mathbf{T}$, which takes a direction $\mathbf{v}$ as input and produces the stress $\mathbf{T}^{(v)}$ on the surface normal to this vector for output, thus expressing a relationship between these two vectors, shown in the figure (right).

In terms of a coordinate basis or fixed frame of reference, a tensor can be represented as an organized multidimensional array of numerical values. The **order** (also *degree* or *rank*) of a tensor is the dimensionality of the array needed to represent it, or equivalently, the number of indices needed to label a component of that array. For example, a linear map is represented by a matrix (a 2-dimensional array) in a basis, and therefore is a 2nd-order tensor. A vector is represented as a 1-dimensional array in a basis, and is a 1st-order tensor. Scalars are single numbers and are thus 0th-order tensors. Because they express a relationship between vectors, tensors themselves must be independent of a particular choice of coordinate system. The coordinate independence of a tensor then takes the form of a covariant and/or contravariant transformation law that relates the array computed in one coordinate system to that computed in another one. The precise form of the transformation law determines the *type* (or *valence*) of the tensor. The tensor type is a pair of natural numbers $(n, m)$, where n is the number of contravariant indices and m is the number of covariant indices. The total order of a tensor is the sum of these two numbers.

Tensors are important in physics because they provide a concise mathematical framework for formulating and solving physics problems in areas such as elasticity, fluid mechanics, and general relativity. Tensors were first conceived by Tullio Levi-Civita and Gregorio Ricci-Curbastro, who continued the earlier work of Bernhard Riemann and Elwin Bruno Christoffel and others, as part of the *absolute differential calculus*. The concept enabled an alternative formulation of the intrinsic differential geometry of a manifold in the form of the Riemann curvature tensor.[2]

## 15.1 Definition

There are several approaches to defining tensors. Although seemingly different, the approaches just describe the same geometric concept using different languages and at different levels of abstraction.

### 15.1.1   As multidimensional arrays

Just as a vector in an n-dimensional space is represented by a one-dimensional array of length n with respect to a given basis, any tensor with respect to a basis is represented by a multidimensional array. For example, a linear operator is represented in a basis as a two-dimensional square $n \times n$ array. The numbers in the multidimensional array are known as the *scalar components* of the tensor or simply its *components*. They are denoted by indices giving their position in the array, as subscripts and superscripts, following the symbolic name of the tensor. For example, the components of an order 2 tensor T could be denoted $T{ij}$, where i and j are indices running from 1 to n, or also by $T^{ij}$. Whether an index is displayed as a superscript or subscript depends on the transformation properties of the tensor, described below. The total number of indices required to uniquely select each component is equal to the dimension of the array, and is called the *order*, *degree* or *rank* of the tensor. However, the term "rank" generally has another meaning in the context of matrices and tensors.

Just as the components of a vector change when we change the basis of the vector space, the components of a tensor also change under such a transformation. Each tensor comes equipped with a *transformation law* that details how the components of the tensor respond to a change of basis. The components of a vector can respond in two distinct ways to a change of basis (see covariance and contravariance of vectors), where the new basis vectors $\hat{\mathbf{e}}_i$ are expressed in terms of the old basis vectors $\mathbf{e}_j$ as,

$$\hat{\mathbf{e}}_i = \sum_{j=1}^{n} \mathbf{e}_j R_i^j = \mathbf{e}_j R_i^j.$$

Here $R^j{}_i$ are the entries of the change of basis matrix, and in the rightmost expression the summation sign was suppressed: this is the Einstein summation convention, which will be used throughout this article.[Note 1] The components $v^j$ of a column vector **v** transform with the inverse of the matrix $R$,

$$\hat{v}^i = (R^{-1})_j^i v^j,$$

where the hat denotes the components in the new basis. This is called a *contravariant* transformation law, because the vector transforms by the *inverse* of the change of basis. In contrast, the components, $w_i$, of a covector (or row vector), w transform with the matrix R itself,

$$\hat{w}_i = w_j R_i^j.$$

This is called a *covariant* transformation law, because the covector transforms by the *same matrix* as the change of basis matrix. The components of a more general tensor transform by some combination of covariant and contravariant transformations, with one transformation law for each index. If the transformation matrix of an index is the inverse matrix of the basis transformation, then the index is called *contravariant* and is traditionally denoted with an upper index (superscript). If the transformation matrix of an index is the basis transformation itself, then the index is called *covariant* and is denoted with a lower index (subscript).

The transformation law for an order $p + q$ tensor with $p$ contravariant indices and $q$ covariant indices is thus given as,

$$\hat{T}_{j_1', \ldots, j_q'}^{i_1', \ldots, i_p'} = (R^{-1})_{i_1}^{i_1'} \cdots (R^{-1})_{i_p}^{i_p'} \; T_{j_1, \ldots, j_q}^{i_1, \ldots, i_p} R_{j_1'}^{j_1} \cdots R_{j_q'}^{j_q}.$$

Here the primed indices denote components in the new coordinates, and the unprimed indices denote the components in the old coordinates. Such a tensor is said to be of order or *type* $(p, q)$. The terms "order", "type", "rank", "valence", and "degree" are all sometimes used for the same concept. Here, the term "order" or "total order" will be used for the total dimension of the array (or its generalisation in other definitions), $p+q$ in the preceding example, and the term "type" for the pair giving the number contravariant and covariant indices. A tensor of type $(p, q)$ is also called as a $(p, q)$-tensor for short.

As an example, the matrix of a linear operator in a basis is a rectangular array $T$ that transforms under a change of basis matrix $R = (R_i^j)$ by $\hat{T} = R^{-1}TR$. In terms of the individual matrix entries, this transformation law has the form $\hat{T}_{j'}^{i'} = (R^{-1})_i^{i'} T_j^i R_{j'}^j$, so the tensor corresponding to the matrix of a linear operator has one covariant and one contravariant index: it is of type (1,1). A linear operator itself does not actually depend on a basis: it is just a linear map that accepts a vector as an argument and produces another vector. The transformation law for the matrix of a linear operator is consistent with the transformation law for a contravariant vector, so that the action of a linear operator on a contravariant vector is represented in coordinates as the matrix product of their respective coordinate representations. That is, the components $(Tv)^i$ are given by $(Tv)^i = T_j^i v^j$. These components transform contravariantly, since

$$(\widehat{Tv})^{i'} = \hat{T}_{j'}^{i'} \hat{v}^{j'} = \left[ (R^{-1})_i^{i'} T_j^i R_{j'}^j \right] \left[ (R^{-1})_j^{j'} v^j) \right]$$
$$= (R^{-1})_i^{i'} (Tv)^i.$$

This discussion motivates the following formal definition:[3]

**Definition.** A tensor of type $(p, q)$ is an assignment of a multidimensional array

$$T^{i_1 \dots i_p}_{j_1 \dots j_q}[\mathbf{f}]$$

to each basis $\mathbf{f} = (\mathbf{e}_1, \dots, \mathbf{e}n)$ of a fixed $n$-dimensional vector space such that, if we apply the change of basis

$$\mathbf{f} \mapsto \mathbf{f} \cdot R = \left( \mathbf{e}_i R^i_1, \dots, \mathbf{e}_i R^i_n \right)$$

then the multidimensional array obeys the transformation law

$$T^{i'_1 \dots i'_p}_{j'_1 \dots j'_q}[\mathbf{f} \cdot R] = (R^{-1})^{i'_1}_{i_1} \cdots (R^{-1})^{i'_p}_{i_p}$$
$$T^{i_1, \dots, i_p}_{j_1, \dots, j_q}[\mathbf{f}] \, R^{j_1}_{j'_1} \cdots R^{j_q}_{j'_q}.$$

The definition of a tensor as a multidimensional array satisfying a transformation law traces back to the work of Ricci.[2] This definition is still used in some physics and engineering text books.[4][5]

**Tensor fields**

Main article: Tensor field

In many applications, especially in differential geometry and physics, it is natural to consider a tensor with components that are functions of the point in a space. This was the setting of Ricci's original work. In modern mathematical terminology such an object is called a tensor field, often referred to simply as a tensor.[2]

In this context, a coordinate basis is often chosen for the tangent vector space. The transformation law may then be expressed in terms of partial derivatives of the coordinate functions,

$$\bar{x}^i(x^1, \dots, x^n),$$

defining a coordinate transformation,[2]

$$\hat{T}^{i'_1 \dots i'_p}_{j'_1 \dots j'_q}(\bar{x}^1, \dots, \bar{x}^n) = \frac{\partial \bar{x}^{i'_1}}{\partial x^{i_1}} \cdots \frac{\partial \bar{x}^{i'_p}}{\partial x^{i_p}} \frac{\partial x^{j_1}}{\partial \bar{x}^{j'_1}}$$

$$\cdots \frac{\partial x^{j_q}}{\partial \bar{x}^{j'_q}} T^{i_1 \dots i_p}_{j_1 \dots j_q}(x^1, \dots, x^n).$$

## 15.1.2    As multilinear maps

A downside to the definition of a tensor using the multidimensional array approach is that it is not apparent from the definition that the defined object is indeed basis independent, as is expected from an intrinsically geometric object. Although it is possible to show that transformation

laws indeed ensure independence from the basis, sometimes a more intrinsic definition is preferred. One approach[6] is to define a tensor as a multilinear map. In that approach a type $(p, q)$ tensor $T$ is defined as a map,

$$T : \underbrace{V^* \times \cdots \times V^*}_{p\text{copies}} \times \underbrace{V \times \cdots \times V}_{q\text{copies}} \to \mathbf{R},$$

where $V$ is a (finite-dimensional) vector space and $V^*$ is the corresponding dual space of covectors, which is linear in each of its arguments.

By applying a multilinear map $T$ of type $(p, q)$ to a basis $\{\mathbf{e}_j\}$ for $V$ and a canonical cobasis $\{\varepsilon^i\}$ for $V^*$,

$$T^{i_1 \dots i_p}_{j_1 \dots j_q} \equiv T(\varepsilon^{i_1}, \dots, \varepsilon^{i_p}, \mathbf{e}_{j_1}, \dots, \mathbf{e}_{j_q}),$$

a $(p+q)$-dimensional array of components can be obtained. A different choice of basis will yield different components. But, because $T$ is linear in all of its arguments, the components satisfy the tensor transformation law used in the multilinear array definition. The multidimensional array of components of $T$ thus form a tensor according to that definition. Moreover, such an array can be realized as the components of some multilinear map $T$. This motivates viewing multilinear maps as the intrinsic objects underlying tensors.

In viewing a tensor as a multilinear map, it is conventional to identify the vector space $V$ with the space of linear functionals on the dual of $V$, the double dual $V^{**}$. There is always a natural linear map from $V$ to its double dual, given by evaluating a linear form in $V^*$ against a vector in $V$. This linear mapping is an isomorphism in finite dimensions, and it is often then expedient to identify $V$ with its double dual.

## 15.1.3    Using tensor products

Main article: Tensor (intrinsic definition)

For some mathematical applications, a more abstract approach is sometimes useful. This can be achieved by defining tensors in terms of elements of tensor products of vector spaces, which in turn are defined through a universal property. A type $(p, q)$ tensor is defined in this context as an element of the tensor product of vector spaces,[7]

$$T \in \underbrace{V \otimes \cdots \otimes V}_{p\text{copies}} \otimes \underbrace{V^* \otimes \cdots \otimes V^*}_{q\text{copies}}.$$

If $vi$ is a basis of $V$ and $wj$ is a basis of $W$, then the tensor product $V \otimes W$ has a natural basis $vi \otimes wj$. The components of a tensor $T$ are the coefficients of the tensor with respect to the basis obtained from a basis $\{e_i\}$ for $V$ and its dual $\{\varepsilon^j\}$, i.e.

$$T = T^{i_1 \ldots i_p}_{j_1 \ldots j_q} \, \mathbf{e}_{i_1} \otimes \cdots \otimes \mathbf{e}_{i_p} \otimes \varepsilon^{j_1} \otimes \cdots \otimes \varepsilon^{j_q}.$$

Using the properties of the tensor product, it can be shown that these components satisfy the transformation law for a type $(p, q)$ tensor. Moreover, the universal property of the tensor product gives a 1-to-1 correspondence between tensors defined in this way and tensors defined as multilinear maps.

Tensor products can be defined in great generality – for example, involving arbitrary modules over a ring. In principle, one could define a "tensor" simply to be an element of any tensor product. However, the mathematics literature usually reserves the term *tensor* for an element of a tensor product of a single vector space $V$ and its dual, as above.

### 15.1.4  Tensors in infinite dimensions

This discussion of tensors so far assumes finite dimensionality of the spaces involved, where the spaces of tensors obtained by each of these constructions are naturally isomorphic.[Note 2] Constructions of spaces of tensors based on the tensor product and multilinear mappings can be generalized, essentially without modification, to vector bundles or coherent sheaves.[8] For infinite-dimensional vector spaces, inequivalent topologies lead to inequivalent notions of tensor, and these various isomorphisms may or may not hold depending on what exactly is meant by a tensor (see topological tensor product). In some applications, it is the tensor product of Hilbert spaces that is intended, whose properties are the most similar to the finite-dimensional case. A more modern view is that it is the tensors' structure as a symmetric monoidal category that encodes their most important properties, rather than the specific models of those categories

## 15.2   Examples

See also: Dyadic tensor

This table shows important examples of tensors, including both tensors on vector spaces and tensor fields on manifolds. The tensors are classified according to their type $(n, m)$, where $n$ is the number of contravariant indices, $m$ is the number of covariant indices, and $n + m$ gives the total order of the tensor. For example, a bilinear form is the same thing as a $(0, 2)$-tensor; an inner product is an example of a $(0,$

2)-tensor, but not all $(0, 2)$-tensors are inner products. In the $(0, M)$-entry of the table, $M$ denotes the dimensionality of the underlying vector space or manifold because for each dimension of the space, a separate index is needed to select that dimension to get a maximally covariant antisymmetric tensor.

Raising an index on an $(n, m)$-tensor produces an $(n + 1, m - 1)$-tensor; this can be visualized as moving diagonally down and to the left on the table. Symmetrically, lowering an index can be visualized as moving diagonally up and to the right on the table. Contraction of an upper with a lower index of an $(n, m)$-tensor produces an $(n - 1, m - 1)$-tensor; this can be visualized as moving diagonally up and to the left on the table.

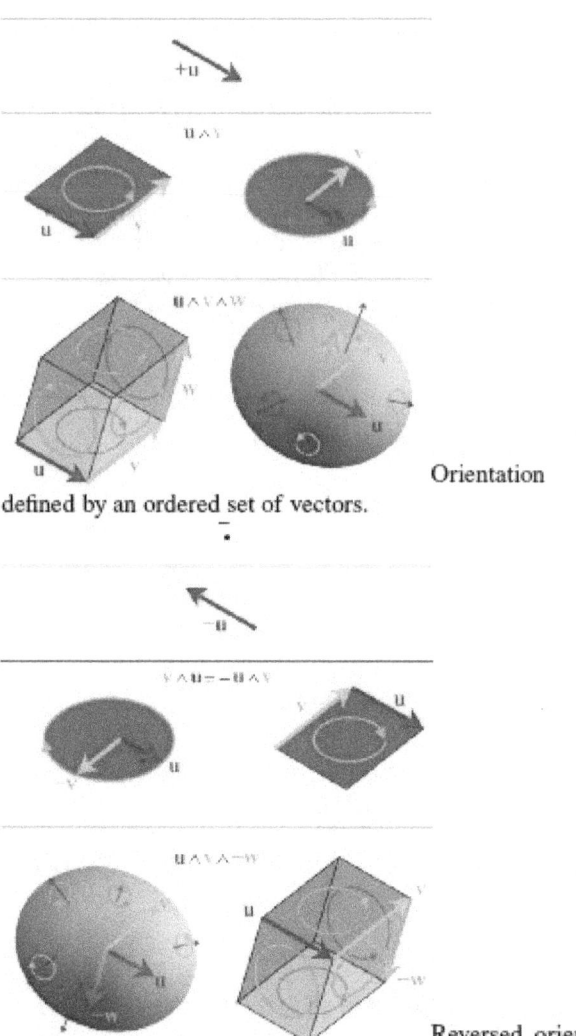

Orientation defined by an ordered set of vectors.

Reversed orientation corresponds to negating the exterior product. Geometric interpretation of grade $n$ elements in a real exterior algebra for $n = 0$ (signed point), 1 (directed line segment, or vector), 2 (oriented plane element), 3 (oriented volume). The exterior product of $n$ vectors can be visualized as any $n$-dimensional shape (e.g. $n$-parallelotope, $n$-ellipsoid); with magnitude (hypervolume), and orientation defined by that on its $n-1$-dimensional boundary and on which side the interior is.[10][11]

## 15.3   Notation

### 15.3.1   Ricci calculus

Ricci calculus is the modern formalism and notation for tensor indices: indicating inner and outer products, covariance and contravariance, summations of tensor components, symmetry and antisymmetry, and partial and covariant derivatives.

### 15.3.2   Einstein summation convention

The Einstein summation convention dispenses with writing summation signs, leaving the summation implicit. Any repeated index symbol is summed over: if the index i is used twice in a given term of a tensor expression, it means that the term is to be summed for all i. Several distinct pairs of indices may be summed this way.

### 15.3.3   Penrose graphical notation

Penrose graphical notation is a diagrammatic notation which replaces the symbols for tensors with shapes, and their indices by lines and curves. It is independent of basis elements, and requires no symbols for the indices.

### 15.3.4   Abstract index notation

The abstract index notation is a way to write tensors such that the indices are no longer thought of as numerical, but rather are indeterminates. This notation captures the expressiveness of indices and the basis-independence of index-free notation.

### 15.3.5   Component-free notation

A component-free treatment of tensors uses notation that emphasises that tensors do not rely on any basis, and is defined in terms of the tensor product of vector spaces.

## 15.4   Operations

There are a number of basic operations that may be conducted on tensors that again produce a tensor. The linear nature of tensor implies that two tensors of the same type may be added together, and that tensors may be multiplied by a scalar with results analogous to the scaling of a vector. On components, these operations are simply performed component for component. These operations do not change the type of the tensor, however there also exist operations that change the type of the tensors.

### 15.4.1   Tensor product

Main article: Tensor product

The tensor product takes two tensors, $S$ and $T$, and produces a new tensor, $S \otimes T$, whose order is the sum of the orders of the original tensors. When described as multilinear maps, the tensor product simply multiplies the two tensors, i.e.

$$(S \otimes T)(v_1, \ldots, v_n, v_{n+1}, \ldots, v_{n+m}) = S($$

$$v_1, \ldots, v_n)T(v_{n+1}, \ldots, v_{n+m}),$$

which again produces a map that is linear in all its arguments. On components the effect similarly is to multiply the components of the two input tensors, i.e.

$$(S \otimes T)^{i_1 \ldots i_l i_{l+1} \ldots i_{l+n}}_{j_1 \ldots j_k j_{k+1} \ldots j_{k+m}} = S^{i_1 \ldots i_l}_{j_1 \ldots j_k} T^{i_{l+1} \ldots i_{l+n}}_{j_{k+1} \ldots j_{k+m}},$$

If S is of type $(l, k)$ and T is of type $(n, m)$, then the tensor product $S \otimes T$ has type $(l+n, k+m)$.

### 15.4.2   Contraction

Main article: Tensor contraction

Tensor contraction is an operation that reduces a type $(n, m)$ tensor to a type $(n-1, m-1)$ tensor. It thereby reduces the total order of a tensor by two. In terms of components, the operation is achieved by summing over one contravariant and one covariant index of tensor. For example, a $(1, 1)$-tensor $T^j_i$ can be contracted to a scalar through

$$T^i_i$$

Where the summation is again implied. When the $(1, 1)$-tensor is interpreted as a linear map, this operation is known as the trace.

The contraction is often used in conjunction with the tensor product to contract an index from each tensor.

The contraction can also be understood in terms of the definition of a tensor as an element of a tensor product of copies of the space $V$ with the space $V^*$ by first decomposing the tensor into a linear combination of simple tensors, and then applying a factor from $V^*$ to a factor from $V$. For example, a tensor

$$T \in V \otimes V \otimes V^*$$

can be written as a linear combination

$$T = v_1 \otimes w_1 \otimes \alpha_1 + v_2 \otimes w_2 \otimes \alpha_2 + \cdots + v_N \otimes w_N \otimes \alpha_N.$$

The contraction of $T$ on the first and last slots is then the vector

$$\alpha_1(v_1)w_1 + \alpha_2(v_2)w_2 + \cdots + \alpha_N(v_N)w_N.$$

In a vector space with an inner product (also known as a metric) $g$, the term contraction is used for removing two contravariant or two covariant indices by forming a trace with the metric tensor or its inverse. For example, a $(2, 0)$-tensor $T^{ij}$ can be contracted to a scalar through

$$T^{ij}g_{ij}$$

(yet again assuming the summation convention).

### 15.4.3  Raising or lowering an index

Main article: Raising and lowering indices

When a vector space is equipped with a nondegenerate bilinear form (or *metric tensor* as it is often called in this context), operations can be defined that convert a contravariant (upper) index into a covariant (lower) index and vice versa. A metric tensor is a (symmetric) $(0, 2)$-tensor, it is thus possible to contract an upper index of a tensor with one of lower indices of the metric tensor in the product. This produces a new tensor with the same index structure as the previous, but with lower index in the position of the contracted upper index. This operation is quite graphically known as *lowering an index*.

Conversely, the inverse operation can be defined, and is called *raising an index*. This is equivalent to a similar contraction on the product with a $(2, 0)$-tensor. This *inverse metric tensor* has components that are the matrix inverse of those if the metric tensor.

## 15.5  Applications

### 15.5.1  Continuum mechanics

Important examples are provided by continuum mechanics. The stresses inside a solid body or fluid are described by a tensor. The stress tensor and strain tensor are both second-order tensors, and are related in a general linear elastic material by a fourth-order elasticity tensor. In detail, the tensor quantifying stress in a 3-dimensional solid object has components that can be conveniently represented as a 3 × 3 array. The three faces of a cube-shaped infinitesimal volume segment of the solid are each subject to some given force. The force's vector components are also three in number. Thus, 3 × 3, or 9 components are required to describe the stress at this cube-shaped infinitesimal segment. Within the bounds of this solid is a whole mass of varying stress quantities, each requiring 9 quantities to describe. Thus, a second-order tensor is needed.

If a particular surface element inside the material is singled out, the material on one side of the surface will apply a force on the other side. In general, this force will not be orthogonal to the surface, but it will depend on the orientation of the surface in a linear manner. This is described by a tensor of type $(2, 0)$, in linear elasticity, or more precisely by a tensor field of type $(2, 0)$, since the stresses may vary from point to point.

### 15.5.2  Other examples from physics

Common applications include

- Electromagnetic tensor (or Faraday's tensor) in electromagnetism

- Finite deformation tensors for describing deformations and strain tensor for strain in continuum mechanics

- Permittivity and electric susceptibility are tensors in anisotropic media

- Four-tensors in general relativity (e.g. stress–energy tensor), used to represent momentum fluxes

- Spherical tensor operators are the eigenfunctions of the quantum angular momentum operator in spherical coordinates

- Diffusion tensors, the basis of diffusion tensor imaging, represent rates of diffusion in biologic environments

- Quantum mechanics and quantum computing utilise tensor products for combination of quantum states

### 15.5.3   Applications of tensors of order > 2

The concept of a tensor of order two is often conflated with that of a matrix. Tensors of higher order do however capture ideas important in science and engineering, as has been shown successively in numerous areas as they develop. This happens, for instance, in the field of computer vision, with the trifocal tensor generalizing the fundamental matrix.

The field of nonlinear optics studies the changes to material polarization density under extreme electric fields. The polarization waves generated are related to the generating electric fields through the nonlinear susceptibility tensor. If the polarization **P** is not linearly proportional to the electric field **E**, the medium is termed *nonlinear*. To a good approximation (for sufficiently weak fields, assuming no permanent dipole moments are present), **P** is given by a Taylor series in **E** whose coefficients are the nonlinear susceptibilities:

$$\frac{P_i}{\varepsilon_0} = \sum_j \chi_{ij}^{(1)} E_j + \sum_{jk} \chi_{ijk}^{(2)} E_j E_k + \sum_{jk\ell} \chi_{ijk\ell}^{(3)} E_j E_k E_\ell + \cdots .$$

Here $\chi^{(1)}$ is the linear susceptibility, $\chi^{(2)}$ gives the Pockels effect and second harmonic generation, and $\chi^{(3)}$ gives the Kerr effect. This expansion shows the way higher-order tensors arise naturally in the subject matter.

## 15.6   Generalizations

### 15.6.1   Tensor products of vector spaces

The vector spaces of a tensor product need not be the same, and sometimes the elements of such a more general tensor product are called "tensors". For example, an element of the tensor product space $V \otimes W$ is a second-order "tensor" in this more general sense,[12] and an order-$d$ tensor may likewise be defined as an element of a tensor product of $d$ different vector spaces.[13] A type $(n, m)$ tensor, in the sense defined previously, is also a tensor of order $n + m$ in this more general sense.

### 15.6.2   Tensors in infinite dimensions

The notion of a tensor can be generalized in a variety of ways to infinite dimensions. One, for instance, is via the tensor product of Hilbert spaces.[14] Another way of generalizing the idea of tensor, common in nonlinear analysis, is via the multilinear maps definition where instead of using finite-dimensional vector spaces and their algebraic duals, one uses infinite-dimensional Banach spaces and their

continuous dual.[15] Tensors thus live naturally on Banach manifolds.[16]

### 15.6.3   Tensor densities

Main article: Tensor density

The concept of a tensor field can be generalized by considering objects that transform differently. An object that transforms as an ordinary tensor field under coordinate transformations, except that it is also multiplied by the determinant of the Jacobian of the inverse coordinate transformation to the $w^{\text{th}}$ power, is called a tensor density with weight $w$ .[17] Invariantly, in the language of multilinear algebra, one can think of tensor densities as multilinear maps taking their values in a density bundle such as the (1-dimensional) space of $n$-forms (where $n$ is the dimension of the space), as opposed to taking their values in just **R**. Higher "weights" then just correspond to taking additional tensor products with this space in the range.

A special case are the scalar densities. Scalar 1-densities are especially important because it makes sense to define their integral over a manifold. They appear, for instance, in the Einstein–Hilbert action in general relativity. The most common example of a scalar 1-density is the volume element, which in the presence of a metric tensor $g$ is the square root of its determinant in coordinates, denoted $\sqrt{\det g}$ . The metric tensor is a covariant tensor of order 2, and so its determinant scales by the square of the coordinate transition:

$$\det(g') = \left( \det \frac{\partial x}{\partial x'} \right)^2 \det(g)$$

which is the transformation law for a scalar density of weight +2.

More generally, any tensor density is the product of an ordinary tensor with a scalar density of the appropriate weight. In the language of vector bundles, the determinant bundle of the tangent bundle is a line bundle that can be used to 'twist' other bundles $w$ times. While locally the more general transformation law can indeed be used to recognise these tensors, there is a global question that arises, reflecting that in the transformation law one may write either the Jacobian determinant, or its absolute value. Non-integral powers of the (positive) transition functions of the bundle of densities make sense, so that the weight of a density, in that sense, is not restricted to integer values. Restricting to changes of coordinates with positive Jacobian determinant is possible on orientable manifolds, because there is a consistent global way to eliminate the minus signs; but otherwise the line bundle of densities and the line bundle of *n*-forms are distinct. For more on the intrinsic meaning, see density on a manifold.

### 15.6.4 Spinors

Main article: Spinor

When changing from one orthonormal basis (called a *frame*) to another by a rotation, the components of a tensor transform by that same rotation. This transformation does not depend on the path taken through the space of frames. However, the space of frames is not simply connected (see orientation entanglement and plate trick): there are continuous paths in the space of frames with the same beginning and ending configurations that are not deformable one into the other. It is possible to attach an additional discrete invariant to each frame called the "spin" that incorporates this path dependence, and which turns out to have values of ±1. A spinor is an object that transforms like a tensor under rotations in the frame, apart from a possible sign that is determined by the spin.

## 15.7 History

The concepts of later tensor analysis arose from the work of Carl Friedrich Gauss in differential geometry, and the formulation was much influenced by the theory of algebraic forms and invariants developed during the middle of the nineteenth century.[18] The word "tensor" itself was introduced in 1846 by William Rowan Hamilton[19] to describe something different from what is now meant by a tensor.[Note 3] The contemporary usage was introduced by Woldemar Voigt in 1898.[20]

Tensor calculus was developed around 1890 by Gregorio Ricci-Curbastro under the title *absolute differential calculus*, and originally presented by Ricci in 1892.[21] It was made accessible to many mathematicians by the publication of Ricci and Tullio Levi-Civita's 1900 classic text *Méthodes de calcul différentiel absolu et leurs applications* (Methods of absolute differential calculus and their applications).[22]

In the 20th century, the subject came to be known as *tensor analysis*, and achieved broader acceptance with the introduction of Einstein's theory of general relativity, around 1915. General relativity is formulated completely in the language of tensors. Einstein had learned about them, with great difficulty, from the geometer Marcel Grossmann.[23] Levi-Civita then initiated a correspondence with Einstein to correct mistakes Einstein had made in his use of tensor analysis. The correspondence lasted 1915–17, and was characterized by mutual respect:

> I admire the elegance of your method of computation; it must be nice to ride through these fields upon the horse of true mathematics while the like of us have to make our way laboriously on foot.
> — Albert Einstein, The Italian Mathematicians of Relativity[24]

Tensors were also found to be useful in other fields such as continuum mechanics. Some well-known examples of tensors in differential geometry are quadratic forms such as metric tensors, and the Riemann curvature tensor. The exterior algebra of Hermann Grassmann, from the middle of the nineteenth century, is itself a tensor theory, and highly geometric, but it was some time before it was seen, with the theory of differential forms, as naturally unified with tensor calculus. The work of Élie Cartan made differential forms one of the basic kinds of tensors used in mathematics.

From about the 1920s onwards, it was realised that tensors play a basic role in algebraic topology (for example in the Künneth theorem).[25] Correspondingly there are types of tensors at work in many branches of abstract algebra, particularly in homological algebra and representation theory. Multilinear algebra can be developed in greater generality than for scalars coming from a field. For example, scalars can come from a ring. But the theory is then less geometric and computations more technical and less algorithmic.[26] Tensors are generalized within category theory by means of the concept of monoidal category, from the 1960s.[27]

## 15.8 See also

### 15.8.1 Foundational

- Cartesian tensor
- Fibre bundle
- Glossary of tensor theory
- Multilinear projection
- One-form
- Tensor product of modules

### 15.8.2 Applications

- Application of tensor theory in engineering
- Covariant derivative

- Curvature

- Diffusion tensor MRI

- Einstein field equations

- Fluid mechanics

- Multilinear subspace learning

- Riemannian geometry

- Structure tensor

- Tensor decomposition

- Tensor derivative

- Tensor software

## 15.9   Notes

[1] The Einstein summation convention, in brief, requires the sum to be taken over all values of the index whenever the same symbol appears as a subscript and superscript in the same term. For example, under this convention $B_i C^i = B_1 C^1 + B_2 C^2 + \cdots B_n C^n$

[2] The double duality isomorphism, for instance, is used to identify $V$ with the double dual space $V^{**}$, which consists of multilinear forms of degree one on $V^*$. It is typical in linear algebra to identify spaces that are naturally isomorphic, treating them as the same space.

[3] Namely, the norm operation in a certain type of algebraic system (now known as a Clifford algebra).

## 15.10   References

### 15.10.1   General

- Bishop, Richard L.; Samuel I. Goldberg (1980) [1968]. *Tensor Analysis on Manifolds*. Dover. ISBN 978-0-486-64039-6.

- Danielson, Donald A. (2003). *Vectors and Tensors in Engineering and Physics* (2/e ed.). Westview (Perseus). ISBN 978-0-8133-4080-7.

- Dimitrienko, Yuriy (2002). *Tensor Analysis and Nonlinear Tensor Functions*. Kluwer Academic Publishers (Springer). ISBN 1-4020-1015-X.

- Jeevanjee, Nadir (2011). *An Introduction to Tensors and Group Theory for Physicists*. Birkhauser. ISBN 978-0-8176-4714-8.

- Lawden, D. F. (2003). *Introduction to Tensor Calculus, Relativity and Cosmology* (3/e ed.). Dover. ISBN 978-0-486-42540-5.

- Lebedev, Leonid P.; Michael J. Cloud (2003). *Tensor Analysis*. World Scientific. ISBN 978-981-238-360-0.

- Lovelock, David; Hanno Rund (1989) [1975]. *Tensors, Differential Forms, and Variational Principles*. Dover. ISBN 978-0-486-65840-7.

- Munkres, James, *Analysis on Manifolds*, Westview Press, 1991. Chapter six gives a "from scratch" introduction to covariant tensors.

- Ricci, Gregorio; Levi-Civita, Tullio (March 1900). "Méthodes de calcul différentiel absolu et leurs applications" (PDF). *Mathematische Annalen* (Springer) **54** (1–2): 125–201. doi:10.1007/BF01454201.

- Kay, David C (1988-04-01). *Schaum's Outline of Tensor Calculus*. McGraw-Hill. ISBN 978-0-07-033484-7.

- Schutz, Bernard, *Geometrical methods of mathematical physics*, Cambridge University Press, 1980.

- Synge J.L., Schild A. (1949). *Tensor Calculus*. first Dover Publications 1978 edition. ISBN 978-0-486-63612-2.

### 15.10.2   Specific

[1] "What is a Tensor?". *Dissemination of IT for the Promotion of Materials Science*. University of Cambridge.

[2] Kline, Morris (1972). *Mathematical thought from ancient to modern times, Vol. 3*. Oxford University Press. pp. 1122–1127. ISBN 0195061373.

[3] Sharpe, R. W. (1997). *Differential Geometry: Cartan's Generalization of Klein's Erlangen Program*. Berlin, New York: Springer-Verlag. p. 194. ISBN 978-0-387-94732-7.

[4] Marion, J.B.; Thornton, S.T. (1995). *Classical Dynamics of Particles and Systems* (4th ed.). Saunders College Publishing. p. 424. ISBN 978-0-03-098967-4.

[5] Griffiths, D.J. (1999). *Introduction to Electrodynamics* (3 ed.). Prentice Hall. pp. 11–12 and 535–. ISBN 978-0-13-805326-0.

[6] For instance, John Lee (2000), *Introduction to smooth manifolds*, Springer, p. 173, ISBN 0-387-95495-3

[7] Hazewinkel, Michiel, ed. (2001), "Affine tensor", *Encyclopedia of Mathematics*, Springer, ISBN 978-1-55608-010-4

[8] Bourbaki, "Algebra", III, where the case of finitely generated projective modules is treated. The global sections of sections of a vector bundle over a compact space form a projective module over the ring of smooth functions. All statements for coherent sheaves are true locally.

[9] Paul Bamberg; Shlomo Sternberg (1991). *A Course in Mathematics for Students of Physics: Volume 2*. Cambridge University Press. p. 669. ISBN 978-0-521-40650-5.

[10] R. Penrose (2007). *The Road to Reality*. Vintage books. ISBN 0-679-77631-1.

[11] J.A. Wheeler, C. Misner, K.S. Thorne (1973). *Gravitation*. W.H. Freeman & Co. p. 83. ISBN 0-7167-0344-0.

[12] M. D. Maia (2011). *Geometry of the Fundamental Interactions: On Riemann's Legacy to High Energy Physics and Cosmology*. Springer Science & Business Media. p. 48. ISBN 978-1-4419-8273-5.

[13] Leslie Hogben, ed. (2013). *Handbook of Linear Algebra, Second Edition* (2nd ed.). CRC Press. p. "15-7". ISBN 978-1-4665-0729-6.

[14] Segal, I. E. (January 1956). "Tensor Algebras Over Hilbert Spaces. I". *Transactions of the American Mathematical Society* (American Mathematical Society) **81** (1): 106–134. doi:10.2307/1992855. JSTOR 1992855.

[15] Abraham, Ralph; Marsden, Jerrold E.; Ratiu, Tudor S. (February 1988) [First Edition 1983]. "Chapter 5 Tensors". *Manifolds, Tensor Analysis and Applications*. Applied Mathematical Sciences, v. 75 75 (2nd ed.). New York: Springer-Verlag. pp. 338–339. ISBN 0-387-96790-7. OCLC 18562688. Elements of $T^r_s$ are called tensors on E, [...].

[16] Lang, Serge (1972). *Differential manifolds*. Reading, Massachusetts: Addison-Wesley Pub. Co. ISBN 0201041669.

[17] Hazewinkel, Michiel, ed. (2001), "Tensor density", *Encyclopedia of Mathematics*, Springer, ISBN 978-1-55608-010-4

[18] Reich, Karin (1994). *Die Entwicklung des Tensorkalküls*. Science networks historical studies, v. 11. Birkhäuser. ISBN 978-3-7643-2814-6. OCLC 31468174.

[19] Hamilton, William Rowan (1854–1855). Wilkins, David R., ed. "On some Extensions of Quaternions" (PDF). *Philosophical Magazine* (7–9): 492–499, 125–137, 261–269, 46–51, 280–290. ISSN 0302-7597. From p. 498: "And if we agree to call the *square root* (taken with a suitable sign) of this scalar product of two conjugate polynomes, P and KP, the common TENSOR of each, ... "

[20] Woldemar Voigt, *Die fundamentalen physikalischen Eigenschaften der Krystalle in elementarer Darstellung* [The fundamental physical properties of crystals in an elementary presentation] (Leipzig, Germany: Veit & Co., 1898), p. 20. From page 20: "*Wir wollen uns deshalb nur darauf*

*stützen, dass Zustände der geschilderten Art bei Spannungen und Dehnungen nicht starrer Körper auftreten, und sie deshalb tensorielle, die für sie charakteristischen physikalischen Grössen aber Tensoren nennen.*" (We therefore want [our presentation] to be based only on [the assumption that] conditions of the type described occur during stresses and strains of non-rigid bodies, and therefore call them "tensorial" but call the characteristic physical quantities for them "tensors".)

[21] Ricci Curbastro, G. (1892). "Résumé de quelques travaux sur les systèmes variables de fonctions associés à une forme différentielle quadratique". *Bulletin des Sciences Mathématiques* 2 (16): 167–189.

[22] (Ricci & Levi-Civita 1900)

[23] Pais, Abraham (2005). *Subtle Is the Lord: The Science and the Life of Albert Einstein*. Oxford University Press. ISBN 978-0-19-280672-7.

[24] Goodstein, Judith R (1982). "The Italian Mathematicians of Relativity". *Centaurus* **26** (3): 241–261. Bibcode:1982Cent...26..241G. doi:10.1111/j.1600-0498.1982.tb00665.x.

[25] Edwin H. Spanier, *Algebraic Topology*, p. 227, "...the Künneth formula expressing the homology of the tensor product...", McGraw Hill, 1966.

[26] Thomas W. Hungerford, *Algebra*, p. 168, "...the classification (up to isomorphism) of modules over an arbitrary ring is quite difficult..." Springer, 1974, ISBN 0387905189.

[27] Saunders Mac Lane, *Categories for the Working Mathematician*, p. 4, "...for example the monoid M ... in the category of abelian groups, × is replaced by the usual tensor product...", Springer, 1971, ISBN 0387900365.

## 15.11 External links

- Weisstein, Eric W., "Tensor", *MathWorld*.

- Introduction to Vectors and Tensors, Vol 1: Linear and Multilinear Algebra by Ray M. Bowen and C. C. Wang.

- Introduction to Vectors and Tensors, Vol 2: Vector and Tensor Analysis by Ray M. Bowen and C. C. Wang.

- An Introduction to Tensors for Students of Physics and Engineering by Joseph C. Kolecki, Glenn Research Center, Cleveland, Ohio, released by NASA

- Foundations of Tensor Analysis for Students of Physics and Engineering With an Introduction to the Theory of Relativity by Joseph C. Kolecki, Glenn Research Center, Cleveland, Ohio, released by NASA

- A discussion of the various approaches to teaching tensors, and recommendations of textbooks

- Introduction to tensors an original approach by S Poirier

- A Quick Introduction to Tensor Analysis by R. A. Sharipov.

- Richard P. Feynman's Lecture on tensors.

# Chapter 16

# Vector space

This article is about linear (vector) spaces. For the structure in incidence geometry, see Linear space (geometry).

A **vector space** (also called a **linear space**) is a collection

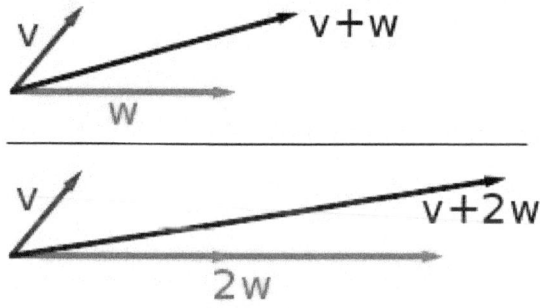

*Vector addition and scalar multiplication: a vector v (blue) is added to another vector w (red, upper illustration). Below, w is stretched by a factor of 2, yielding the sum v + 2w.*

of objects called **vectors**, which may be added together and multiplied ("scaled") by numbers, called *scalars* in this context. Scalars are often taken to be real numbers, but there are also vector spaces with scalar multiplication by complex numbers, rational numbers, or generally any field. The operations of vector addition and scalar multiplication must satisfy certain requirements, called *axioms*, listed below.

Euclidean vectors are an example of a vector space. They represent physical quantities such as forces: any two forces (of the same type) can be added to yield a third, and the multiplication of a force vector by a real multiplier is another force vector. In the same vein, but in a more geometric sense, vectors representing displacements in the plane or in three-dimensional space also form vector spaces. Vectors in vector spaces do not necessarily have to be arrow-like objects as they appear in the mentioned examples: vectors are regarded as abstract mathematical objects with particular properties, which in some cases can be visualized as arrows.

Vector spaces are the subject of linear algebra and are well understood from this point of view since vector spaces are

characterized by their dimension, which, roughly speaking, specifies the number of independent directions in the space. A vector space may be endowed with additional structure, such as a norm or inner product. Such spaces arise naturally in mathematical analysis, notably in the guise of infinite-dimensional function spaces whose vectors are functions. Analytical problems call for the ability to decide whether a sequence of vectors converges to a given vector. This is accomplished by considering vector spaces with additional structure, mostly spaces endowed with a suitable topology, thus allowing the consideration of proximity and continuity issues. These topological vector spaces, in particular Banach spaces and Hilbert spaces, have a richer theory.

Historically, the first ideas leading to vector spaces can be traced back as far as the 17th century's analytic geometry, matrices, systems of linear equations, and Euclidean vectors. The modern, more abstract treatment, first formulated by Giuseppe Peano in 1888, encompasses more general objects than Euclidean space, but much of the theory can be seen as an extension of classical geometric ideas like lines, planes and their higher-dimensional analogs.

Today, vector spaces are applied throughout mathematics, science and engineering. They are the appropriate linear-algebraic notion to deal with systems of linear equations; offer a framework for Fourier expansion, which is employed in image compression routines; or provide an environment that can be used for solution techniques for partial differential equations. Furthermore, vector spaces furnish an abstract, coordinate-free way of dealing with geometrical and physical objects such as tensors. This in turn allows the examination of local properties of manifolds by linearization techniques. Vector spaces may be generalized in several ways, leading to more advanced notions in geometry and abstract algebra.

## 16.1    Introduction and definition

The concept of vector space will first be explained by describing two particular examples:

### 16.1.1    First example: arrows in the plane

The first example of a vector space consists of arrows in a fixed plane, starting at one fixed point. This is used in physics to describe forces or velocities. Given any two such arrows, **v** and **w**, the parallelogram spanned by these two arrows contains one diagonal arrow that starts at the origin, too. This new arrow is called the *sum* of the two arrows and is denoted **v** + **w**. In the special case of two arrows on the same line, their sum is the arrow on this line whose length is the sum or the difference of the lengths, depending on whether the arrows have the same direction. Another operation that can be done with arrows is scaling: given any positive real number $a$, the arrow that has the same direction as **v**, but is dilated or shrunk by multiplying its length by $a$, is called *multiplication* of **v** by $a$. It is denoted $a$**v**. When $a$ is negative, $a$**v** is defined as the arrow pointing in the opposite direction, instead.

The following shows a few examples: if $a = 2$, the resulting vector $a$**w** has the same direction as **w**, but is stretched to the double length of **w** (right image below). Equivalently 2**w** is the sum **w** + **w**. Moreover, $(-1)$**v** = $-$**v** has the opposite direction and the same length as **v** (blue vector pointing down in the right image).

### 16.1.2    Second example:  ordered pairs of numbers

A second key example of a vector space is provided by pairs of real numbers $x$ and $y$. (The order of the components $x$ and $y$ is significant, so such a pair is also called an ordered pair.) Such a pair is written as $(x, y)$. The sum of two such pairs and multiplication of a pair with a number is defined as follows:

$$(x_1, y_1) + (x_2, y_2) = (x_1 + x_2, y_1 + y_2)$$

and

$$a\,(x, y) = (ax, ay).$$

The first example above reduces to this one if the arrows are represented by the pair of Cartesian coordinates of their end points.

### 16.1.3    Definition

A vector space over a field $F$ is a set $V$ together with two operations that satisfy the eight axioms listed below. Elements of $V$ are commonly called *vectors*. Elements of $F$ are commonly called *scalars*. The first operation, called *vector addition* or simply *addition*, takes any two vectors **v** and **w** and assigns to them a third vector which is commonly written as **v** + **w**, and called the sum of these two vectors. The second operation, called *scalar multiplication* takes any scalar $a$ and any vector **v** and gives another vector $a$**v**.

In this article, vectors are distinguished from scalars by boldface.[nb 1] In the two examples above, the field is the field of the real numbers and the set of the vectors consists of the planar arrows with fixed starting point and of pairs of real numbers, respectively.

To qualify as a vector space, the set $V$ and the operations of addition and multiplication must adhere to a number of requirements called axioms.[1] In the list below, let **u**, **v** and **w** be arbitrary vectors in $V$, and $a$ and $b$ scalars in $F$.

These axioms generalize properties of the vectors introduced in the above examples. Indeed, the result of addition of two ordered pairs (as in the second example above) does not depend on the order of the summands:

$$(x_v, y_v) + (x_w, y_w) = (x_w, y_w) + (x_v, y_v).$$

Likewise, in the geometric example of vectors as arrows, **v** + **w** = **w** + **v** since the parallelogram defining the sum of the vectors is independent of the order of the vectors. All other axioms can be checked in a similar manner in both examples. Thus, by disregarding the concrete nature of the particular type of vectors, the definition incorporates these two and many more examples in one notion of vector space.

Subtraction of two vectors and division by a (non-zero) scalar can be defined as

$$\mathbf{v} - \mathbf{w} = \mathbf{v} + (-\mathbf{w}),$$
$$\mathbf{v}/a = (1/a)\mathbf{v}.$$

When the scalar field $F$ is the real numbers **R**, the vector space is called a *real vector space*. When the scalar field is the complex numbers, it is called a *complex vector space*. These two cases are the ones used most often in engineering. The general definition of a vector space allows scalars to be elements of any fixed field $F$. The notion is then known as an *F-vector spaces* or a *vector space over F*. A field is, essentially, a set of numbers possessing addition, subtraction, multiplication and division operations.[nb 3] For example, rational numbers also form a field.

In contrast to the intuition stemming from vectors in the plane and higher-dimensional cases, there is, in general vector spaces, no notion of nearness, angles or distances. To deal with such matters, particular types of vector spaces are introduced; see below.

### 16.1.4 Alternative formulations and elementary consequences

The requirement that vector addition and scalar multiplication be binary operations includes (by definition of binary operations) a property called closure: that $u + v$ and $av$ are in $V$ for all $a$ in $F$, and $u$, $v$ in $V$. Some older sources mention these properties as separate axioms.[2]

In the parlance of abstract algebra, the first four axioms can be subsumed by requiring the set of vectors to be an abelian group under addition. The remaining axioms give this group an $F$-module structure. In other words, there is a ring homomorphism $f$ from the field $F$ into the endomorphism ring of the group of vectors. Then scalar multiplication $av$ is defined as $(f(a))(v)$.[3]

There are a number of direct consequences of the vector space axioms. Some of them derive from elementary group theory, applied to the additive group of vectors: for example the zero vector $0$ of $V$ and the additive inverse $-v$ of any vector $v$ are unique. Other properties follow from the distributive law, for example $av$ equals $0$ if and only if $a$ equals 0 or $v$ equals $0$.

## 16.2 History

Vector spaces stem from affine geometry via the introduction of coordinates in the plane or three-dimensional space. Around 1636, Descartes and Fermat founded analytic geometry by equating solutions to an equation of two variables with points on a plane curve.[4] In 1804, to achieve geometric solutions without using coordinates, Bolzano introduced certain operations on points, lines and planes, which are predecessors of vectors.[5] His work was then used in the conception of barycentric coordinates by Möbius in 1827.[6] The definition of vectors was founded on Bellavitis' notion of the bipoint, an oriented segment of which one end is the origin and the other a target, then further elaborated with the presentation of complex numbers by Argand and Hamilton and the introduction of quaternions and biquaternions by the latter.[7] They are elements in $\mathbf{R}^2$, $\mathbf{R}^4$, and $\mathbf{R}^8$; their treatment as linear combinations can be traced back to Laguerre in 1867, who also defined systems of linear equations.

In 1857, Cayley introduced matrix notation, which allows for a harmonization and simplification of linear maps. Around the same time, Grassmann studied the barycen-

tric calculus initiated by Möbius. He envisaged sets of abstract objects endowed with operations.[8] In his work, the concepts of linear independence and dimension, as well as scalar products, are present. In fact, Grassmann's 1844 work exceeds the framework of vector spaces, since his consideration of multiplication led him to what are today called algebras. Peano was the first to give the modern definition of vector spaces and linear maps in 1888.[9]

An important development of vector spaces is due to the construction of function spaces by Lebesgue. This was later formalized by Banach and Hilbert, around 1920.[10] At that time, algebra and the new field of functional analysis began to interact, notably with key concepts such as spaces of $p$-integrable functions and Hilbert spaces.[11] Vector spaces, including infinite-dimensional ones, then became a firmly established notion, and many mathematical branches started making use of this concept.

## 16.3 Examples

Main article: Examples of vector spaces

### 16.3.1 Coordinate spaces

Main article: Coordinate space

The most simple example of a vector space over a field $F$ is the field itself, equipped with its standard addition and multiplication. More generally, a vector space can be composed of $n$-tuples (sequences of length $n$) of elements of $F$, such as

$$(a_1, a_2, ..., an),\text{ where each } ai \text{ is an element of } F.\text{[12]}$$

A vector space composed of all the $n$-tuples of a field $F$ is known as a *coordinate space*, usually denoted $F^n$. The case $n = 1$ is the above-mentioned simplest example, in which the field $F$ is also regarded as a vector space over itself. The case $F = \mathbf{R}$ and $n = 2$ was discussed in the introduction above.

### 16.3.2 Complex numbers and other field extensions

The set of complex numbers $\mathbf{C}$, i.e., numbers that can be written in the form $x + iy$ for real numbers $x$ and $y$ where $i$ is the imaginary unit, form a vector space over the reals with the usual addition and multiplication: $(x + iy) + (a + ib) = (x + a) + i(y + b)$ and $c \cdot (x + iy) = (c \cdot x) + i(c \cdot y)$

for real numbers $x$, $y$, $a$, $b$ and $c$. The various axioms of a vector space follow from the fact that the same rules hold for complex number arithmetic.

In fact, the example of complex numbers is essentially the same (i.e., it is *isomorphic*) to the vector space of ordered pairs of real numbers mentioned above: if we think of the complex number $x + i\,y$ as representing the ordered pair $(x, y)$ in the complex plane then we see that the rules for sum and scalar product correspond exactly to those in the earlier example.

More generally, field extensions provide another class of examples of vector spaces, particularly in algebra and algebraic number theory: a field $F$ containing a smaller field $E$ is an $E$-vector space, by the given multiplication and addition operations of $F$.[13] For example, the complex numbers are a vector space over $\mathbf{R}$, and the field extension $\mathbf{Q}(i\sqrt{5})$ is a vector space over $\mathbf{Q}$.

### 16.3.3   Function spaces

Functions from any fixed set $\Omega$ to a field $F$ also form vector spaces, by performing addition and scalar multiplication pointwise. That is, the sum of two functions $f$ and $g$ is the function $(f + g)$ given by

$$(f + g)(w) = f(w) + g(w),$$

and similarly for multiplication. Such function spaces occur in many geometric situations, when $\Omega$ is the real line or an interval, or other subsets of $\mathbf{R}$. Many notions in topology and analysis, such as continuity, integrability or differentiability are well-behaved with respect to linearity: sums and scalar multiples of functions possessing such a property still have that property.[14] Therefore, the set of such functions are vector spaces. They are studied in greater detail using the methods of functional analysis, see below. Algebraic constraints also yield vector spaces: the vector space $F[\mathbf{x}]$ is given by polynomial functions:

$f(x) = r_0 + r_1 x + \ldots + r_{n-1}x^{n-1} + r_n x^n$, where the coefficients $r_0, \ldots, r_n$ are in $F$.[15]

### 16.3.4   Linear equations

Main articles: Linear equation, Linear differential equation and Systems of linear equations

Systems of homogeneous linear equations are closely tied to vector spaces.[16] For example, the solutions of

are given by triples with arbitrary $a$, $b = a/2$, and $c = -5a/2$. They form a vector space: sums and scalar multiples of such triples still satisfy the same ratios of the three variables; thus they are solutions, too. Matrices can be used to condense multiple linear equations as above into one vector equation, namely

$$A\mathbf{x} = \mathbf{0},$$

where $A = \begin{bmatrix} 1 & 3 & 1 \\ 4 & 2 & 2 \end{bmatrix}$ is the matrix containing the coefficients of the given equations, $\mathbf{x}$ is the vector $(a, b, c)$, $A\mathbf{x}$ denotes the matrix product, and $\mathbf{0} = (0, 0)$ is the zero vector. In a similar vein, the solutions of homogeneous *linear differential equations* form vector spaces. For example,

$$f''(x) + 2f'(x) + f(x) = 0$$

yields $f(x) = a\,e^{-x} + bx\,e^{-x}$, where $a$ and $b$ are arbitrary constants, and $e^x$ is the natural exponential function.

## 16.4   Basis and dimension

Main articles: Basis and Dimension
*Bases* allow to represent vectors by a sequence of scalars

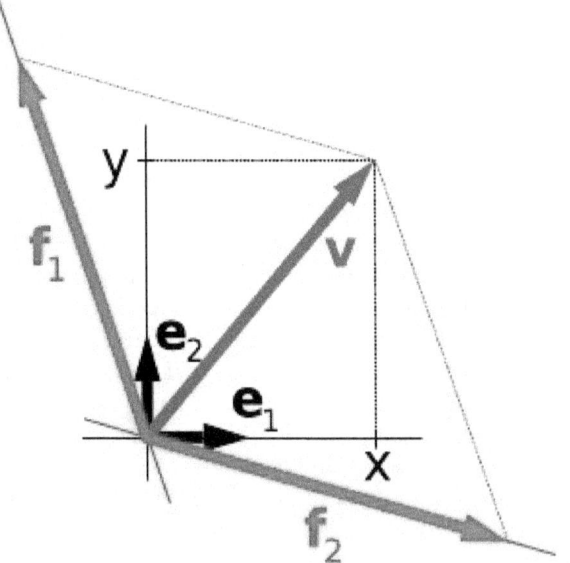

*A vector v in $\mathbf{R}^2$ (blue) expressed in terms of different bases: using the standard basis of $\mathbf{R}^2$ $v = x e_1 + y e_2$ (black), and using a different, non-orthogonal basis: $v = f_1 + f_2$ (red).*

called *coordinates* or *components*. A basis is a (finite or infinite) set $B = \{b_i\}i \in I$ of vectors $b_i$, for convenience often indexed by some index set $I$, that spans the whole space and is linearly independent. "Spanning the whole space" means that any vector **v** can be expressed as a finite sum (called a *linear combination*) of the basis elements:

where the *ak* are scalars, called the coordinates (or the components) of the vector **v** with respect to the basis *B*, and b*ik* ($k = 1, ..., n$) elements of *B*. Linear independence means that the coordinates *ak* are uniquely determined for any vector in the vector space.

For example, the coordinate vectors $e_1 = (1, 0, ..., 0)$, $e_2 = (0, 1, 0, ..., 0)$, to *en* $= (0, 0, ..., 0, 1)$, form a basis of $F^n$, called the standard basis, since any vector $(x_1, x_2, ..., xn)$ can be uniquely expressed as a linear combination of these vectors:

$$(x_1, x_2, ..., xn) = x_1(1, 0, ..., 0) + x_2(0, 1, 0, ..., 0)$$
$$+ ... + xn(0, ..., 0, 1) = x_1e_1 + x_2e_2 + ... + xnen.$$

The corresponding coordinates $x_1$, $x_2$, ..., *xn* are just the Cartesian coordinates of the vector.

Every vector space has a basis. This follows from Zorn's lemma, an equivalent formulation of the Axiom of Choice.[17] Given the other axioms of Zermelo–Fraenkel set theory, the existence of bases is equivalent to the axiom of choice.[18] The ultrafilter lemma, which is weaker than the axiom of choice, implies that all bases of a given vector space have the same number of elements, or cardinality (cf. *Dimension theorem for vector spaces*).[19] It is called the *dimension* of the vector space, denoted dim *V*. If the space is spanned by finitely many vectors, the above statements can be proven without such fundamental input from set theory.[20]

The dimension of the coordinate space $F^n$ is *n*, by the basis exhibited above. The dimension of the polynomial ring $F[x]$ introduced above is countably infinite, a basis is given by $1, x, x^2, ...$ A fortiori, the dimension of more general function spaces, such as the space of functions on some (bounded or unbounded) interval, is infinite.[nb 4] Under suitable regularity assumptions on the coefficients involved, the dimension of the solution space of a homogeneous ordinary differential equation equals the degree of the equation.[21] For example, the solution space for the above equation is generated by $e^{-x}$ and $xe^{-x}$. These two functions are linearly independent over **R**, so the dimension of this space is two, as is the degree of the equation.

A field extension over the rationals **Q** can be thought of as a vector space over **Q** (by defining vector addition as field addition, defining scalar multiplication as field multiplication by elements of **Q**, and otherwise ignoring the field multiplication). The dimension (or degree) of the field extension

$\mathbf{Q}(\alpha)$ over **Q** depends on $\alpha$. If $\alpha$ satisfies some polynomial equation

$$qn\alpha^n + qn_{-1}\alpha^{n-1} + ... + q_0 = 0, \text{ with rational co-}$$
efficients *qn*, ..., $q_0$.

("$\alpha$ is algebraic"), the dimension is finite. More precisely, it equals the degree of the minimal polynomial having $\alpha$ as a root.[22] For example, the complex numbers **C** are a two-dimensional real vector space, generated by 1 and the imaginary unit *i*. The latter satisfies $i^2 + 1 = 0$, an equation of degree two. Thus, **C** is a two-dimensional **R**-vector space (and, as any field, one-dimensional as a vector space over itself, **C**). If $\alpha$ is not algebraic, the dimension of $\mathbf{Q}(\alpha)$ over **Q** is infinite. For instance, for $\alpha = \pi$ there is no such equation, in other words $\pi$ is transcendental.[23]

## 16.5    Linear maps and matrices

Main article: Linear map

The relation of two vector spaces can be expressed by *linear map* or *linear transformation*. They are functions that reflect the vector space structure— i.e., they preserve sums and scalar multiplication:

$$f(\mathbf{x} + \mathbf{y}) = f(\mathbf{x}) + f(\mathbf{y}) \text{ and } f(a \cdot \mathbf{x}) = a \cdot f(\mathbf{x}) \text{ for}$$
all **x** and **y** in *V*, all *a* in $F$.[24]

An *isomorphism* is a linear map $f : V \rightarrow W$ such that there exists an inverse map $g : W \rightarrow V$, which is a map such that the two possible compositions $f \circ g : W \rightarrow W$ and $g \circ f : V \rightarrow V$ are identity maps. Equivalently, *f* is both one-to-one (injective) and onto (surjective).[25] If there exists an isomorphism between *V* and *W*, the two spaces are said to be *isomorphic*; they are then essentially identical as vector spaces, since all identities holding in *V* are, via *f*, transported to similar ones in *W*, and vice versa via *g*.

For example, the "arrows in the plane" and "ordered pairs of numbers" vector spaces in the introduction are isomorphic: a planar arrow **v** departing at the origin of some (fixed) coordinate system can be expressed as an ordered pair by considering the *x*- and *y*-component of the arrow, as shown in the image at the right. Conversely, given a pair $(x, y)$, the arrow going by *x* to the right (or to the left, if *x* is negative), and *y* up (down, if *y* is negative) turns back the arrow **v**.

Linear maps $V \rightarrow W$ between two vector spaces form a vector space Hom$F(V, W)$, also denoted L$(V, W)$.[26] The space of linear maps from *V* to *F* is called the *dual vector space*, denoted $V^*$.[27] Via the injective natural map $V \rightarrow V^{**}$, any vector space can be embedded into its *bidual*;

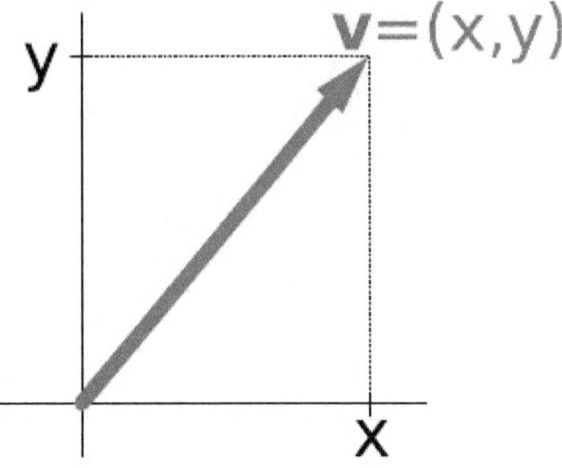

*Describing an arrow vector v by its coordinates* x *and* y *yields an isomorphism of vector spaces.*

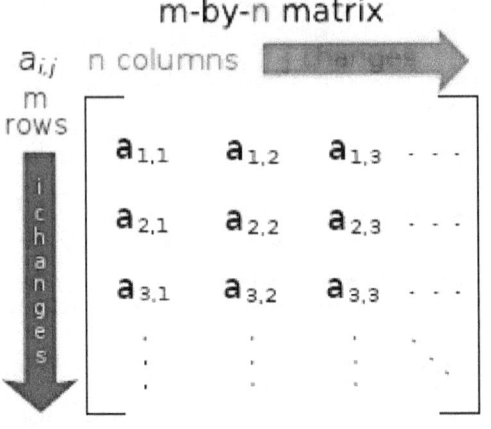

*A typical matrix*

the map is an isomorphism if and only if the space is finite-dimensional.[28]

Once a basis of $V$ is chosen, linear maps $f : V \to W$ are completely determined by specifying the images of the basis vectors, because any element of $V$ is expressed uniquely as a linear combination of them.[29] If dim $V$ = dim $W$, a 1-to-1 correspondence between fixed bases of $V$ and $W$ gives rise to a linear map that maps any basis element of $V$ to the corresponding basis element of $W$. It is an isomorphism, by its very definition.[30] Therefore, two vector spaces are isomorphic if their dimensions agree and vice versa. Another way to express this is that any vector space is *completely classified* (up to isomorphism) by its dimension, a single number. In particular, any $n$-dimensional $F$-vector space $V$ is isomorphic to $F^n$. There is, however, no "canonical" or preferred isomorphism; actually an isomorphism $\varphi : F^n \to V$ is equivalent to the choice of a basis of $V$, by mapping the standard basis of $F^n$ to $V$, via $\varphi$. The freedom of choosing a convenient basis is particularly useful in the infinite-dimensional context, see below.

### 16.5.1   Matrices

Main articles: Matrix and Determinant
*Matrices* are a useful notion to encode linear maps.[31] They are written as a rectangular array of scalars as in the image at the right. Any $m$-by-$n$ matrix $A$ gives rise to a linear map from $F^n$ to $F^m$, by the following

$$\mathbf{x} = (x_1, x_2, \cdots, x_n) \mapsto \left( \sum_{j=1}^{n} a_{1j}x_j, \sum_{j=1}^{n} a_{2j}x_j, \cdots, \sum_{j=1}^{n} a_{mj}x_j \right)$$
, where $\sum$ denotes summation,

or, using the matrix multiplication of the matrix $A$ with the coordinate vector $\mathbf{x}$:

$$\mathbf{x} \mapsto A\mathbf{x}.$$

Moreover, after choosing bases of $V$ and $W$, *any* linear map $f : V \to W$ is uniquely represented by a matrix via this assignment.[32]

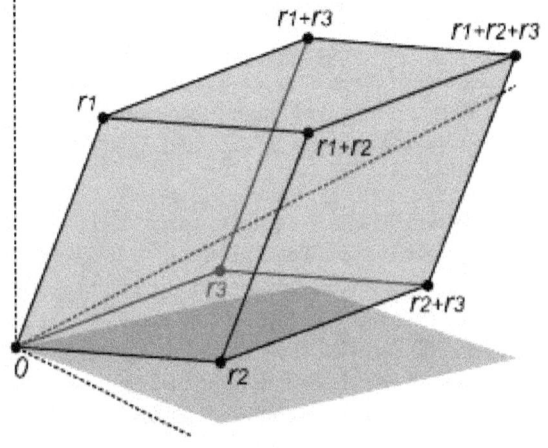

*The volume of this parallelepiped is the absolute value of the determinant of the 3-by-3 matrix formed by the vectors* $\mathbf{r}_1$, $\mathbf{r}_2$, *and* $\mathbf{r}_3$.

The determinant det $(A)$ of a square matrix $A$ is a scalar that tells whether the associated map is an isomorphism or not: to be so it is sufficient and necessary that the determinant is nonzero.[33] The linear transformation of $\mathbf{R}^n$ corresponding to a real $n$-by-$n$ matrix is orientation preserving if and only if its determinant is positive.

### 16.5.2 Eigenvalues and eigenvectors

Main article: Eigenvalues and eigenvectors

Endomorphisms, linear maps $f : V \to V$, are particularly important since in this case vectors $\mathbf{v}$ can be compared with their image under $f$, $f(\mathbf{v})$. Any nonzero vector $\mathbf{v}$ satisfying $\lambda \mathbf{v} = f(\mathbf{v})$, where $\lambda$ is a scalar, is called an *eigenvector* of $f$ with *eigenvalue* $\lambda$.[nb 5][34] Equivalently, $\mathbf{v}$ is an element of the kernel of the difference $f - \lambda \cdot \mathrm{Id}$ (where Id is the identity map $V \to V$). If $V$ is finite-dimensional, this can be rephrased using determinants: $f$ having eigenvalue $\lambda$ is equivalent to

$$\det(f - \lambda \cdot \mathrm{Id}) = 0.$$

By spelling out the definition of the determinant, the expression on the left hand side can be seen to be a polynomial function in $\lambda$, called the characteristic polynomial of $f$.[35] If the field $F$ is large enough to contain a zero of this polynomial (which automatically happens for $F$ algebraically closed, such as $F = \mathbf{C}$) any linear map has at least one eigenvector. The vector space $V$ may or may not possess an eigenbasis, a basis consisting of eigenvectors. This phenomenon is governed by the Jordan canonical form of the map.[nb 6] The set of all eigenvectors corresponding to a particular eigenvalue of $f$ forms a vector space known as the *eigenspace* corresponding to the eigenvalue (and $f$) in question. To achieve the spectral theorem, the corresponding statement in the infinite-dimensional case, the machinery of functional analysis is needed, see below.

# 16.6  Basic constructions

In addition to the above concrete examples, there are a number of standard linear algebraic constructions that yield vector spaces related to given ones. In addition to the definitions given below, they are also characterized by universal properties, which determine an object $X$ by specifying the linear maps from $X$ to any other vector space.

### 16.6.1  Subspaces and quotient spaces

Main articles: Linear subspace and Quotient vector space
A nonempty subset $W$ of a vector space $V$ that is closed under addition and scalar multiplication (and therefore contains the $\mathbf{0}$-vector of $V$) is called a *linear subspace* of $V$, or simply a *subspace* of $V$, when the ambient space is unambiguously a vector space.[36][nb 7] Subspaces of $V$ are vector spaces (over the same field) in their own right. The intersection of all subspaces containing a given set $S$ of vectors

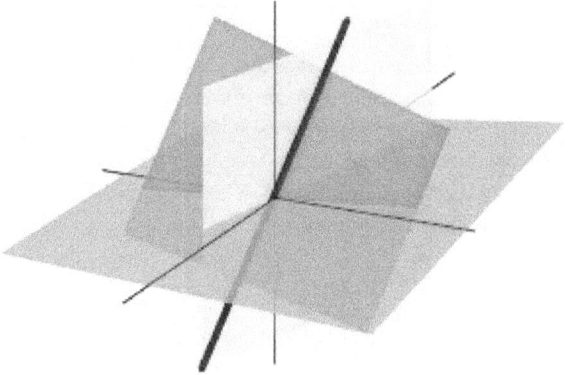

*A line passing through the origin (blue, thick) in $\mathbf{R}^3$ is a linear subspace. It is the intersection of two planes (green and yellow).*

is called its span, and it is the smallest subspace of $V$ containing the set $S$. Expressed in terms of elements, the span is the subspace consisting of all the linear combinations of elements of $S$.[37]

A linear subspace of dimension 1 is a **vector line**. A linear subspace of dimension 2 is a **vector plane**. A linear subspace that contains all elements but one of a basis of the ambient space is a **vector hyperplane**. In a vector space of finite dimension $n$, a vector hyperplane is thus a subspace of dimension $n - 1$.

The counterpart to subspaces are *quotient vector spaces*.[38] Given any subspace $W \subset V$, the quotient space $V/W$ ("$V$ modulo $W$") is defined as follows: as a set, it consists of $\mathbf{v} + W = \{\mathbf{v} + \mathbf{w} : \mathbf{w} \in W\}$, where $\mathbf{v}$ is an arbitrary vector in $V$. The sum of two such elements $\mathbf{v}_1 + W$ and $\mathbf{v}_2 + W$ is $(\mathbf{v}_1 + \mathbf{v}_2) + W$, and scalar multiplication is given by $a \cdot (\mathbf{v} + W) = (a \cdot \mathbf{v}) + W$. The key point in this definition is that $\mathbf{v}_1 + W = \mathbf{v}_2 + W$ if and only if the difference of $\mathbf{v}_1$ and $\mathbf{v}_2$ lies in $W$.[nb 8] This way, the quotient space "forgets" information that is contained in the subspace $W$.

The kernel $\ker(f)$ of a linear map $f : V \to W$ consists of vectors $\mathbf{v}$ that are mapped to $\mathbf{0}$ in $W$.[39] Both kernel and image $\mathrm{im}(f) = \{f(\mathbf{v}) : \mathbf{v} \in V\}$ are subspaces of $V$ and $W$, respectively.[40] The existence of kernels and images is part of the statement that the category of vector spaces (over a fixed field $F$) is an abelian category, i.e. a corpus of mathematical objects and structure-preserving maps between them (a category) that behaves much like the category of abelian groups.[41] Because of this, many statements such as the first isomorphism theorem (also called rank–nullity theorem in matrix-related terms)

$$V / \ker(f) \equiv \mathrm{im}(f).$$

and the second and third isomorphism theorem can be formulated and proven in a way very similar to the corresponding statements for groups.

An important example is the kernel of a linear map $\mathbf{x} \mapsto A\mathbf{x}$ for some fixed matrix $A$, as above.  The kernel of this map is the subspace of vectors $\mathbf{x}$ such that $A\mathbf{x} = 0$, which is precisely the set of solutions to the system of homogeneous linear equations belonging to $A$. This concept also extends to linear differential equations

$$a_0 f + a_1 \frac{df}{dx} + a_2 \frac{d^2 f}{dx^2} + \cdots + a_n \frac{d^n f}{dx^n} = 0 \,,\text{ where}$$
the coefficients $a_i$ are functions in $x$, too.

In the corresponding map

$$f \mapsto D(f) = \sum_{i=0}^{n} a_i \frac{d^i f}{dx^i}$$

the derivatives of the function $f$ appear linearly (as opposed to $f''(x)^2$, for example).  Since differentiation is a linear procedure (i.e., $(f + g)' = f' + g'$ and $(c \cdot f)' = c \cdot f'$ for a constant $c$) this assignment is linear, called a linear differential operator.  In particular, the solutions to the differential equation $D(f) = 0$ form a vector space (over $\mathbf{R}$ or $\mathbf{C}$).

### 16.6.2  Direct product and direct sum

Main articles: Direct product and Direct sum of modules

The *direct product* of vector spaces and the *direct sum* of vector spaces are two ways of combining an indexed family of vector spaces into a new vector space.

The *direct product* $\prod_{i \in I} V_i$ of a family of vector spaces $V_i$ consists of the set of all tuples $(v_i)i \in I$, which specify for each index $i$ in some index set $I$ an element $v_i$ of $V_i$.[42] Addition and scalar multiplication is performed componentwise. A variant of this construction is the *direct sum* $\oplus_{i \in I} V_i$ (also called coproduct and denoted $\coprod_{i \in I} V_i$ ), where only tuples with finitely many nonzero vectors are allowed. If the index set $I$ is finite, the two constructions agree, but in general they are different.

### 16.6.3  Tensor product

Main article: Tensor product of vector spaces

The *tensor product* $V \otimes_F W$, or simply $V \otimes W$, of two vector spaces $V$ and $W$ is one of the central notions of multilinear algebra which deals with extending notions such as linear maps to several variables. A map $g : V \times W \to X$ is called

bilinear if $g$ is linear in both variables $\mathbf{v}$ and $\mathbf{w}$.  That is to say, for fixed $\mathbf{w}$ the map $\mathbf{v} \mapsto g(\mathbf{v}, \mathbf{w})$ is linear in the sense above and likewise for fixed $\mathbf{v}$.

The tensor product is a particular vector space that is a *universal* recipient of bilinear maps $g$, as follows. It is defined as the vector space consisting of finite (formal) sums of symbols called tensors

$$\mathbf{v}_1 \otimes \mathbf{w}_1 + \mathbf{v}_2 \otimes \mathbf{w}_2 + \ldots + \mathbf{v}n \otimes \mathbf{w}n,$$

subject to the rules

$a \cdot (\mathbf{v} \otimes \mathbf{w}) = (a \cdot \mathbf{v}) \otimes \mathbf{w} = \mathbf{v} \otimes (a \cdot \mathbf{w})$, where $a$ is a scalar,

$(\mathbf{v}_1 + \mathbf{v}_2) \otimes \mathbf{w} = \mathbf{v}_1 \otimes \mathbf{w} + \mathbf{v}_2 \otimes \mathbf{w}$, and

$\mathbf{v} \otimes (\mathbf{w}_1 + \mathbf{w}_2) = \mathbf{v} \otimes \mathbf{w}_1 + \mathbf{v} \otimes \mathbf{w}_2$.[43]

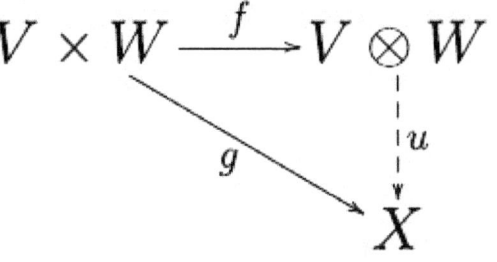

*Commutative diagram depicting the universal property of the tensor product.*

These rules ensure that the map $f$ from the $V \times W$ to $V \otimes W$ that maps a tuple $(\mathbf{v}, \mathbf{w})$ to $\mathbf{v} \otimes \mathbf{w}$ is bilinear. The universality states that given *any* vector space $X$ and *any* bilinear map $g : V \times W \to X$, there exists a unique map $u$, shown in the diagram with a dotted arrow, whose composition with $f$ equals $g$: $u(\mathbf{v} \otimes \mathbf{w}) = g(\mathbf{v}, \mathbf{w})$.[44] This is called the universal property of the tensor product, an instance of the method—much used in advanced abstract algebra—to indirectly define objects by specifying maps from or to this object.

## 16.7  Vector spaces with additional structure

From the point of view of linear algebra, vector spaces are completely understood insofar as any vector space is characterized, up to isomorphism, by its dimension. However, vector spaces *per se* do not offer a framework to deal with the question—crucial to analysis—whether a sequence of functions converges to another function. Likewise, linear algebra is not adapted to deal with infinite series, since the addition operation allows only finitely many terms to be added. Therefore, the needs of functional analysis require considering additional structures.

A vector space may be given a partial order ≤, under which some vectors can be compared.[45] For example, $n$-dimensional real space $\mathbf{R}^n$ can be ordered by comparing its vectors componentwise. Ordered vector spaces, for example Riesz spaces, are fundamental to Lebesgue integration, which relies on the ability to express a function as a difference of two positive functions

$$f = f^+ - f^-,$$

where $f^+$ denotes the positive part of $f$ and $f^-$ the negative part.[46]

### 16.7.1 Normed vector spaces and inner product spaces

Main articles: Normed vector space and Inner product space

"Measuring" vectors is done by specifying a norm, a datum which measures lengths of vectors, or by an inner product, which measures angles between vectors. Norms and inner products are denoted $|\mathbf{v}|$ and $\langle \mathbf{v}, \mathbf{w} \rangle$, respectively. The datum of an inner product entails that lengths of vectors can be defined too, by defining the associated norm $|\mathbf{v}| := \sqrt{\langle \mathbf{v}, \mathbf{v} \rangle}$. Vector spaces endowed with such data are known as *normed vector spaces* and *inner product spaces*, respectively.[47]

Coordinate space $F^n$ can be equipped with the standard dot product:

$$\langle \mathbf{x}, \mathbf{y} \rangle = \mathbf{x} \cdot \mathbf{y} = x_1 y_1 + \cdots + x_n y_n.$$

In $\mathbf{R}^2$, this reflects the common notion of the angle between two vectors $\mathbf{x}$ and $\mathbf{y}$, by the law of cosines:

$$\mathbf{x} \cdot \mathbf{y} = \cos\left(\angle(\mathbf{x}, \mathbf{y})\right) \cdot |\mathbf{x}| \cdot |\mathbf{y}|.$$

Because of this, two vectors satisfying $\langle \mathbf{x}, \mathbf{y} \rangle = 0$ are called orthogonal. An important variant of the standard dot product is used in Minkowski space: $\mathbf{R}^4$ endowed with the Lorentz product

$$\langle \mathbf{x} | \mathbf{y} \rangle = x_1 y_1 + x_2 y_2 + x_3 y_3 - x_4 y_4. \text{ [48]}$$

In contrast to the standard dot product, it is not positive definite: $\langle \mathbf{x} | \mathbf{x} \rangle$ also takes negative values, for example for $\mathbf{x} = (0, 0, 0, 1)$. Singling out the fourth coordinate—corresponding to time, as opposed to three space-dimensions—makes it useful for the mathematical treatment of special relativity.

### 16.7.2 Topological vector spaces

Main article: Topological vector space

Convergence questions are treated by considering vector spaces $V$ carrying a compatible topology, a structure that allows one to talk about elements being close to each other.[49][50] Compatible here means that addition and scalar multiplication have to be continuous maps. Roughly, if $\mathbf{x}$ and $\mathbf{y}$ in $V$, and $a$ in $F$ vary by a bounded amount, then so do $\mathbf{x} + \mathbf{y}$ and $a\mathbf{x}$.[nb 9] To make sense of specifying the amount a scalar changes, the field $F$ also has to carry a topology in this context; a common choice are the reals or the complex numbers.

In such *topological vector spaces* one can consider series of vectors. The infinite sum

$$\sum_{i=0}^{\infty} f_i$$

denotes the limit of the corresponding finite partial sums of the sequence $(f_i)_{i \in \mathbf{N}}$ of elements of $V$. For example, the $f_i$ could be (real or complex) functions belonging to some function space $V$, in which case the series is a function series. The mode of convergence of the series depends on the topology imposed on the function space. In such cases, pointwise convergence and uniform convergence are two prominent examples.

A way to ensure the existence of limits of certain infinite series is to restrict attention to spaces where any Cauchy sequence has a limit; such a vector space is called complete. Roughly, a vector space is complete provided that it contains all necessary limits. For example, the vector space of polynomials on the unit interval [0,1], equipped with the topology of uniform convergence is not complete because any continuous function on [0,1] can be uniformly approximated by a sequence of polynomials, by the Weierstrass approximation theorem.[51] In contrast, the space of *all* continuous functions on [0,1] with the same topology is complete.[52] A norm gives rise to a topology by defining that a sequence of vectors $\mathbf{v}_n$ converges to $\mathbf{v}$ if and only if

$$\lim_{n \to \infty} |\mathbf{v}_n - \mathbf{v}| = 0.$$

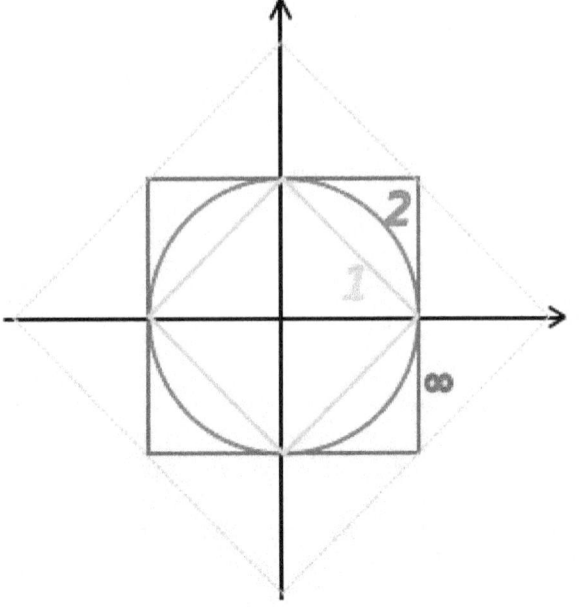

*Unit "spheres" in $\mathbf{R}^2$ consist of plane vectors of norm 1. Depicted are the unit spheres in different p-norms, for p = 1, 2, and ∞. The bigger diamond depicts points of 1-norm equal to $\sqrt{2}$ .*

Banach and Hilbert spaces are complete topological vector spaces whose topologies are given, respectively, by a norm and an inner product. Their study—a key piece of functional analysis—focusses on infinite-dimensional vector spaces, since all norms on finite-dimensional topological vector spaces give rise to the same notion of convergence.[53] The image at the right shows the equivalence of the 1-norm and ∞-norm on $\mathbf{R}^2$: as the unit "balls" enclose each other, a sequence converges to zero in one norm if and only if it so does in the other norm. In the infinite-dimensional case, however, there will generally be inequivalent topologies, which makes the study of topological vector spaces richer than that of vector spaces without additional data.

From a conceptual point of view, all notions related to topological vector spaces should match the topology. For example, instead of considering all linear maps (also called functionals) $V \to W$, maps between topological vector spaces are required to be continuous.[54] In particular, the (topological) dual space $V^*$ consists of continuous functionals $V \to \mathbf{R}$ (or to $\mathbf{C}$). The fundamental Hahn–Banach theorem is concerned with separating subspaces of appropriate topological vector spaces by continuous functionals.[55]

### Banach spaces

Main article: Banach space

*Banach spaces*, introduced by Stefan Banach, are complete normed vector spaces.[56] A first example is the vector space $\ell^p$ consisting of infinite vectors with real entries $\mathbf{x} = (x_1, x_2, ...)$ whose *p*-norm ($1 \le p \le \infty$) given by

$$|\mathbf{x}|_p := \left(\sum_i |x_i|^p\right)^{1/p} \text{ for } p < \infty \text{ and } |\mathbf{x}|_\infty := \sup_i |x_i|$$

is finite. The topologies on the infinite-dimensional space $\ell^p$ are inequivalent for different *p*. E.g. the sequence of vectors $\mathbf{x}n = (2^{-n}, 2^{-n}, ..., 2^{-n}, 0, 0, ...)$, i.e. the first $2^n$ components are $2^{-n}$, the following ones are 0, converges to the zero vector for $p = \infty$, but does not for $p = 1$:

$$|x_n|_\infty = \sup(2^{-n}, 0) = 2^{-n} \to 0 \text{ , but } |x_n|_1 = \sum_{i=1}^{2^n} 2^{-n} = 2^n \cdot 2^{-n} = 1.$$

More generally than sequences of real numbers, functions $f: \Omega \to \mathbf{R}$ are endowed with a norm that replaces the above sum by the Lebesgue integral

$$|f|_p := \left(\int_\Omega |f(x)|^p \, dx\right)^{1/p}.$$

The space of integrable functions on a given domain $\Omega$ (for example an interval) satisfying $|f|p < \infty$, and equipped with this norm are called Lebesgue spaces, denoted $L^p(\Omega)$.[nb 10] These spaces are complete.[57] (If one uses the Riemann integral instead, the space is *not* complete, which may be seen as a justification for Lebesgue's integration theory.[nb 11]) Concretely this means that for any sequence of Lebesgue-integrable functions $f_1, f_2, ...$ with $|fn|p < \infty$, satisfying the condition

$$\lim_{k, n \to \infty} \int_\Omega |f_k(x) - f_n(x)|^p \, dx = 0$$

there exists a function $f(x)$ belonging to the vector space $L^p(\Omega)$ such that

$$\lim_{k \to \infty} \int_\Omega |f(x) - f_k(x)|^p \, dx = 0.$$

Imposing boundedness conditions not only on the function, but also on its derivatives leads to Sobolev spaces.[58]

### Hilbert spaces

Main article: Hilbert space

Complete inner product spaces are known as *Hilbert spaces*,

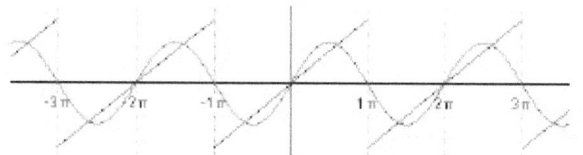

The succeeding snapshots show summation of 1 to 5 terms in approximating a periodic function (blue) by finite sum of sine functions (red).

in honor of David Hilbert.[59] The Hilbert space $L^2(\Omega)$, with inner product given by

$$\langle f , g \rangle = \int_\Omega f(x)\overline{g(x)}\,dx,$$

where $\overline{g(x)}$ denotes the complex conjugate of $g(x)$,[60][nb 12] is a key case.

By definition, in a Hilbert space any Cauchy sequence converges to a limit. Conversely, finding a sequence of functions $fn$ with desirable properties that approximates a given limit function, is equally crucial. Early analysis, in the guise of the Taylor approximation, established an approximation of differentiable functions $f$ by polynomials.[61] By the Stone–Weierstrass theorem, every continuous function on $[a, b]$ can be approximated as closely as desired by a polynomial.[62] A similar approximation technique by trigonometric functions is commonly called Fourier expansion, and is much applied in engineering, see below. More generally, and more conceptually, the theorem yields a simple description of what "basic functions", or, in abstract Hilbert spaces, what basic vectors suffice to generate a Hilbert space $H$, in the sense that the *closure* of their span (i.e., finite linear combinations and limits of those) is the whole space. Such a set of functions is called a *basis* of $H$, its cardinality is known as the Hilbert space dimension.[nb 13] Not only does the theorem exhibit suitable basis functions as sufficient for approximation purposes, but together with the Gram-Schmidt process, it enables one to construct a basis of orthogonal vectors.[63] Such orthogonal bases are the Hilbert space generalization of the coordinate axes in finite-dimensional Euclidean space.

The solutions to various differential equations can be interpreted in terms of Hilbert spaces. For example, a great many fields in physics and engineering lead to such equations and frequently solutions with particular physical properties are used as basis functions, often orthogonal.[64] As an example from physics, the time-dependent Schrödinger equation in quantum mechanics describes the change of physical properties in time by means of a partial differential equation, whose solutions are called wavefunctions.[65] Definite values for physical properties such as energy, or momentum, correspond to eigenvalues of a certain (linear) differential operator and the associated wavefunctions are called eigenstates. The spectral theorem decomposes a linear compact operator acting on functions in terms of these eigenfunctions and their eigenvalues.[66]

### 16.7.3 Algebras over fields

Main articles: Algebra over a field and Lie algebra

General vector spaces do not possess a multiplication be-

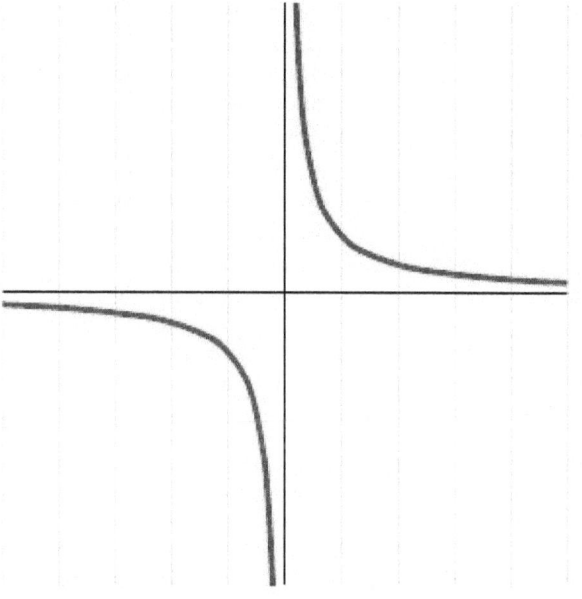

A hyperbola, given by the equation x · y = 1. The coordinate ring of functions on this hyperbola is given by $R[x, y]/(x \cdot y - 1)$, an infinite-dimensional vector space over $R$.

tween vectors. A vector space equipped with an additional bilinear operator defining the multiplication of two vectors is an *algebra over a field*.[67] Many algebras stem from functions on some geometrical object: since functions with values in a given field can be multiplied pointwise, these entities form algebras. The Stone–Weierstrass theorem mentioned above, for example, relies on Banach algebras which are both Banach spaces and algebras.

Commutative algebra makes great use of rings of polynomials in one or several variables, introduced above. Their multiplication is both commutative and associative. These rings and their quotients form the basis of algebraic geometry, because they are rings of functions of algebraic geometric objects.[68]

Another crucial example are *Lie algebras*, which are neither commutative nor associative, but the failure to be so is limited by the constraints ([x, y] denotes the product of x and y):

- $[x, y] = -[y, x]$ (anticommutativity), and

- $[x, [y, z]] + [y, [z, x]] + [z, [x, y]] = 0$ (Jacobi identity).[69]

Examples include the vector space of $n$-by-$n$ matrices, with $[x, y] = xy - yx$, the commutator of two matrices, and $\mathbf{R}^3$, endowed with the cross product.

The tensor algebra $T(V)$ is a formal way of adding products to any vector space $V$ to obtain an algebra.[70] As a vector space, it is spanned by symbols, called simple tensors

$$\mathbf{v}_1 \otimes \mathbf{v}_2 \otimes ... \otimes \mathbf{v}n, \text{ where the degree } n \text{ varies.}$$

The multiplication is given by concatenating such symbols, imposing the distributive law under addition, and requiring that scalar multiplication commute with the tensor product $\otimes$, much the same way as with the tensor product of two vector spaces introduced above. In general, there are no relations between $\mathbf{v}_1 \otimes \mathbf{v}_2$ and $\mathbf{v}_2 \otimes \mathbf{v}_1$. Forcing two such elements to be equal leads to the symmetric algebra, whereas forcing $\mathbf{v}_1 \otimes \mathbf{v}_2 = - \mathbf{v}_2 \otimes \mathbf{v}_1$ yields the exterior algebra.[71]

When a field, $F$ is explicitly stated, a common term used is $F$-algebra.

## 16.8   Applications

Vector spaces have manifold applications as they occur in many circumstances, namely wherever functions with values in some field are involved. They provide a framework to deal with analytical and geometrical problems, or are used in the Fourier transform. This list is not exhaustive: many more applications exist, for example in optimization. The minimax theorem of game theory stating the existence of a unique payoff when all players play optimally can be formulated and proven using vector spaces methods.[72] Representation theory fruitfully transfers the good understanding of linear algebra and vector spaces to other mathematical domains such as group theory.[73]

### 16.8.1   Distributions

Main article: Distribution

A *distribution* (or *generalized function*) is a linear map assigning a number to each "test" function, typically a smooth function with compact support, in a continuous way: in the above terminology the space of distributions is the (continuous) dual of the test function space.[74] The latter space is endowed with a topology that takes into account not only $f$ itself, but also all its higher derivatives. A standard example is the result of integrating a test function $f$ over some domain $\Omega$:

$$I(f) = \int_{\Omega} f(x) \, dx.$$

When $\Omega = \{p\}$, the set consisting of a single point, this reduces to the Dirac distribution, denoted by $\delta$, which associates to a test function $f$ its value at the $p$: $\delta(f) = f(p)$. Distributions are a powerful instrument to solve differential equations. Since all standard analytic notions such as derivatives are linear, they extend naturally to the space of distributions. Therefore, the equation in question can be transferred to a distribution space, which is bigger than the underlying function space, so that more flexible methods are available for solving the equation. For example, Green's functions and fundamental solutions are usually distributions rather than proper functions, and can then be used to find solutions of the equation with prescribed boundary conditions. The found solution can then in some cases be proven to be actually a true function, and a solution to the original equation (e.g., using the Lax–Milgram theorem, a consequence of the Riesz representation theorem).[75]

### 16.8.2   Fourier analysis

Main article: Fourier analysis
Resolving a periodic function into a sum of trigonometric

*The heat equation describes the dissipation of physical properties over time, such as the decline of the temperature of a hot body placed in a colder environment (yellow depicts colder regions than red).*

functions forms a *Fourier series*, a technique much used in physics and engineering.[nb 14][76] The underlying vector space is usually the Hilbert space $L^2(0, 2\pi)$, for which the functions $\sin mx$ and $\cos mx$ ($m$ an integer) form an orthogonal basis.[77] The Fourier expansion of an $L^2$ function $f$ is

$$\frac{a_0}{2} + \sum_{m=1}^{\infty} [a_m \cos(mx) + b_m \sin(mx)].$$

The coefficients $a_m$ and $b_m$ are called Fourier coefficients of $f$, and are calculated by the formulas[78]

$$a_m = \frac{1}{\pi} \int_0^{2\pi} f(t) \cos(mt)\, dt \ , \quad b_m = \frac{1}{\pi} \int_0^{2\pi} f(t) \sin(mt)\, dt.$$

In physical terms the function is represented as a superposition of sine waves and the coefficients give information about the function's frequency spectrum.[79] A complex-number form of Fourier series is also commonly used.[78] The concrete formulae above are consequences of a more general mathematical duality called Pontryagin duality.[80] Applied to the group **R**, it yields the classical Fourier transform; an application in physics are reciprocal lattices, where the underlying group is a finite-dimensional real vector space endowed with the additional datum of a lattice encoding positions of atoms in crystals.[81]

Fourier series are used to solve boundary value problems in partial differential equations.[82] In 1822, Fourier first used this technique to solve the heat equation.[83] A discrete version of the Fourier series can be used in sampling applications where the function value is known only at a finite number of equally spaced points. In this case the Fourier series is finite and its value is equal to the sampled values at all points.[84] The set of coefficients is known as the discrete Fourier transform (DFT) of the given sample sequence. The DFT is one of the key tools of digital signal processing, a field whose applications include radar, speech encoding, image compression.[85] The JPEG image format is an application of the closely related discrete cosine transform.[86]

The fast Fourier transform is an algorithm for rapidly computing the discrete Fourier transform.[87] It is used not only for calculating the Fourier coefficients but, using the convolution theorem, also for computing the convolution of two finite sequences.[88] They in turn are applied in digital filters[89] and as a rapid multiplication algorithm for polynomials and large integers (Schönhage-Strassen algorithm).[90][91]

### 16.8.3 Differential geometry

Main article: Tangent space
The tangent plane to a surface at a point is naturally a vector space whose origin is identified with the point of contact. The tangent plane is the best linear approximation, or linearization, of a surface at a point.[nb 15] Even in a three-dimensional Euclidean space, there is typically no natural

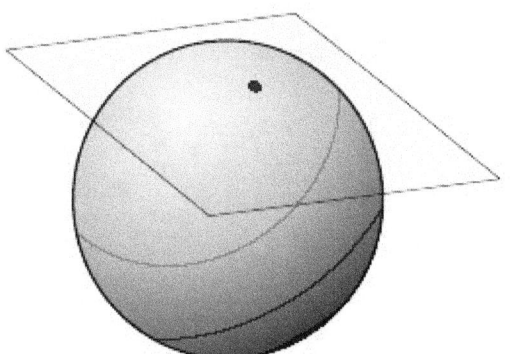

*The tangent space to the 2-sphere at some point is the infinite plane touching the sphere in this point.*

way to prescribe a basis of the tangent plane, and so it is conceived of as an abstract vector space rather than a real coordinate space. The *tangent space* is the generalization to higher-dimensional differentiable manifolds.[92]

Riemannian manifolds are manifolds whose tangent spaces are endowed with a suitable inner product.[93] Derived therefrom, the Riemann curvature tensor encodes all curvatures of a manifold in one object, which finds applications in general relativity, for example, where the Einstein curvature tensor describes the matter and energy content of space-time.[94][95] The tangent space of a Lie group can be given naturally the structure of a Lie algebra and can be used to classify compact Lie groups.[96]

## 16.9 Generalizations

### 16.9.1 Vector bundles

Main articles: Vector bundle and Tangent bundle
A *vector bundle* is a family of vector spaces parametrized continuously by a topological space $X$.[92] More precisely, a vector bundle over $X$ is a topological space $E$ equipped with a continuous map

$$\pi : E \to X$$

such that for every $x$ in $X$, the fiber $\pi^{-1}(x)$ is a vector space. The case dim $V = 1$ is called a line bundle. For any vector space $V$, the projection $X \times V \to X$ makes the product $X \times V$ into a "trivial" vector bundle. Vector bundles over $X$ are required to be locally a product of $X$ and some (fixed) vector space $V$: for every $x$ in $X$, there is a neighborhood $U$ of $x$ such that the restriction of $\pi$ to $\pi^{-1}(U)$ is isomorphic[nb 16] to the trivial bundle $U \times V \to U$. Despite their locally trivial character, vector bundles may (depending on the shape of the underlying space $X$) be "twisted" in the large (i.e.,

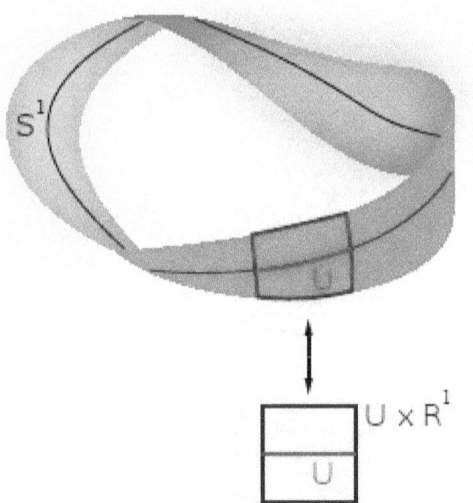

*A Möbius strip. Locally, it looks like* U × **R**.

the bundle need not be (globally isomorphic to) the trivial bundle $X \times V$). For example, the Möbius strip can be seen as a line bundle over the circle $S^1$ (by identifying open intervals with the real line). It is, however, different from the cylinder $S^1 \times \mathbf{R}$, because the latter is orientable whereas the former is not.[97]

Properties of certain vector bundles provide information about the underlying topological space. For example, the tangent bundle consists of the collection of tangent spaces parametrized by the points of a differentiable manifold. The tangent bundle of the circle $S^1$ is globally isomorphic to $S^1 \times \mathbf{R}$, since there is a global nonzero vector field on $S^1$.[nb 17] In contrast, by the hairy ball theorem, there is no (tangent) vector field on the 2-sphere $S^2$ which is everywhere nonzero.[98] K-theory studies the isomorphism classes of all vector bundles over some topological space.[99] In addition to deepening topological and geometrical insight, it has purely algebraic consequences, such as the classification of finite-dimensional real division algebras: $\mathbf{R}, \mathbf{C}$, the quaternions $\mathbf{H}$ and the octonions $\mathbf{O}$.

The cotangent bundle of a differentiable manifold consists, at every point of the manifold, of the dual of the tangent space, the cotangent space. Sections of that bundle are known as differential one-forms.

### 16.9.2   Modules

Main article: Module

*Modules* are to rings what vector spaces are to fields: the same axioms, applied to a ring $R$ instead of a field $F$, yield modules.[100] The theory of modules, compared to that of vector spaces, is complicated by the presence of ring elements that do not have multiplicative inverses. For example, modules need not have bases, as the **Z**-module (i.e., abelian group) **Z**/2**Z** shows; those modules that do (including all vector spaces) are known as free modules. Nevertheless, a vector space can be compactly defined as a module over a ring which is a field with the elements being called vectors. Some authors use the term *vector space* to mean modules over a division ring.[101] The algebro-geometric interpretation of commutative rings via their spectrum allows the development of concepts such as locally free modules, the algebraic counterpart to vector bundles.

### 16.9.3   Affine and projective spaces

Main articles: Affine space and Projective space

Roughly, *affine spaces* are vector spaces whose origins are

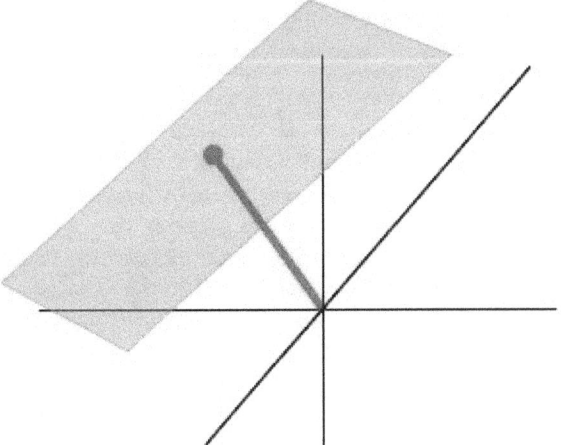

*An affine plane (light blue) in* $\mathbf{R}^3$. *It is a two-dimensional subspace shifted by a vector* **x** *(red).*

not specified.[102] More precisely, an affine space is a set with a free transitive vector space action. In particular, a vector space is an affine space over itself, by the map

$$V \times V \to V, (\mathbf{v}, \mathbf{a}) \mapsto \mathbf{a} + \mathbf{v}.$$

If $W$ is a vector space, then an affine subspace is a subset of $W$ obtained by translating a linear subspace $V$ by a fixed vector $\mathbf{x} \in W$; this space is denoted by $\mathbf{x} + V$ (it is a coset of $V$ in $W$) and consists of all vectors of the form $\mathbf{x} + \mathbf{v}$ for $\mathbf{v} \in V$. An important example is the space of solutions of a system of inhomogeneous linear equations

$$A\mathbf{x} = \mathbf{b}$$

generalizing the homogeneous case $\mathbf{b} = 0$ above.[103] The space of solutions is the affine subspace $\mathbf{x} + V$ where $\mathbf{x}$ is a particular solution of the equation, and $V$ is the space of solutions of the homogeneous equation (the nullspace of $A$).

The set of one-dimensional subspaces of a fixed finite-dimensional vector space $V$ is known as *projective space*; it may be used to formalize the idea of parallel lines intersecting at infinity.[104] Grassmannians and flag manifolds generalize this by parametrizing linear subspaces of fixed dimension $k$ and flags of subspaces, respectively.

## 16.10    See also

- Vector (mathematics and physics), for a list of various kinds of vectors

## 16.11    Notes

[1] It is also common, especially in physics, to denote vectors with an arrow on top: $\vec{v}$.

[2] This axiom refers to two different operations: scalar multiplication: $bv$; and field multiplication: $ab$. It does not assert the associativity of either operation. More formally, scalar multiplication is the *semigroup action* of the scalars on the vector space. Combined with the axiom of the identity element of scalar multiplication, it is a *monoid action*.

[3] Some authors (such as Brown 1991) restrict attention to the fields $\mathbf{R}$ or $\mathbf{C}$, but most of the theory is unchanged for an arbitrary field.

[4] The indicator functions of intervals (of which there are infinitely many) are linearly independent, for example.

[5] The nomenclature derives from German "eigen", which means own or proper.

[6] Roman 2005, ch. 8, p. 140. See also Jordan–Chevalley decomposition.

[7] This is typically the case when a vector space is also considered as an affine space. In this case, a linear subspace contains the zero vector, while an affine subspace does not necessarily contain it.

[8] Some authors (such as Roman 2005) choose to start with this equivalence relation and derive the concrete shape of $V/W$ from this.

[9] This requirement implies that the topology gives rise to a uniform structure, Bourbaki 1989, ch. II

[10] The triangle inequality for |–|p is provided by the Minkowski inequality. For technical reasons, in the context of functions one has to identify functions that agree almost everywhere to get a norm, and not only a seminorm.

[11] "Many functions in $L^2$ of Lebesgue measure, being unbounded, cannot be integrated with the classical Riemann integral. So spaces of Riemann integrable functions would not be complete in the $L^2$ norm, and the orthogonal decomposition would not apply to them. This shows one of the advantages of Lebesgue integration.", Dudley 1989, §5.3, p. 125

[12] For $p \neq 2$, $L^p(\Omega)$ is not a Hilbert space.

[13] A basis of a Hilbert space is not the same thing as a basis in the sense of linear algebra above. For distinction, the latter is then called a Hamel basis.

[14] Although the Fourier series is periodic, the technique can be applied to any $L^2$ function on an interval by considering the function to be continued periodically outside the interval. See Kreyszig 1988, p. 601

[15] That is to say (BSE-3 2001), the plane passing through the point of contact $P$ such that the distance from a point $P_1$ on the surface to the plane is infinitesimally small compared to the distance from $P_1$ to $P$ in the limit as $P_1$ approaches $P$ along the surface.

[16] That is, there is a homeomorphism from $\pi^{-1}(U)$ to $V \times U$ which restricts to linear isomorphisms between fibers.

[17] A line bundle, such as the tangent bundle of $S^1$ is trivial if and only if there is a section that vanishes nowhere, see Husemoller 1994, Corollary 8.3. The sections of the tangent bundle are just vector fields.

## 16.12    Footnotes

[1] Roman 2005, ch. 1, p. 27

[2] van der Waerden 1993, Ch. 19

[3] Bourbaki 1998, §II.1.1. Bourbaki calls the group homomorphisms $f(a)$ *homotheties*.

[4] Bourbaki 1969, ch. "Algèbre linéaire et algèbre multilinéaire", pp. 78–91

[5] Bolzano 1804

[6] Möbius 1827

[7] Hamilton 1853

[8] Grassmann 2000

[9] Peano 1888, ch. IX

[10] Banach 1922

[11] Dorier 1995, Moore 1995

[12] Lang 1987, ch. I.1

[13] Lang 2002, ch. V.1

[14]  e.g. Lang 1993, ch. XII.3., p. 335

[15]  Lang 1987, ch. IX.1

[16]  Lang 1987, ch. VI.3.

[17]  Roman 2005, Theorem 1.9, p. 43

[18]  Blass 1984

[19]  Halpern 1966, pp. 670–673

[20]  Artin 1991, Theorem 3.3.13

[21]  Braun 1993, Th. 3.4.5, p. 291

[22]  Stewart 1975, Proposition 4.3, p. 52

[23]  Stewart 1975, Theorem 6.5, p. 74

[24]  Roman 2005, ch. 2, p. 45

[25]  Lang 1987, ch. IV.4, Corollary, p. 106

[26]  Lang 1987, Example IV.2.6

[27]  Lang 1987, ch. VI.6

[28]  Halmos 1974, p. 28, Ex. 9

[29]  Lang 1987, Theorem IV.2.1, p. 95

[30]  Roman 2005, Th. 2.5 and 2.6, p. 49

[31]  Lang 1987, ch. V.1

[32]  Lang 1987, ch. V.3., Corollary, p. 106

[33]  Lang 1987, Theorem VII.9.8, p. 198

[34]  Roman 2005, ch. 8, p. 135–156

[35]  Lang 1987, ch. IX.4

[36]  Roman 2005, ch. 1, p. 29

[37]  Roman 2005, ch. 1, p. 35

[38]  Roman 2005, ch. 3, p. 64

[39]  Lang 1987, ch. IV.3.

[40]  Roman 2005, ch. 2, p. 48

[41]  Mac Lane 1998

[42]  Roman 2005, ch. 1, pp. 31–32

[43]  Lang 2002, ch. XVI.1

[44]  Roman 2005, Th. 14.3. See also Yoneda lemma.

[45]  Schaefer & Wolff 1999, pp. 204–205

[46]  Bourbaki 2004, ch. 2, p. 48

[47]  Roman 2005, ch. 9

[48]  Naber 2003, ch. 1.2

[49]  Treves 1967

[50]  Bourbaki 1987

[51]  Kreyszig 1989, §4.11-5

[52]  Kreyszig 1989, §1.5-5

[53]  Choquet 1966, Proposition III.7.2

[54]  Treves 1967, p. 34–36

[55]  Lang 1983, Cor. 4.1.2, p. 69

[56]  Treves 1967, ch. 11

[57]  Treves 1967, Theorem 11.2, p. 102

[58]  Evans 1998, ch. 5

[59]  Treves 1967, ch. 12

[60]  Dennery 1996, p.190

[61]  Lang 1993, Th. XIII.6, p. 349

[62]  Lang 1993, Th. III.1.1

[63]  Choquet 1966, Lemma III.16.11

[64]  Kreyszig 1999, Chapter 11

[65]  Griffiths 1995, Chapter 1

[66]  Lang 1993, ch. XVII.3

[67]  Lang 2002, ch. III.1, p. 121

[68]  Eisenbud 1995, ch. 1.6

[69]  Varadarajan 1974

[70]  Lang 2002, ch. XVI.7

[71]  Lang 2002, ch. XVI.8

[72]  Luenberger 1997, §7.13

[73]  See representation theory and group representation.

[74]  Lang 1993, Ch. XI.1

[75]  Evans 1998, Th. 6.2.1

[76]  Folland 1992, p. 349 *ff*

[77]  Gasquet & Witomski 1999, p. 150

[78]  Gasquet & Witomski 1999, §4.5

[79]  Gasquet & Witomski 1999, p. 57

[80]  Loomis 1953, Ch. VII

[81]  Ashcroft & Mermin 1976, Ch. 5

[82]  Kreyszig 1988, p. 667

[83]  Fourier 1822

[84] Gasquet & Witomski 1999, p. 67

[85] Ifeachor & Jervis 2002, pp. 3–4, 11

[86] Wallace Feb 1992

[87] Ifeachor & Jervis 2002, p. 132

[88] Gasquet & Witomski 1999, §10.2

[89] Ifeachor & Jervis 2002, pp. 307–310

[90] Gasquet & Witomski 1999, §10.3

[91] Schönhage & Strassen 1971

[92] Spivak 1999, ch. 3

[93] Jost 2005. See also Lorentzian manifold.

[94] Misner, Thorne & Wheeler 1973, ch. 1.8.7, p. 222 and ch. 2.13.5, p. 325

[95] Jost 2005, ch. 3.1

[96] Varadarajan 1974, ch. 4.3, Theorem 4.3.27

[97] Kreyszig 1991, §34, p. 108

[98] Eisenberg & Guy 1979

[99] Atiyah 1989

[100] Artin 1991, ch. 12

[101] Grillet, Pierre Antoine. Abstract algebra. Vol. 242. Springer Science & Business Media, 2007.

[102] Meyer 2000, Example 5.13.5, p. 436

[103] Meyer 2000, Exercise 5.13.15–17, p. 442

[104] Coxeter 1987

# 16.13  References

## 16.13.1  Algebra

- Artin, Michael (1991), *Algebra*, Prentice Hall, ISBN 978-0-89871-510-1

- Blass, Andreas (1984), "Existence of bases implies the axiom of choice", *Axiomatic set theory (Boulder, Colorado, 1983)*, Contemporary Mathematics **31**, Providence, R.I.: American Mathematical Society, pp. 31–33, MR 763890

- Brown, William A. (1991), *Matrices and vector spaces*, New York: M. Dekker, ISBN 978-0-8247-8419-5

- Lang, Serge (1987), *Linear algebra*, Berlin, New York: Springer-Verlag, ISBN 978-0-387-96412-6

- Lang, Serge (2002), *Algebra*, Graduate Texts in Mathematics **211** (Revised third ed.), New York: Springer-Verlag, ISBN 978-0-387-95385-4, MR 1878556

- Mac Lane, Saunders (1999), *Algebra* (3rd ed.), pp. 193–222, ISBN 0-8218-1646-2

- Meyer, Carl D. (2000), *Matrix Analysis and Applied Linear Algebra*, SIAM, ISBN 978-0-89871-454-8

- Roman, Steven (2005), *Advanced Linear Algebra*, Graduate Texts in Mathematics **135** (2nd ed.), Berlin, New York: Springer-Verlag, ISBN 978-0-387-24766-3

- Spindler, Karlheinz (1993), *Abstract Algebra with Applications: Volume 1: Vector spaces and groups*, CRC, ISBN 978-0-8247-9144-5

- van der Waerden, Bartel Leendert (1993), *Algebra* (in German) (9th ed.), Berlin, New York: Springer-Verlag, ISBN 978-3-540-56799-8

## 16.13.2  Analysis

- Bourbaki, Nicolas (1987), *Topological vector spaces*, Elements of mathematics, Berlin, New York: Springer-Verlag, ISBN 978-3-540-13627-9

- Bourbaki, Nicolas (2004), *Integration I*, Berlin, New York: Springer-Verlag, ISBN 978-3-540-41129-1

- Braun, Martin (1993), *Differential equations and their applications: an introduction to applied mathematics*, Berlin, New York: Springer-Verlag, ISBN 978-0-387-97894-9

- BSE-3 (2001), "Tangent plane", in Hazewinkel, Michiel, *Encyclopedia of Mathematics*, Springer, ISBN 978-1-55608-010-4

- Choquet, Gustave (1966), *Topology*, Boston, MA: Academic Press

- Dennery, Philippe; Krzywicki, Andre (1996), *Mathematics for Physicists*, Courier Dover Publications, ISBN 978-0-486-69193-0

- Dudley, Richard M. (1989), *Real analysis and probability*, The Wadsworth & Brooks/Cole Mathematics Series, Pacific Grove, CA: Wadsworth & Brooks/Cole Advanced Books & Software, ISBN 978-0-534-10050-6

- Dunham, William (2005), *The Calculus Gallery*, Princeton University Press, ISBN 978-0-691-09565-3

- Evans, Lawrence C. (1998), *Partial differential equations*, Providence, R.I.: American Mathematical Society, ISBN 978-0-8218-0772-9

- Folland, Gerald B. (1992), *Fourier Analysis and Its Applications*, Brooks-Cole, ISBN 978-0-534-17094-3

- Gasquet, Claude; Witomski, Patrick (1999), *Fourier Analysis and Applications: Filtering, Numerical Computation, Wavelets*, Texts in Applied Mathematics, New York: Springer-Verlag, ISBN 0-387-98485-2

- Ifeachor, Emmanuel C.; Jervis, Barrie W. (2001), *Digital Signal Processing: A Practical Approach* (2nd ed.), Harlow, Essex, England: Prentice-Hall (published 2002), ISBN 0-201-59619-9

- Krantz, Steven G. (1999), *A Panorama of Harmonic Analysis*, Carus Mathematical Monographs, Washington, DC: Mathematical Association of America, ISBN 0-88385-031-1

- Kreyszig, Erwin (1988), *Advanced Engineering Mathematics* (6th ed.), New York: John Wiley & Sons, ISBN 0-471-85824-2

- Kreyszig, Erwin (1989), *Introductory functional analysis with applications*, Wiley Classics Library, New York: John Wiley & Sons, ISBN 978-0-471-50459-7, MR 992618

- Lang, Serge (1983), *Real analysis*, Addison-Wesley, ISBN 978-0-201-14179-5

- Lang, Serge (1993), *Real and functional analysis*, Berlin, New York: Springer-Verlag, ISBN 978-0-387-94001-4

- Loomis, Lynn H. (1953), *An introduction to abstract harmonic analysis*, Toronto-New York–London: D. Van Nostrand Company, Inc., pp. x+190

- Schaefer, Helmut H.; Wolff, M.P. (1999), *Topological vector spaces* (2nd ed.), Berlin, New York: Springer-Verlag, ISBN 978-0-387-98726-2

- Treves, François (1967), *Topological vector spaces, distributions and kernels*, Boston, MA: Academic Press

### 16.13.3    Historical references

- Banach, Stefan (1922), "Sur les opérations dans les ensembles abstraits et leur application aux équations intégrales (On operations in abstract sets and their application to integral equations)" (PDF), *Fundamenta Mathematicae* (in French) 3, ISSN 0016-2736

- Bolzano, Bernard (1804), *Betrachtungen über einige Gegenstände der Elementargeometrie (Considerations of some aspects of elementary geometry)* (in German)

- Bourbaki, Nicolas (1969), *Éléments d'histoire des mathématiques (Elements of history of mathematics)* (in French), Paris: Hermann

- Dorier, Jean-Luc (1995), "A general outline of the genesis of vector space theory", *Historia Mathematica* **22** (3): 227–261, doi:10.1006/hmat.1995.1024, MR 1347828

- Fourier, Jean Baptiste Joseph (1822), *Théorie analytique de la chaleur* (in French), Chez Firmin Didot, père et fils

- Grassmann, Hermann (1844), *Die Lineale Ausdehnungslehre - Ein neuer Zweig der Mathematik* (in German), O. Wigand, reprint: Hermann Grassmann. Translated by Lloyd C. Kannenberg. (2000), Kannenberg, L.C., ed., *Extension Theory*, Providence, R.I.: American Mathematical Society, ISBN 978-0-8218-2031-5

- Hamilton, William Rowan (1853), *Lectures on Quaternions*, Royal Irish Academy

- Möbius, August Ferdinand (1827), *Der Barycentrische Calcul : ein neues Hülfsmittel zur analytischen Behandlung der Geometrie (Barycentric calculus: a new utility for an analytic treatment of geometry)* (in German)

- Moore, Gregory H. (1995), "The axiomatization of linear algebra: 1875–1940", *Historia Mathematica* **22** (3): 262–303, doi:10.1006/hmat.1995.1025

- Peano, Giuseppe (1888), *Calcolo Geometrico secondo l'Ausdehnungslehre di H. Grassmann preceduto dalle Operazioni della Logica Deduttiva* (in Italian), Turin

### 16.13.4    Further references

- Ashcroft, Neil; Mermin, N. David (1976), *Solid State Physics*, Toronto: Thomson Learning, ISBN 978-0-03-083993-1

- Atiyah, Michael Francis (1989), *K-theory*, Advanced Book Classics (2nd ed.), Addison-Wesley, ISBN 978-0-201-09394-0, MR 1043170

- Bourbaki, Nicolas (1998), *Elements of Mathematics : Algebra I Chapters 1-3*, Berlin, New York: Springer-Verlag, ISBN 978-3-540-64243-5

- Bourbaki, Nicolas (1989), *General Topology. Chapters 1-4*, Berlin, New York: Springer-Verlag, ISBN 978-3-540-64241-1

- Coxeter, Harold Scott MacDonald (1987), *Projective Geometry* (2nd ed.), Berlin, New York: Springer-Verlag, ISBN 978-0-387-96532-1

- Eisenberg, Murray; Guy, Robert (1979), "A proof of the hairy ball theorem", *The American Mathematical Monthly* (Mathematical Association of America) **86** (7): 572–574, doi:10.2307/2320587, JSTOR 2320587

- Eisenbud, David (1995), *Commutative algebra*, Graduate Texts in Mathematics **150**, Berlin, New York: Springer-Verlag, ISBN 978-0-387-94269-8, MR 1322960

- Goldrei, Derek (1996), *Classic Set Theory: A guided independent study* (1st ed.), London: Chapman and Hall, ISBN 0-412-60610-0

- Griffiths, David J. (1995), *Introduction to Quantum Mechanics*, Upper Saddle River, NJ: Prentice Hall, ISBN 0-13-124405-1

- Halmos, Paul R. (1974), *Finite-dimensional vector spaces*, Berlin, New York: Springer-Verlag, ISBN 978-0-387-90093-3

- Halpern, James D. (Jun 1966), "Bases in Vector Spaces and the Axiom of Choice", *Proceedings of the American Mathematical Society* (American Mathematical Society) **17** (3): 670–673, doi:10.2307/2035388, JSTOR 2035388

- Husemoller, Dale (1994), *Fibre Bundles* (3rd ed.), Berlin, New York: Springer-Verlag, ISBN 978-0-387-94087-8

- Jost, Jürgen (2005), *Riemannian Geometry and Geometric Analysis* (4th ed.), Berlin, New York: Springer-Verlag, ISBN 978-3-540-25907-7

- Kreyszig, Erwin (1991), *Differential geometry*, New York: Dover Publications, pp. xiv+352, ISBN 978-0-486-66721-8

- Kreyszig, Erwin (1999), *Advanced Engineering Mathematics* (8th ed.), New York: John Wiley & Sons, ISBN 0-471-15496-2

- Luenberger, David (1997), *Optimization by vector space methods*, New York: John Wiley & Sons, ISBN 978-0-471-18117-0

- Mac Lane, Saunders (1998), *Categories for the Working Mathematician* (2nd ed.), Berlin, New York: Springer-Verlag, ISBN 978-0-387-98403-2

- Misner, Charles W.; Thorne, Kip; Wheeler, John Archibald (1973), *Gravitation*, W. H. Freeman, ISBN 978-0-7167-0344-0

- Naber, Gregory L. (2003), *The geometry of Minkowski spacetime*, New York: Dover Publications, ISBN 978-0-486-43235-9, MR 2044239

- Schönhage, A.; Strassen, Volker (1971), "Schnelle Multiplikation großer Zahlen (Fast multiplication of big numbers)" (PDF), *Computing* (in German) **7**: 281–292, doi:10.1007/bf02242355, ISSN 0010-485X

- Spivak, Michael (1999), *A Comprehensive Introduction to Differential Geometry (Volume Two)*, Houston, TX: Publish or Perish

- Stewart, Ian (1975), *Galois Theory*, Chapman and Hall Mathematics Series, London: Chapman and Hall, ISBN 0-412-10800-3

- Varadarajan, V. S. (1974), *Lie groups, Lie algebras, and their representations*, Prentice Hall, ISBN 978-0-13-535732-3

- Wallace, G.K. (Feb 1992), "The JPEG still picture compression standard", *IEEE Transactions on Consumer Electronics* **38** (1): xviii–xxxiv, doi:10.1109/30.125072, ISSN 0098-3063

- Weibel, Charles A. (1994), *An introduction to homological algebra*, Cambridge Studies in Advanced Mathematics **38**, Cambridge University Press, ISBN 978-0-521-55987-4, OCLC 36131259, MR 1269324

## 16.14   External links

- Hazewinkel, Michiel, ed. (2001), "Vector space", *Encyclopedia of Mathematics*, Springer, ISBN 978-1-55608-010-4

- A lecture about fundamental concepts related to vector spaces (given at MIT)

- A graphical simulator for the concepts of span, linear dependency, base and dimension

# Chapter 17

# Symmetric bilinear form

A **symmetric bilinear form** on a vector space is a linear map from two copies of the vector space to the field of scalars such that the order of the two vectors does not affect the value of the map. In other words, it is a bilinear function $B$ that maps every pair $(u, v)$ of elements of the vector space $V$ to the underlying field such that $B(u, v) = B(v, u)$ for every $u$ and $v$ in $V$. They are also referred to more briefly as just **symmetric forms** when "bilinear" is understood.

Symmetric bilinear forms on finite-dimensional vector spaces precisely correspond to symmetric matrices given a basis for $V$. Among bilinear forms, the symmetric ones are important because they are the ones for which the vector space admits a particularly simple kind of basis known as an orthogonal basis (at least when the characteristic of the field is not 2).

Given a symmetric bilinear form $B$, the function $q(x) = B(x, x)$ is the associated quadratic form on the vector space. Moreover, if the characteristic of the field is not 2, $B$ is the unique symmetric bilinear form associated with $q$.

## 17.1  Formal definition

Let $V$ be a vector space of dimension $n$ over a field $K$. A map $B : V \times V \to K$ is a symmetric bilinear form on the space if:

- $B(u, v) = B(v, u)$    $\forall u, v \in V$

- $B(u + v, w) = B(u, w) + B(v, w)$    $\forall u, v, w \in V$

- $B(\lambda v, w) = \lambda B(v, w)$    $\forall \lambda \in K, \forall v, w \in V$

The last two axioms only imply linearity in the first argument, but the first axiom then immediately implies linearity in the second argument as well.

## 17.2  Examples

Let $V = \mathbf{R}^n$, the $n$ dimensional real vector space. Then the standard dot product is a symmetric bilinear form, $B(x, y) = x \cdot y$. The matrix corresponding to this bilinear form (see below) on a standard basis is the identity matrix.

Let $V$ be any vector space (including possibly infinite-dimensional), and assume $T$ is a linear function from $V$ to the field. Then the function defined by $B(x, y) = T(x)T(y)$ is a symmetric bilinear form.

Let $V$ be the vector space of continuous single-variable real functions. For $f, g \in V$ one can define $B(f, g) = \int_0^1 f(t)g(t)dt$. By the properties of definite integrals, this defines a symmetric bilinear form on $V$. This is an example of a symmetric bilinear form which is not associated to any symmetric matrix (since the vector space is infinite-dimensional).

## 17.3  Matrix representation

Let $C = \{e_1, \ldots, e_n\}$ be a basis for $V$. Define the $n \times n$ matrix $A$ by $A_{ij} = B(e_i, e_j)$. The matrix $A$ is a symmetric matrix exactly due to symmetry of the bilinear form. If the $n \times 1$ matrix $x$ represents a vector $v$ with respect to this basis, and analogously, $y$ represents $w$, then $B(v, w)$ is given by :

$$x^\mathsf{T} A y = y^\mathsf{T} A x.$$

Suppose $C'$ is another basis for $V$, with : $\begin{bmatrix} e'_1 & \cdots & e'_n \end{bmatrix} = \begin{bmatrix} e_1 & \cdots & e_n \end{bmatrix} S$ with $S$ an invertible $n \times n$ matrix. Now the new matrix representation for the symmetric bilinear form is given by

$$A' = S^\mathsf{T} A S.$$

## 17.4 Orthogonality and singularity

A symmetric bilinear form is always reflexive. Two vectors $v$ and $w$ are defined to be orthogonal with respect to the bilinear form $B$ if $B(v, w) = 0$, which is, due to reflexivity, equivalent to $B(w, v) = 0$.

The **radical** of a bilinear form $B$ is the set of vectors orthogonal with every vector in $V$. That this is a subspace of $V$ follows from the linearity of $B$ in each of its arguments. When working with a matrix representation $A$ with respect to a certain basis, $v$, represented by $x$, is in the radical if and only if

$$Ax = 0 \iff x^\mathsf{T} A = 0.$$

The matrix $A$ is singular if and only if the radical is nontrivial.

If $W$ is a subset of $V$, then its *orthogonal complement* $W^\perp$ is the set of all vectors in $V$ that are orthogonal to every vector in $W$; it is a subspace of $V$. When $B$ is non-degenerate, the radical of $B$ is trivial and the dimension of $W^\perp$ is $\dim(W^\perp) = \dim(V) - \dim(W)$.

## 17.5 Orthogonal basis

A basis $C = \{e_1, \ldots, e_n\}$ is orthogonal with respect to $B$ if and only if :

$$B(e_i, e_j) = 0 \ \forall i \neq j.$$

When the characteristic of the field is not two, $V$ always has an orthogonal basis. This can be proven by induction.

A basis $C$ is orthogonal if and only if the matrix representation $A$ is a diagonal matrix.

### 17.5.1 Signature and Sylvester's law of inertia

In a more general form, Sylvester's law of inertia says that, when working over an ordered field, the numbers of diagonal elements in the diagonalized form of a matrix that are positive, negative and zero respectively are independent of the chosen orthogonal basis. These three numbers form the *signature* of the bilinear form.

### 17.5.2 Real case

When working in a space over the reals, one can go a bit a further. Let $C = \{e_1, \ldots, e_n\}$ be an orthogonal basis.

We define a new basis $C' = \{e'_1, \ldots, e'_n\}$

$$e'_i = \begin{cases} e_i & \text{if } B(e_i, e_i) = 0 \\ \frac{e_i}{\sqrt{B(e_i, e_i)}} & \text{if } B(e_i, e_i) > 0 \\ \frac{e_i}{\sqrt{-B(e_i, e_i)}} & \text{if } B(e_i, e_i) < 0 \end{cases}$$

Now, the new matrix representation $A$ will be a diagonal matrix with only 0, 1 and −1 on the diagonal. Zeroes will appear if and only if the radical is nontrivial.

### 17.5.3 Complex case

When working in a space over the complex numbers, one can go further as well and it is even easier. Let $C = \{e_1, \ldots, e_n\}$ be an orthogonal basis.

We define a new basis $C' = \{e'_1, \ldots, e'_n\}$ :

$$e'_i = \begin{cases} e_i & \text{if } B(e_i, e_i) = 0 \\ e_i / \sqrt{B(e_i, e_i)} & \text{if } B(e_i, e_i) \neq 0 \end{cases}$$

Now the new matrix representation $A$ will be a diagonal matrix with only 0 and 1 on the diagonal. Zeroes will appear if and only if the radical is nontrivial.

## 17.6 Orthogonal polarities

Let $B$ be a symmetric bilinear form with a trivial radical on the space $V$ over the field $K$ with characteristic not 2. One can now define a map from $D(V)$, the set of all subspaces of $V$, to itself:

$$\alpha : D(V) \to D(V) : W \mapsto W^\perp.$$

This map is an **orthogonal polarity** on the projective space $PG(W)$. Conversely, one can prove all orthogonal polarities are induced in this way, and that two symmetric bilinear forms with trivial radical induce the same polarity if and only if they are equal up to scalar multiplication.

## 17.7 References

- Adkins, William A.; Weintraub, Steven H. (1992). *Algebra: An Approach via Module Theory*. Graduate Texts in Mathematics **136**. Springer-Verlag. ISBN 3-540-97839-9. Zbl 0768.00003.

- Milnor, J.; Husemoller, D. (1973). *Symmetric Bilinear Forms*. Ergebnisse der Mathematik und ihrer Grenzgebiete **73**. Springer-Verlag. ISBN 3-540-06009-X. Zbl 0292.10016.

- Weisstein, Eric W., "Symmetric Bilinear Form", *MathWorld*.

# Chapter 18

# Metric signature

The **signature** $(p, q, r)$ of a metric tensor $g$ (or equivalently, a real quadratic form thought of as a real symmetric bilinear form on a finite-dimensional vector space) is the number (counted with multiplicity) of positive, negative and zero eigenvalues of the real symmetric matrix $gab$ of the metric tensor with respect to a basis. Alternatively, it can be defined as the dimensions of a maximal positive, negative and null subspace. By Sylvester's law of inertia these numbers do not depend on the choice of basis. The signature thus classifies the metric up to a choice of basis. The signature is often denoted by a pair of integers $(p, q)$ implying $r = 0$ or as an explicit list of signs of eigenvalues such as $(+, -, -, -)$ or $(-, +, +, +)$ for the signature $(1, 3)$ resp. $(3, 1)$.[1]

The signature is said to be **indefinite** or **mixed** if both $p$ and $q$ are nonzero, and degenerate if $r$ is nonzero. A Riemannian metric is a metric with a (positive) definite signature. A Lorentzian metric is one with signature $(p, 1)$, or $(1, q)$.

There is another notion of **signature** of a nondegenerate metric tensor given by a single number $s$ defined as $p - q$, where $p$ and $q$ are as above, which is equivalent to the above definition when the dimension $n = p + q$ is given or implicit. For example, $s = 1 - 3 = -2$ for $(+, -, -, -)$ and $s = 3 - 1 = +2$ for $(-, +, +, +)$.

## 18.1 Definition

The signature of a metric tensor is defined as the signature of the corresponding quadratic form.[2] It is the number $(p, q, r)$ of positive, negative and zero eigenvalues of any matrix (i.e. in any basis for the underlying vector space) representing the form, counted with their algebraic multiplicity. Usually, $q = 0$ is required, which is the same as saying a metric tensor must be nondegenerate, i.e. no nonzero vector is orthogonal to all vectors.

By Sylvester's law of inertia, the numbers $(p, q, r)$ are basis independent.

## 18.2 Properties

### 18.2.1 Signature and dimension

By the spectral theorem a symmetric $n \times n$ matrix over the reals is always diagonalizable, and has therefore exactly $n$ real eigenvalues (counted with algebraic multiplicity). Thus $p + q + r = n = \dim(V)$.

### 18.2.2 Sylvester's law of inertia: independence of basis choice and existence of orthonormal basis

According to Sylvester's law of inertia, the signature of the scalar product (a.k.a. real symmetric bilinear form), $g$ does not depend on the choice of basis. Moreover, for every metric $g$ of signature $(p, q, r)$ there exists a basis such that $gab = +1$ for $a = b = 1, ..., p$, $gab = -1$ for $a = b = p + 1, ..., p + q$ and $gab = 0$ otherwise. It follows that there exists an isometry $(V_1, g_1) \to (V_2, g_2)$ if and only if the signatures of $g_1$ and $g_2$ are equal. Likewise the signature is equal for two congruent matrices and classifies a matrix up to congruency. Equivalently, the signature is constant on the orbits of the general linear group $GL(V)$ on the space of symmetric rank 2 contravariant tensors $S^2 V^*$ and classifies each orbit.

### 18.2.3 Geometrical interpretation of the indices

The number $p$ (resp. $q$) is the maximal dimension of a vector subspace on which the scalar product $g$ is positive-definite (resp. negative-definite), and $r$ is the dimension of the radical of the scalar product $g$ or the null subspace of symmetric matrix $gab$ of the scalar product. Thus a nondegenerate scalar product has signature $(p, q, 0)$, with $p + q = n$. The special cases $(n, 0, 0)$ and $(0, n, 0)$ correspond to positive-definite and negative-definite scalar products which can be transformed into each other by negation.

## 18.3 Examples

### 18.3.1 Matrices

The signature of the $n \times n$ identity matrix is (n, 0, 0). The signature of a diagonal matrix is the number of positive, negative and zero numbers on its main diagonal.

The following matrices have both the same signature (1, 1, 0), therefore they are congruent because of Sylvester's law of inertia:

$$\begin{pmatrix} 1 & 0 \\ 0 & -1 \end{pmatrix}, \quad \begin{pmatrix} 0 & 1 \\ 1 & 0 \end{pmatrix}.$$

### 18.3.2 Scalar products

The standard scalar product defined on $\mathbb{R}^n$ has the signature (n, 0, 0). A scalar product has this signature if and only if it is a positive definite scalar product.

A negative definite scalar product has the signature (0, n, 0). A positive semi-definite scalar product has a signature (p, 0, r), where $p + r = n$.

The Minkowski space is $\mathbb{R}^4$ and has a scalar product defined by the matrix

$$\begin{pmatrix} -1 & 0 & 0 & 0 \\ 0 & 1 & 0 & 0 \\ 0 & 0 & 1 & 0 \\ 0 & 0 & 0 & 1 \end{pmatrix}$$

and has signature (3, 1, 0), which is known as space-positive.

Sometimes it is used with the opposite signs, thus obtaining the signature (1, 3, 0), which is known as time-positive.

$$\begin{pmatrix} 1 & 0 & 0 & 0 \\ 0 & -1 & 0 & 0 \\ 0 & 0 & -1 & 0 \\ 0 & 0 & 0 & -1 \end{pmatrix}$$

## 18.4 How to compute the signature

There are some methods for computing the signature of a matrix.

- For any nondegenerate symmetric matrix of $n \times n$, diagonalize it (or find all of eigenvalues of it) and count the number of positive and negative signs.

- For a symmetric matrix, the characteristic polynomial will have all real roots whose signs may in some cases be completely determined by Descartes' rule of signs.

- Lagrange algorithm gives a way to compute an orthogonal basis, and thus compute a diagonal matrix congruent (thus, with the same signature) to the other one: the signature of a diagonal matrix is the number of positive, negative and zero elements on its diagonal.

- According to Jacobi's criterion, a symmetric matrix is positive-definite if and only if all the determinants of its main minors are positive.

## 18.5 Signature in physics

In mathematics, the usual convention for any Riemannian manifold is to use a positive-definite metric tensor (meaning that after diagonalization, elements on the diagonal are all positive).

In theoretical physics, spacetime is modeled by a pseudo-Riemannian manifold. The signature counts how many time-like or space-like characters are in the spacetime, in the sense defined by special relativity: as used in particle physics, the metric is positive definite on the time-like subspace, and negative definite on the space-like subspace. In the specific case of the Minkowski metric,

$$ds^2 = c^2 dt^2 - dx^2 - dy^2 - dz^2$$

the metric signature is (1, 3, 0), since it is positive definite in the time direction, and negative definite in the three spatial directions $x$, $y$ and $z$. (Sometimes the opposite sign convention is used, but with the one given here $s$ directly measures proper time.)

## 18.6 Signature change

If a metric is regular everywhere then the signature of the metric is constant. However if one allows for metrics that are degenerate or discontinuous on some hypersurfaces, then signature of the metric may change at these surfaces.[3] Such signature changing metrics may possibly have applications in cosmology and quantum gravity.

## 18.7 See also

- pseudo-Riemannian manifold

- Sign convention

# 18.8 Notes

[1] Rowland, Todd. "Matrix Signature." From MathWorld--A Wolfram Web Resource, created by Eric W. Weisstein. http://mathworld.wolfram.com/MatrixSignature.html

[2] Landau, L.D.; Lifshitz, E.M. (2002) [1939]. *The Classical Theory of Fields*. Course of Theoretical Physics 2 (4th ed.). Butterworth–Heinemann. pp. 245–246. ISBN 0 7506 2768 9.

[3] Dray, Tevian; Ellis, George; Hellaby, Charles; Manogue, Corinne A. (1997). "Gravity and signature change". *General Relativity and Gravitation* **29**: 591–597. arXiv:gr-qc/9610063. Bibcode:1997GReGr..29..591D. doi:10.1023/A:1018895302693.

# Chapter 19

# Metric tensor (general relativity)

This article is about metrics in general relativity. For a discussion of metrics in general, see metric tensor.

In general relativity, the **metric tensor** (or simply, the **metric**) is the fundamental object of study. It may loosely be thought of as a generalization of the gravitational potential familiar from Newtonian gravitation. The metric captures all the geometric and causal structure of spacetime, being used to define notions such as time, distance, volume, curvature, angle, and separating the future and the past.

## 19.1 Notation and conventions

Throughout this article we work with a metric signature that is mostly positive $(- + + +)$; see sign convention. The gravitation constant $G$ will be kept explicit. The summation convention, where repeated indices are automatically summed over, is employed.

## 19.2 Definition

Mathematically, spacetime is represented by a 4-dimensional differentiable manifold $M$ and the metric is given as a covariant, second-order, symmetric tensor on $M$, conventionally denoted by $g$. Moreover, the metric is required to be nondegenerate with signature (-+++). A manifold $M$ equipped with such a metric is a type of Lorentzian manifold.

Explicitly, the metric is a symmetric bilinear form on each tangent space of $M$ which varies in a smooth (or differentiable) manner from point to point. Given two tangent vectors $u$ and $v$ at a point $x$ in $M$, the metric can be evaluated on $u$ and $v$ to give a real number:

$$g_x(u, v) = g_x(v, u) \in \mathbb{R}.$$

This can be thought of as a generalization of the dot product in ordinary Euclidean space. This analogy is not exact, however. Unlike Euclidean space — where the dot product is positive definite — the metric gives each tangent space the structure of Minkowski space.

## 19.3 Local coordinates and matrix representations

Physicists usually work in local coordinates (i.e. coordinates defined on some local patch of $M$). In local coordinates $x^\mu$ (where $\mu$ is an index which runs from 0 to 3) the metric can be written in the form

$$g = g_{\mu\nu} dx^\mu \otimes dx^\nu.$$

The factors $dx^\mu$ are one-form gradients of the scalar coordinate fields $x^\mu$. The metric is thus a linear combination of tensor products of one-form gradients of coordinates. The coefficients $g_{\mu\nu}$ are a set of 16 real-valued functions (since the tensor $g$ is actually a *tensor field*, which is defined at all points of a spacetime manifold). In order for the metric to be symmetric we must have

$$g_{\mu\nu} = g_{\nu\mu},$$

giving 10 independent coefficients. If we denote the symmetric tensor product by juxtaposition (so that $dx^\mu dx^\nu = dx^\nu dx^\mu$) we can write the metric in the form

$$g = g_{\mu\nu} dx^\mu dx^\nu.$$

If the local coordinates are specified, or understood from context, the metric can be written as a 4 × 4 symmetric matrix with entries $g_{\mu\nu}$. The nondegeneracy of $g_{\mu\nu}$ means that this matrix is non-singular (i.e. has non-vanishing determinant), while the Lorentzian signature of $g$ implies that the matrix has one negative and three positive eigenvalues. Note that physicists often refer to this matrix or the coordinates $g_{\mu\nu}$ themselves as the metric (see, however, abstract index notation).

With the quantities $dx^\mu$ being regarded as the components of an infinitesimal coordinate displacement four-vector, the metric determines the invariant square of an infinitesimal line element, often referred to as an *interval*. The interval is often denoted

$$ds^2 = g_{\mu\nu}dx^\mu dx^\nu.$$

The interval $ds^2$ imparts information about the causal structure of spacetime. When $ds^2 < 0$, the interval is timelike and the square root of the absolute value of $ds^2$ is an incremental proper time. Only timelike intervals can be physically traversed by a massive object. When $ds^2 = 0$, the interval is lightlike, and can only be traversed by light. When $ds^2 > 0$, the interval is spacelike and the square root of $ds^2$ acts as an incremental proper length. Spacelike intervals cannot be traversed, since they connect events that are outside each other's light cones. Events can be causally related only if they are within each other's light cones.

The components of the metric depend on the choice of local coordinate system. Under a change of coordinates $x^\mu \rightarrow x^{\bar\mu}$, the metric components transform as

$$g_{\bar\mu\bar\nu} = \frac{\partial x^\rho}{\partial x^{\bar\mu}} \frac{\partial x^\sigma}{\partial x^{\bar\nu}} g_{\rho\sigma} = \Lambda^\rho{}_{\bar\mu} \Lambda^\sigma{}_{\bar\nu} g_{\rho\sigma}.$$

## 19.4 Examples

### 19.4.1 Flat spacetime

The simplest example of a Lorentzian manifold is flat spacetime, which can be given as $\mathbf{R}^4$ with coordinates $(t, x, y, z)$ and the metric

$$ds^2 = -c^2 dt^2 + dx^2 + dy^2 + dz^2 = \eta_{\mu\nu}dx^\mu dx^\nu.$$

Note that these coordinates actually cover all of $\mathbf{R}^4$. The flat space metric (or Minkowski metric) is often denoted by the symbol $\eta$ and is the metric used in special relativity. In the above coordinates, the matrix representation of $\eta$ is

$$\eta = \begin{pmatrix} -c^2 & 0 & 0 & 0 \\ 0 & 1 & 0 & 0 \\ 0 & 0 & 1 & 0 \\ 0 & 0 & 0 & 1 \end{pmatrix}$$

(An alternative convention replaces coordinate $t$ by $ct$, and defines $\eta$ as in Minkowski space#Standard basis.)

In spherical coordinates $(t, r, \theta, \phi)$, the flat space metric takes the form

$$ds^2 = -c^2 dt^2 + dr^2 + r^2 d\Omega^2$$

where

$$d\Omega^2 = d\theta^2 + \sin^2\theta\, d\phi^2$$

is the standard metric on the 2-sphere.

### 19.4.2 Schwarzschild metric

Besides the flat space metric the most important metric in general relativity is the Schwarzschild metric which can be given in one set of local coordinates by

$$ds^2 = -\left(1 - \frac{2GM}{rc^2}\right) c^2 dt^2 + \left(1 - \frac{2GM}{rc^2}\right)^{-1}$$
$$\text{x}\ \ (dr^2 + r^2 d\Omega^2)$$

where, again, $d\Omega^2$ is the standard metric on the 2-sphere. Here $G$ is the gravitation constant and $M$ is a constant with the dimensions of mass. Its derivation can be found here. The Schwarzschild metric approaches the Minkowski metric as $M$ approaches zero (except at the origin where it is undefined). Similarly, when $r$ goes to infinity, the Schwarzschild metric approaches the Minkowski metric.

### 19.4.3 Other metrics

Other notable metrics are:

- Alcubierre metric,
- Bondi metric,
- de Sitter/anti-de Sitter metrics,
- Eddington–Finkelstein coordinates,
- Friedmann–Lemaître–Robertson–Walker metric,
- Gullstrand–Painlevé coordinates,
- Isotropic coordinates,
- Kerr metric,
- Kerr–Newman metric,

- Kruskal–Szekeres coordinates,

- Lemaître coordinates,

- Lemaître–Tolman metric,

- Peres metric,

- Reissner–Nordström metric,

- Rindler coordinates,

- Weyl–Lewis–Papapetrou coordinates.

Some of them are without the event horizon or can be without the gravitational singularity.

## 19.5   Volume

The metric $g$ induces a natural volume form (up to a sign), which can be used to integrate over a region of a manifold. Given local coordinates $x^\mu$ for the manifold, the volume form can be written

$$\mathrm{vol}_g = \pm\sqrt{|\det[g_{\mu\nu}]|}\, dx^0 \wedge dx^1 \wedge dx^2 \wedge dx^3$$

where $\det[g\mu\nu]$ is the determinant of the matrix of components of the metric tensor for the given coordinate system.

## 19.6   Curvature

The metric $g$ completely determines the curvature of spacetime. According to the fundamental theorem of Riemannian geometry, there is a unique connection $\nabla$ on any semi-Riemannian manifold that is compatible with the metric and torsion-free. This connection is called the Levi-Civita connection. The Christoffel symbols of this connection are given in terms of partial derivatives of the metric in local coordinates $x^\mu$ by the formula

$$\Gamma^\lambda{}_{\mu\nu} = \frac{1}{2} g^{\lambda\rho} \left( \frac{\partial g_{\rho\mu}}{\partial x^\nu} + \frac{\partial g_{\rho\nu}}{\partial x^\mu} - \frac{\partial g_{\mu\nu}}{\partial x^\rho} \right)$$

The curvature of spacetime is then given by the Riemann curvature tensor which is defined in terms of the Levi-Civita connection $\nabla$. In local coordinates this tensor is given by:

$$R^\rho{}_{\sigma\mu\nu} = \partial_\mu \Gamma^\rho{}_{\nu\sigma} - \partial_\nu \Gamma^\rho{}_{\mu\sigma} + \Gamma^\rho{}_{\mu\lambda}\Gamma^\lambda{}_{\nu\sigma} - \Gamma^\rho{}_{\nu\lambda}\Gamma^\lambda{}_{\mu\sigma}.$$

The curvature is then expressible purely in terms of the metric $g$ and its derivatives.

## 19.7   Einstein's equations

One of the core ideas of general relativity is that the metric (and the associated geometry of spacetime) is determined by the matter and energy content of spacetime. Einstein's field equations:

$$R_{\mu\nu} - \frac{1}{2} R g_{\mu\nu} = \frac{8\pi G}{c^4} T_{\mu\nu}$$

where the Ricci curvature tensor

$$R_{\nu\rho} \overset{\text{def}}{=} R^\mu{}_{\nu\mu\rho}$$

and the scalar curvature

$$R \overset{\text{def}}{=} g^{\mu\nu} R_{\mu\nu}$$

relate the metric (and the associated curvature tensors) to the stress–energy tensor $T_{\mu\nu}$ . This tensor equation is a complicated set of nonlinear partial differential equations for the metric components. Exact solutions of Einstein's field equations are very difficult to find.

## 19.8   See also

- Alternatives to general relativity

- Basic introduction to the mathematics of curved spacetime

- Mathematics of general relativity

- Ricci calculus

## 19.9   References

See general relativity resources for a list of references.

## 19.10   External links

- Caltech Tutorial on Relativity — A simple introduction to the basics of metrics in the context of relativity.

# Chapter 20

# Event (relativity)

In physics, and in particular relativity, an **event** is a point in spacetime (that is, a specific place and time) and the physical situation or occurrence associated with it. For example, a glass breaking on the floor is an event; it occurs at a unique place and a unique time.[1] Strictly speaking, the notion of an event is an idealization, in the sense that it specifies a definite time and place, whereas any actual event is bound to have a finite extent, both in time and in space.[2]

Upon choosing a frame of reference, one can assign coordinates to the event: three spatial coordinates $\vec{x} = (x, y, z)$ to describe the location and one time coordinate $t$ to specify the moment at which the event occurs. These four coordinates $(\vec{x}, t)$ together form a four-vector associated to the event.

One of the goals of relativity is to specify the possibility of one event influencing another. This is done by means of the metric tensor, which allows for determining the causal structure of spacetime. The difference (or interval) between two events can be classified into spacelike, lightlike and timelike separations. Only if two events are separated by a lightlike or timelike interval can one influence the other.

## 20.1 References

[1] A.P. French (1968), Special Relativity, MIT Introductory Physics Series, CRC Press, ISBN 0-7487-6422-4, p 86

[2] Leo Sartori (1996), Understanding Relativity: a simplified approach to Einstein's theories, University of California Press, ISBN 0-520-20029-2, p 9

# Chapter 21

# Sylvester's law of inertia

**Sylvester's law of inertia** is a theorem in matrix algebra about certain properties of the coefficient matrix of a real quadratic form that remain invariant under a change of basis. Namely, if $A$ is the symmetric matrix that defines the quadratic form, and $S$ is any invertible matrix such that $D = SAS^T$ is diagonal, then the number of negative elements in the diagonal of $D$ is always the same, for all such $S$; and the same goes for the number of positive elements.

This property is named after J. J. Sylvester who published its proof in 1852.[1][2]

## 21.1   Statement of the theorem

Let $A$ be a symmetric square matrix of order $n$ with real entries. Any non-singular matrix $S$ of the same size is said to transform $A$ into another symmetric matrix $B = SAS^T$, also of order $n$, where $S^T$ is the transpose of $S$. It is also said that matrices $A$ and $B$ are congruent. If $A$ is the coefficient matrix of some quadratic form of $\mathbf{R}^n$, then $B$ is the matrix for the same form after the change of basis defined by $S$.

A symmetric matrix $A$ can always be transformed in this way into a diagonal matrix $D$ which has only entries 0, +1 and −1 along the diagonal. Sylvester's law of inertia states that the number of diagonal entries of each kind is an invariant of $A$, i.e. it does not depend on the matrix $S$ used.

The number of +1s, denoted $n_+$, is called the **positive index of inertia** of $A$, and the number of −1s, denoted $n_-$, is called the **negative index of inertia**. The number of 0s, denoted $n_0$, is the dimension of the kernel of $A$, and also the corank of $A$. These numbers satisfy an obvious relation

$$n_0 + n_+ + n_- = n.$$

The difference sign$(A) = n_- - n_+$ is usually called the **signature** of $A$. (However, some authors use that term for the triple $(n_0, n_+, n_-)$ consisting of the corank and the positive and negative indices of inertia of $A$; for a non-degenerate

form of a given dimension these are equivalent data, but in general the triple yields more data.)

If the matrix $A$ has the property that every principal upper left $k \times k$ minor $\Delta k$ is non-zero then the negative index of inertia is equal to the number of sign changes in the sequence

$$\Delta_0 = 1, \Delta_1, \ldots, \Delta_n = \det A.$$

## 21.2   Statement in terms of eigenvalues

The positive and negative indices of a symmetric matrix $A$ are also the number of positive and negative eigenvalues of $A$. Any symmetric real matrix $A$ has an eigendecomposition of the form $QEQ^T$ where $E$ is a diagonal matrix containing the eigenvalues of $A$, and $Q$ is an orthonormal square matrix containing the eigenvectors. The matrix $E$ can be written $E = WDW^T$ where $D$ is diagonal with entries 0, +1, or −1, and $W$ is diagonal with $Wii = \sqrt{|Eii|}$. The matrix $S = QW$ transforms $D$ to $A$.

## 21.3   Law of inertia for quadratic forms

In the context of quadratic forms, a real quadratic form $Q$ in $n$ variables (or on an $n$-dimensional real vector space) can by a suitable change of basis (by non-singular linear transformation from x to y) be brought to the diagonal form

$$Q(x_1, x_2, \ldots, x_n) = \sum_{i=1}^{n} a_i x_i^2$$

with each $ai \in \{0, 1, -1\}$. Sylvester's law of inertia states that the number of coefficients of a given sign is an invariant

146

of $Q$, i.e. does not depend on a particular choice of diagonalizing basis. Expressed geometrically, the law of inertia says that all maximal subspaces on which the restriction of the quadratic form is positive definite (respectively, negative definite) have the same dimension. These dimensions are the positive and negative indices of inertia.

## 21.4 Generalizations

Sylvester's law of inertia is also valid if $A$ and $B$ have complex entries. In this case, it is said that $A$ and $B$ are *-congruent if and only if there exists a non-singular complex matrix $S$ such that $B = SAS^*$.

In the complex scenario, a way to state Sylvester's law of inertia is that if $A$ and $B$ are Hermitian matrices, then $A$ and $B$ are *-congruent if and only if they have the same inertia. A theorem due to Ikramov[3] generalizes the law of inertia to any normal matrices $A$ and $B$:

If $A$ and $B$ are normal matrices, then $A$ and $B$ are congruent if and only if they have the same number of eigenvalues on each open ray from the origin in the complex plane.

## 21.5 See also

* Metric signature

* Morse theory

* Cholesky decomposition

* Haynsworth inertia additivity formula

## 21.6 References

[1] Sylvester, J J (1852). "A demonstration of the theorem that every homogeneous quadratic polynomial is reducible by real orthogonal substitutions to the form of a sum of positive and negative squares" (PDF). *Philosophical Magazine (Ser. 4)* **4** (23): 138–142. doi:10.1080/14786445208647087. Retrieved 2008-06-27.

[2] Norman, C.W. (1986). *Undergraduate algebra*. Oxford University Press. pp. 360–361. ISBN 0-19-853248-2.

[3] Ikramov, Kh. D. (2001). "On the inertia law for normal matrices". *Doklady Math.* **64**: 141–142.

* Garling, D. J. H. (2011). *Clifford algebras. An introduction*. London Mathematical Society Student Texts **78**. Cambridge: Cambridge University Press. ISBN 978-1-107-09638-7. Zbl 1235.15025.

## 21.7 External links

* Sylvester's law on PlanetMath.

* Sylvester's law of inertia and *-congruence

# Chapter 22

# Postulates of special relativity

*See also: Special relativity*

## 22.1 Postulates of special relativity

1. First postulate (principle of relativity)

> The laws by which the states of physical systems undergo change are not affected, whether these changes of state be referred to the one or the other of two systems of coordinates in uniform translatory motion. OR: The laws of physics are the same in all inertial frames of reference.

2. Second postulate (invariance of $c$)

> As measured in any inertial frame of reference, light is always propagated in empty space with a definite velocity $c$ that is independent of the state of motion of the emitting body. OR: The speed of light in free space has the same value $c$ in all inertial frames of reference.

The two-postulate basis for special relativity is the one historically used by Einstein, and it remains the starting point today. As Einstein himself later acknowledged, the derivation of the Lorentz transformation tacitly makes use of some additional assumptions, including spatial homogeneity, isotropy, and memorylessness.[1] Also Hermann Minkowski implicitly used both postulates when he introduced the Minkowski space formulation, even though he showed that $c$ can be seen as a space-time constant, and the identification with the speed of light is derived from optics.[2]

## 22.2 Alternative derivations of special relativity

Historically, Hendrik Lorentz and Henri Poincaré (1892–1905) derived the Lorentz transformation from Maxwell's equations, which served to explain the negative result of all aether drift measurements. By that the luminiferous aether becomes undetectable in agreement with what Poincaré called the principle of relativity (see History of Lorentz transformations and Lorentz ether theory). A more modern example of deriving the Lorentz transformation from electrodynamics (without using the historical aether concept at all), was given by Richard Feynman.[3]

Following Einstein's original derivation and the group theoretical presentation by Minkowski, many alternative derivations have been proposed, based on various sets of assumptions. It has often been argued (such as by Vladimir Ignatowski in 1910,[4][5][6] or Philipp Frank and Hermann Rothe in 1911,[7][8] and many others in subsequent years[9]) that a formula equivalent to the Lorentz transformation, up to a nonnegative free parameter, follows from just the relativity postulate itself, without first postulating the universal light speed. (Also these formulations rely on the aforementioned various assumptions such as isotropy). The numerical value of the parameter in these transformations can then be determined by experiment, just as the numerical values of the parameter pair $c$ and the Vacuum permittivity are left to be determined by experiment even when using Einstein's original postulates. Experiment rules out the validity of the Galilean transformations. When the numerical values in both Einstein's and other approaches have been found then these different approaches result in the same theory.

## 22.3 Mathematical formulation of the postulates

In the rigorous mathematical formulation of special relativity, we suppose that the universe exists on a four-dimensional spacetime $M$. Individual points in spacetime are known as events; physical objects in spacetime are described by worldlines (if the object is a point particle) or worldsheets (if the object is larger than a point). The worldline or worldsheet only describes the motion of the object; the object may also have several other physical characteristics such as energy-momentum, mass, charge, etc.

In addition to events and physical objects, there are a class of inertial frames of reference. Each inertial frame of reference provides a coordinate system $(x_1, x_2, x_3, t)$ for events in the spacetime $M$. Furthermore, this frame of reference also gives coordinates to all other physical characteristics of objects in the spacetime, for instance it will provide coordinates $(p_1, p_2, p_3, E)$ for the momentum and energy of an object, coordinates $(E_1, E_2, E_3, B_1, B_2, B_3)$ for an electromagnetic field, and so forth.

We assume that given any two inertial frames of reference, there exists a coordinate transformation that converts the coordinates from one frame of reference to the coordinates in another frame of reference. This transformation not only provides a conversion for spacetime coordinates $(x_1, x_2, x_3, t)$ , but will also provide a conversion for all other physical coordinates, such as a conversion law for momentum and energy $(p_1, p_2, p_3, E)$ , etc. (In practice, these conversion laws can be efficiently handled using the mathematics of tensors).

We also assume that the universe obeys a number of physical laws. Mathematically, each physical law can be expressed with respect to the coordinates given by an inertial frame of reference by a mathematical equation (for instance, a differential equation) which relates the various coordinates of the various objects in the spacetime. A typical example is Maxwell's equations. Another is Newton's first law.

1. First Postulate (Principle of relativity)

> Under transitions between inertial reference frames, the equations of all fundamental laws of physics stay form-invariant, while all the numerical constants entering these equations preserve their values. Thus, if a fundamental physical law is expressed with a mathematical equation in one inertial frame, it must be expressed by an identical equation in any other inertial frame, provided both frames are parameterised with charts of the same type. (The caveat on charts is relaxed, if we employ connections to write the law in a covari-

ant form.)

2. Second Postulate (Invariance of $c$)

> There exists an absolute constant $0 < c < \infty$ with the following property. If $A$, $B$ are two events which have coordinates $(x_1, x_2, x_3, t)$ and $(y_1, y_2, y_3, s)$ in one inertial frame $F$ , and have coordinates $(x_1', x_2', x_3', t')$ and $(y_1', y_2', y_3', s')$ in another inertial frame $F'$ , then
>
> $$\sqrt{(x_1 - y_1)^2 + (x_2 - y_2)^2 + (x_3 - y_3)^2} = c(s - t) \quad \text{if and only if}$$
> $$\sqrt{(x_1' - y_1')^2 + (x_2' - y_2')^2 + (x_3' - y_3')^2} = c(s' - t') .$$

Informally, the Second Postulate asserts that objects travelling at speed $c$ in one reference frame will necessarily travel at speed $c$ in all reference frames. This postulate is a subset of the postulates that underlie Maxwell's equations in the interpretation given to them in the context of special relativity. However, Maxwell's equations rely on several other postulates, some of which are now known to be false (e.g., Maxwell's equations cannot account for the quantum attributes of electromagnetic radiation).

The second postulate can be used to imply a stronger version of itself, namely that the spacetime interval is invariant under changes of inertial reference frame. In the above notation, this means that

$$c^2(s - t)^2 - (x_1 - y_1)^2 - (x_2 - y_2)^2 - (x_3 - y_3)^2$$
$$= c^2(s' - t')^2 - (x_1' - y_1')^2 - (x_2' - y_2')^2 - (x_3' - y_3')^2$$

for any two events $A$, $B$. This can in turn be used to deduce the transformation laws between reference frames; see Lorentz transformation.

The postulates of special relativity can be expressed very succinctly using the mathematical language of pseudo-Riemannian manifolds. The second postulate is then an assertion that the four-dimensional spacetime $M$ is a pseudo-Riemannian manifold equipped with a metric $g$ of signature $(1,3)$, which is given by the Minkowski metric when measured in each inertial reference frame. This metric is viewed as one of the physical quantities of the theory, thus it transforms in a certain manner when the frame of reference is changed, and it can be legitimately used in describing the laws of physics. The first postulate is an assertion that the laws of physics are invariant when represented in any frame of reference for which $g$ is given by the Minkowski metric. One advantage of this formulation is that it is now easy to compare special relativity with general relativity, in which the same two postulates hold but the assumption that the metric is required to be Minkowski is dropped.

The theory of Galilean relativity is the limiting case of special relativity in the limit $c \to \infty$ (which is sometimes referred to as the non-relativistic limit). In this theory, the first postulate remains unchanged, but the second postulate is modified to:

> If $A$, $B$ are two events which have coordinates $(x_1, x_2, x_3, t)$ and $(y_1, y_2, y_3, s)$ in one inertial frame $F$, and have coordinates $(x_1', x_2', x_3', t')$ and $(y_1', y_2', y_3', s')$ in another inertial frame $F'$, then $s - t = s' - t'$. Furthermore, if $s - t = s' - t' = 0$, then
>
> $$\sqrt{(x_1 - y_1)^2 + (x_2 - y_2)^2 + (x_3 - y_3)^2}$$
> $$= \sqrt{(x_1' - y_1')^2 + (x_2' - y_2')^2 + (x_3' - y_3')^2}$$

The physical theory given by classical mechanics, and Newtonian gravity is consistent with Galilean relativity, but not special relativity. Conversely, Maxwell's equations are not consistent with Galilean relativity unless one postulates the existence of a physical aether. In a surprising number of cases, the laws of physics in special relativity (such as the famous equation $E = mc^2$) can be deduced by combining the postulates of special relativity with the hypothesis that the laws of special relativity approach the laws of classical mechanics in the non-relativistic limit.

## 22.4   Notes

[1]  Albert Einstein, Morgan document, 1921

[2]  Minkowski, Hermann (1909), "Raum und Zeit", *Physikalische Zeitschrift* **10**: 75–88

   - Various English translations on Wikisource: Space and Time.

[3]  Feynman, R.P. (1970), "21–6. The potentials for a charge moving with constant velocity; the Lorentz formula", *The Feynman Lectures on Physics* 2, Reading: Addison Wesley Longman, ISBN 0-201-02115-3

[4]  Ignatowsky, W. v. (1910). "Einige allgemeine Bemerkungen über das Relativitätsprinzip". *Physikalische Zeitschrift* **11**: 972–976.

   - English Wikisource translation: Some General Remarks on the Relativity Principle

[5]  Ignatowsky, W. v. (1911). "Das Relativitätsprinzip". *Archiv der Mathematik und Physik* **18**: 17–40.

[6]  Ignatowsky, W. v. (1911). "Eine Bemerkung zu meiner Arbeit: "Einige allgemeine Bemerkungen zum Relativitätsprinzip"". *Physikalische Zeitschrift* **12**: 779.

[7]  Frank, Philipp & Rothe, Hermann (1910), "Über die Transformation der Raum-Zeitkoordinaten von ruhenden auf bewegte Systeme", *Annalen der Physik* **339** (5): 825–855, Bibcode:1911AnP...339..825F, doi:10.1002/andp.19113390502

   - English translation: On the Transformation of Space-Time Coordinates from Stationary to Moving Systems

[8]  Frank, Philipp & Rothe, Hermann (1912). "Zur Herleitung der Lorentztransformation". *Physikalische Zeitschrift* **13**: 750–753.

[9]  Baccetti, Valentina; Tate, Kyle; Visser, Matt (2012), "Inertial frames without the relativity principle", *Journal of High Energy Physics*: 119, arXiv:1112.1466, Bibcode:2012JHEP...05..119B, doi:10.1007/JHEP05(2012)119; See references 5–25 therein.

# Chapter 23

# Lorentz group

*Hendrik Antoon Lorentz (1853–1928), after whom the Lorentz group is named.*

In physics and mathematics, the **Lorentz group** is the group of all Lorentz transformations of Minkowski spacetime, the classical setting for all (nongravitational) physical phenomena. The Lorentz group is named for the Dutch physicist Hendrik Lorentz.

Under the Lorentz transformations, these laws and equations are invariant:

- The kinematical laws of special relativity

- Maxwell's field equations in the theory of electromagnetism

- The Dirac equation in the theory of the electron

Therefore, the Lorentz group expresses the fundamental symmetry of many known fundamental laws of nature.

## 23.1   Basic properties

The Lorentz group is a subgroup of the Poincaré group—the group of all isometries of Minkowski spacetime. Lorentz transformations are, precisely, isometries that leave the origin fixed. Thus, the Lorentz group is an isotropy subgroup of the isometry group of Minkowski spacetime. For this reason, the Lorentz group is sometimes called the **homogeneous Lorentz group** while the Poincaré group is sometimes called the *inhomogeneous Lorentz group*. Lorentz transformations are examples of linear transformations; general isometries of Minkowski spacetime are affine transformations. Mathematically, the Lorentz group may be described as the generalized orthogonal group O(1,3), the matrix Lie group that preserves the quadratic form

$$(t, x, y, z) \mapsto t^2 - x^2 - y^2 - z^2$$

on $\mathbf{R}^4$. This quadratic form is, when put on matrix form (see classical orthogonal group), interpreted in physics as the metric tensor of Minkowski spacetime.

The Lorentz group is a six-dimensional noncompact nonabelian real Lie group that is not connected. All four of its connected components are not simply connected. The identity component (i.e., the component containing the identity element) of the Lorentz group is itself a group, and is often called the **restricted Lorentz group**, and is denoted SO⁺(1,3). The restricted Lorentz group consists of those Lorentz transformations that preserve the orientation of space and direction of time. The restricted Lorentz group has often been presented through a facility of biquaternion algebra.

151

The restricted Lorentz group arises in other ways in pure mathematics. For example, it arises as the point symmetry group of a certain ordinary differential equation. This fact also has physical significance.

### 23.1.1   Connected components

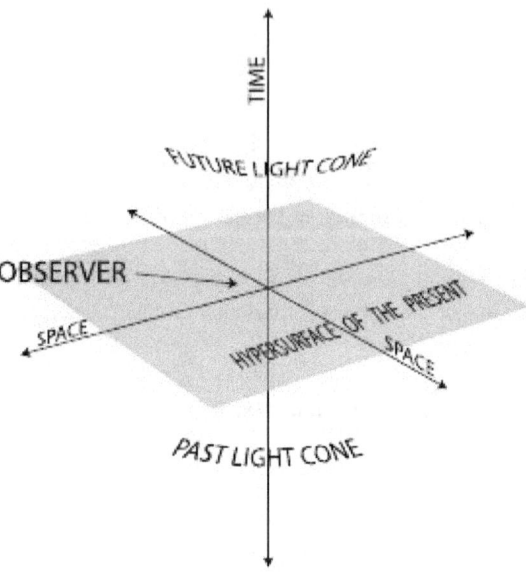

*Light cone in 2D space plus a time dimension.*

Because it is a Lie group, the Lorentz group O(1,3) is both a group and admits a topological description as a smooth manifold. As a manifold, it has four connected components. Intuitively, this means that it consists of four topologically separated pieces.

The four connected components can be categorized by two transformational properties its elements have:

- some elements are reversed under time-inverting Lorentz transformations, for example, a future-pointing timelike vector would be invert to a past-pointing vector

- some elements have orientation reversed by **improper Lorentz transformations**, for example, certain vierbein (tetrads)

Lorentz transformations that preserve the direction of time are called **orthochronous**. The subgroup of orthochronous transformations is often denoted O$^+$(1,3). Those that preserve orientation are called **proper**, and as linear transformations they have determinant +1. (The improper Lorentz transformations have determinant −1.) The subgroup of proper Lorentz transformations is denoted SO(1,3).

The subgroup of all Lorentz transformations preserving both orientation and direction of time is called the **proper, orthochronous Lorentz group** or **restricted Lorentz group**, and is denoted by SO$^+$(1, 3). (Note that some authors refer to SO(1,3) or even O(1,3) when they actually mean SO$^+$(1, 3).)

The set of the four connected components can be given a group structure as the quotient group O(1,3)/SO$^+$(1,3), which is isomorphic to the Klein four-group. Every element in O(1,3) can be written as the semidirect product of a proper, orthochronous transformation and an element of the discrete group

$$\{1, P, T, PT\}$$

where $P$ and $T$ are the space inversion and time reversal operators:

$$P = \mathrm{diag}(1, -1, -1, -1)$$
$$T = \mathrm{diag}(-1, 1, 1, 1).$$

Thus an arbitrary Lorentz transformation can be specified as a proper, orthochronous Lorentz transformation along with a further two bits of information, which pick out one of the four connected components. This pattern is typical of finite-dimensional Lie groups.

## 23.2   Restricted Lorentz group

The restricted Lorentz group is the identity component of the Lorentz group, which means that it consists of all Lorentz transformations that can be connected to the identity by a continuous curve lying in the group. The restricted Lorentz group is a connected normal subgroup of the full Lorentz group with the same dimension, in this case with dimension six.

The restricted Lorentz group is generated by ordinary spatial rotations and Lorentz boosts (which can be thought of as hyperbolic rotations in a plane that includes a timelike direction). Since every proper, orthochronous Lorentz transformation can be written as a product of a rotation (specified by 3 real parameters) and a boost (also specified by 3 real parameters), it takes 6 real parameters to specify an arbitrary proper orthochronous Lorentz transformation. This is one way to understand why the restricted Lorentz group is six-dimensional. (See also the Lie algebra of the Lorentz group.)

The set of all rotations forms a Lie subgroup isomorphic to the ordinary rotation group SO(3). The set of all boosts, however, does *not* form a subgroup, since composing two boosts does not, in general, result in another boost. (Rather, a pair of non-colinear boosts is equivalent to a boost and a rotation, and this relates to Thomas rotation.) A boost in some direction, or a rotation about some axis, generates a one-parameter subgroup.

### 23.2.1 Surfaces of transitivity

If a group $G$ acts on a space $V$, then a surface $S \subset V$ is a **surface of transitivity** if $S$ is invariant under $G$, i.e., $gs \in S$ $\forall g \in G$, $\forall s \in S$, and for any two points $s_1$, $s_2 \in S$ there is a $g \in G$ such that $gs_1 = s_2$. By definition of the Lorentz group, it preserves the quadratic form

$$Q(x) = x_0^2 - x_1^2 - x_2^2 - x_3^2.$$

The surfaces of transitivity of the orthochronous Lorentz group $O^+(1, 3)$, $Q(x) = $ const. of spacetime are the following:[1]

- $Q(x) > 0$, $x_0 > 0$ is the upper branch of a hyperboloid of two sheets.

- $Q(x) > 0$, $x_0 < 0$ is the lower branch of this hyperboloid.

- $Q(x) = 0$, $x_0 > 0$ is the upper branch of the light cone.

- $Q(x) = 0$, $x_0 < 0$ is the lower branch of the light cone.

- $Q(x) < 0$ is a hyperboloid of one sheet.

- The origin $x_0 = x_1 = x_2 = x_3 = 0$.

These surfaces are 3-dimensional, so the images are not faithful, but they are faithful for the corresponding facts about $O^+(1, 2)$. For the full Lorentz group, the surfaces of transitivity are only four since the transformation $T$ takes an upper branch of a hyperboloid (cone) to a lower one and vice versa.

These observations constitute a good starting point for finding all infinite-dimensional unitary representations of the Lorentz group, in fact, of the Poincaré group, using the method of induced representations.[2] One begins with a "standard vector", one for each surface of transitivity, and then ask which subgroup preserves these vectors. These subgroups are called little groups by physicists. The problem is then essentially reduced to the easier problem of finding representations of the little groups. For example, a standard vector in one of the hyperbolas of two sheets could be suitably chosen as $(m, 0, 0, 0)$. For each $m \neq 0$, the vector pierces exactly one sheet. In this case the little group is $SO(3)$, the rotation group, all of whose representations are known. The precise infinite-dimensional unitary

representation under which a particle transform is part of its classification. Not all representations can correspond to physical particles (as far as is known). Standard vectors on the one-sheeted hyperbolas would correspond to tachyons. Particles on the light cone are photons, and more hypothetically, gravitons. The "particle" corresponding to the origin is the vacuum.

### 23.2.2 Relation to the Möbius group

See also: Algebra of physical space

The restricted Lorentz group $SO^+(1, 3)$ is isomorphic to the projective special linear group $PSL(2,\mathbf{C})$, which is in turn isomorphic to the Möbius group, the symmetry group of conformal geometry on the Riemann sphere. (This observation was utilized by Roger Penrose as the starting point of twistor theory.)

This may be shown by constructing a surjective homomorphism of Lie groups from $SL(2,\mathbf{C})$ to $SO^+(1,3)$, which we will call the **spinor map**. This proceeds as follows:

We can define an action of $SL(2,\mathbf{C})$ on Minkowski spacetime by writing a point of spacetime as a two-by-two Hermitian matrix in the form

$$X = \begin{bmatrix} t + z & x - iy \\ x + iy & t - z \end{bmatrix}.$$

This presentation has the pleasant feature that

$$\det X = t^2 - x^2 - y^2 - z^2.$$

Therefore, we have identified the space of Hermitian matrices (which is four-dimensional, as a *real* vector space) with Minkowski spacetime in such a way that the determinant of a Hermitian matrix is the squared length of the corresponding vector in Minkowski spacetime. $SL(2,\mathbf{C})$ acts on the space of Hermitian matrices via

$$X \mapsto PXP^*$$

where $P^*$ is the Hermitian transpose of $P$, and this action preserves the determinant. Therefore, $SL(2,\mathbf{C})$ acts on Minkowski spacetime by (linear) isometries. This defines a map from $SL(2,\mathbf{C})$ to the Lorentz group $SO^+(1,3)$, and the map is evidently a homomorphism. This is the spinor map.

The kernel of the spinor map is the two element subgroup $\pm I$, and it happens that the map is surjective. By the first isomorphism theorem, the quotient group $PSL(2,\mathbf{C}) = SL(2,\mathbf{C}) / \{\pm I\}$ is isomorphic to $SO^+(1,3)$.

### Appearance of the night sky

This isomorphism has the consequence that Möbius transformations of the Riemann sphere represent the way that Lorentz transformations change the appearance of the night sky, as seen by an observer who is maneuvering at relativistic velocities relative to the "fixed stars".

Suppose the "fixed stars" live in Minkowski spacetime and are modeled by points on the celestial sphere. Then a given point on the celestial sphere can be associated with $\xi = u + iv$, a complex number that corresponds to the point on the Riemann sphere, and can be identified with a null vector (a light-like vector) in Minkowski space

$$\begin{bmatrix} u^2 + v^2 + 1 \\ 2u \\ -2v \\ u^2 + v^2 - 1 \end{bmatrix}$$

or the Hermitian matrix

$$N = 2 \begin{bmatrix} u^2 + v^2 & u + iv \\ u - iv & 1 \end{bmatrix}.$$

The set of real scalar multiples of this null vector, called a *null line* through the origin, represents a *line of sight* from an observer at a particular place and time (an arbitrary event we can identify with the origin of Minkowski spacetime) to various distant objects, such as stars. Then the points of the celestial sphere (equivalently, lines of sight) are identified with certain Hermitian matrices.

### 23.2.3   Conjugacy classes

Because the restricted Lorentz group $SO^+(1, 3)$ is isomorphic to the Möbius group $PSL(2,\mathbf{C})$, its conjugacy classes also fall into five classes:

- **Elliptic** transformations

- **Hyperbolic** transformations

- **Loxodromic** transformations

- **Parabolic** transformations

- The trivial **identity** transformation

In the article on Möbius transformations, it is explained how this classification arises by considering the fixed points of Möbius transformations in their action on the Riemann sphere, which corresponds here to null eigenspaces

of restricted Lorentz transformations in their action on Minkowski spacetime.

An example of each type is given in the subsections below, along with the effect of the one-parameter subgroup it generates (e.g., on the appearance of the night sky).

The Möbius transformations are the conformal transformations of the Riemann sphere (or celestial sphere). Then conjugating with an arbitrary element of $SL(2,\mathbf{C})$ obtains the following examples of arbitrary elliptic, hyperbolic, loxodromic, and parabolic (restricted) Lorentz transformations, respectively. The effect on the **flow lines** of the corresponding one-parameter subgroups is to transform the pattern seen in the examples by some conformal transformation. For example, an elliptic Lorentz transformation can have any two distinct fixed points on the celestial sphere, but points still flow along circular arcs from one fixed point toward the other. The other cases are similar.

### Elliptic

An elliptic element of $SL(2,\mathbf{C})$ is

$$P_1 = \begin{bmatrix} \exp(i\theta/2) & 0 \\ 0 & \exp(-i\theta/2) \end{bmatrix}$$

and has fixed points $\xi = 0, \infty$. Writing the action as $X \mapsto P_1 X P_1{}^*$ and collecting terms, the spinor map converts this to the (restricted) Lorentz transformation

$$Q_1 = \begin{bmatrix} 1 & 0 & 0 & 0 \\ 0 & \cos(\theta) & -\sin(\theta) & 0 \\ 0 & \sin(\theta) & \cos(\theta) & 0 \\ 0 & 0 & 0 & 1 \end{bmatrix} = \exp\left( \theta \begin{bmatrix} 0 & 0 & 0 & 0 \\ 0 & 0 & -1 & 0 \\ 0 & 1 & 0 & 0 \\ 0 & 0 & 0 & 0 \end{bmatrix} \right).$$

This transformation then represents a rotation about the z axis, $\exp(i\theta J z)$. The one-parameter subgroup it generates is obtained by taking $\theta$ to be a real variable, the rotation angle, instead of a constant.

The corresponding continuous transformations of the celestial sphere (except for the identity) all share the same two fixed points, the North and South poles. The transformations move all other points around latitude circles so that this group yields a continuous counterclockwise rotation about the z axis as $\theta$ increases. The *angle doubling* evident in the spinor map is a characteristic feature of *spinorial double coverings*.

### Hyperbolic

A hyperbolic element of $SL(2,\mathbf{C})$ is

$$P_2 = \begin{bmatrix} \exp(\beta/2) & 0 \\ 0 & \exp(-\beta/2) \end{bmatrix}$$

and has fixed points $\xi = 0, \infty$. Under stereographic projection from the Riemann sphere to the Euclidean plane, the effect of this Möbius transformation is a dilation from the origin.

The spinor map converts this to the Lorentz transformation

$$Q_2 = \begin{bmatrix} \cosh(\beta) & 0 & 0 & \sinh(\beta) \\ 0 & 1 & 0 & 0 \\ 0 & 0 & 1 & 0 \\ \sinh(\beta) & 0 & 0 & \cosh(\beta) \end{bmatrix}$$

$$= \exp\left( \beta \begin{bmatrix} 0 & 0 & 0 & 1 \\ 0 & 0 & 0 & 0 \\ 0 & 0 & 0 & 0 \\ 1 & 0 & 0 & 0 \end{bmatrix} \right).$$

This transformation represents a boost along the z axis with rapidity $\beta$. The one-parameter subgroup it generates is obtained by taking $\beta$ to be a real variable, instead of a constant. The corresponding continuous transformations of the celestial sphere (except for the identity) all share the same fixed points (the North and South poles), and they move all other points along longitudes away from the South pole and toward the North pole.

### Loxodromic

A loxodromic element of SL(2,**C**) is

$$P_3 = P_2 P_1 = P_1 P_2 = \begin{bmatrix} \exp\left((\beta + i\theta)/2\right) & 0 \\ 0 & \exp\left(-(\beta + i\theta)/2\right) \end{bmatrix}$$

and has fixed points $\xi = 0, \infty$. The spinor map converts this to the Lorentz transformation

$$Q_3 = Q_2 Q_1 = Q_1 Q_2.$$

The one-parameter subgroup this generates is obtained by replacing $\beta + i\theta$ with any real multiple of this complex constant. (If $\beta$, $\theta$ vary independently, then a *two-dimensional* abelian subgroup is obtained, consisting of simultaneous rotations about the z axis and boosts along the z-axis; in contrast, the *one-dimensional* subgroup discussed here consists of those elements of this two-dimensional subgroup such that the **rapidity** of the boost and **angle** of the rotation have a *fixed ratio*.)

The corresponding continuous transformations of the celestial sphere (excepting the identity) all share the same two fixed points (the North and South poles). They move all other points away from the South pole and toward the North pole (or vice versa), along a family of curves called **loxodromes**. Each loxodrome spirals infinitely often around each pole.

### Parabolic

A parabolic element of SL(2,**C**) is

$$P_4 = \begin{bmatrix} 1 & \alpha \\ 0 & 1 \end{bmatrix}$$

and has the single fixed point $\xi = \infty$ on the Riemann sphere. Under stereographic projection, it appears as an ordinary translation along the real axis.

The spinor map converts this to the matrix (representing a Lorentz transformation)

$$Q_4 = \begin{bmatrix} 1 + |\alpha|^2/2 & \operatorname{Re}(\alpha) & \operatorname{Im}(\alpha) & -|\alpha|^2/2 \\ \operatorname{Re}(\alpha) & 1 & 0 & -\operatorname{Re}(\alpha) \\ -\operatorname{Im}(\alpha) & 0 & 1 & \operatorname{Im}(\alpha) \\ |\alpha|^2/2 & \operatorname{Re}(\alpha) & \operatorname{Im}(\alpha) & 1 - |\alpha|^2/2 \end{bmatrix}$$

$$= \exp \begin{bmatrix} 0 & \operatorname{Re}(\alpha) & \operatorname{Im}(\alpha) & 0 \\ \operatorname{Re}(\alpha) & 0 & 0 & -\operatorname{Re}(\alpha) \\ -\operatorname{Im}(\alpha) & 0 & 0 & \operatorname{Im}(\alpha) \\ 0 & \operatorname{Re}(\alpha) & \operatorname{Im}(\alpha) & 0 \end{bmatrix}.$$

This generates a two-parameter abelian subgroup, which is obtained by considering $\alpha$ a complex variable rather than a constant. The corresponding continuous transformations of the celestial sphere (except for the identity transformation) move points along a family of circles that are all tangent at the North pole to a certain great circle. All points other than the North pole itself move along these circles.

Parabolic Lorentz transformations are often called **null rotations**, since they preserve null vectors, just as rotations preserve timelike vectors and boosts preserve spacelike vectors. Since these are likely to be the least familiar of the four types of nonidentity Lorentz transformations (elliptic, hyperbolic, loxodromic, parabolic), it is illustrated here how to determine the effect of an example of a parabolic Lorentz transformation on Minkowski spacetime.

The matrix given above yields the transformation

$$\begin{bmatrix} t \\ x \\ y \\ z \end{bmatrix} \rightarrow \begin{bmatrix} t \\ x \\ y \\ z \end{bmatrix} + \operatorname{Re}(\alpha) \begin{bmatrix} x \\ t - z \\ 0 \\ x \end{bmatrix} + \operatorname{Im}(\alpha) \begin{bmatrix} y \\ 0 \\ z - t \\ y \end{bmatrix} + \frac{|\alpha|^2}{2} \begin{bmatrix} t - z \\ 0 \\ 0 \\ t - z \end{bmatrix}.$$

Now, without loss of generality, pick $Im(\alpha)=0$. Differentiating this transformation with respect to the now real group parameter $\alpha$ and evaluating at $\alpha=0$ produces the corresponding vector field (first order linear partial differential operator),

$x\,(\partial_t + \partial_z) + (t - z)\,\partial_x.$

Apply this to a function $f(t,x,y,z)$, and demand that it stays invariant, i.e., it is annihilated by this transformation. The solution of the resulting first order linear partial differential equation can be expressed in the form

$$f(t, x, y, z) = F(y,\ t - z,\ t^2 - x^2 - z^2),$$

where F is an *arbitrary* smooth function. The arguments of F give three *rational invariants* describing how points (events) move under this parabolic transformation, as they themselves do not move,

$$y = c_1, \quad t - z = c_2, \quad t^2 - x^2 - z^2 = c_3.$$

Choosing real values for the constants on the right hand sides yields three conditions, and thus specifies a curve in Minkowski spacetime. This curve is an orbit of the transformation.

The form of the rational invariants shows that these flow-lines (orbits) have a simple description: suppressing the inessential coordinate y, each orbit is the intersection of a *null plane*, $t = z + c_2$, with a *hyperboloid*, $t^2 - x^2 - z^2 = c_3$. The case $c_3 = 0$ has the hyperboloid degenerate to a light cone with the orbits becoming parabolas lying in corresponding null planes.

A particular null line lying on the light cone is left *invariant*; this corresponds to the unique (double) fixed point on the Riemann sphere mentioned above. The other null lines through the origin are "swung around the cone" by the transformation. Following the motion of one such null line as $\alpha$ increases corresponds to following the motion of a point along one of the circular flow lines on the celestial sphere, as described above.

A choice $Re(\alpha)=0$ instead, produces similar orbits, now with the roles of x and y interchanged.

Parabolic transformations lead to the gauge symmetry of massless particles (like photons) with helicity $|h| \geq 1$. In the above explicit example, a massless particle moving in the z direction, so with 4-momentum $P=(p,0,0,p)$, is not affected at all by the x-boost and y-rotation combination $Kx-Jy$ displayed above, in the "little group" of its motion. This is evident from the explicit transformation law discussed: like any light-like vector, $P$ itself is now invariant, i.e., all traces or effects of $\alpha$ have disappeared. $c_1 = c_2 = c_3 = 0$, in the special case discussed. (The other similar generator, $Ky+Jx$ as well as it and $Jz$ comprise altogether the little group of the lightlike vector, isomorphic to E(2).)

## 23.3  Lie algebra

As with any Lie group, the best way to study many aspects of the Lorentz group is via its Lie algebra. The Lorentz group is a subgroup of the diffeomorphism group of $\mathbf{R}^4$ and therefore its Lie algebra can be identified with vector fields on $\mathbf{R}^4$. In particular, the vectors that generate isometries on a space are its Killing vectors, which provides a convenient alternative to the left-invariant vector field for calculating the Lie algebra. We can write down a set of six generators:

- vector fields on $\mathbf{R}^4$ generating three rotations $i\,J$,

$$-y\partial_x + x\partial_y \equiv iJ_z, \qquad -z\partial_y + y\partial_z \equiv iJ_x, \qquad -x\partial_z + z\partial_x \equiv iJ_y;$$

- vector fields on $\mathbf{R}^4$ generating three boosts $i\,K$,

$$x\partial_t + t\partial_x \equiv iK_x, \qquad y\partial_t + t\partial_y \equiv iK_y, \qquad z\partial_t + t\partial_z \equiv iK_z.$$

It may be helpful to briefly recall here how to obtain a one-parameter group from a vector field, written in the form of a first order linear partial differential operator such as

$$-y\partial_x + x\partial_y.$$

The corresponding initial value problem is

$$\frac{\overline{\partial x}}{\partial \lambda} = -y,\quad \frac{\overline{\partial y}}{\partial \lambda} = x,\ x(0) = x_0,\ y(0) = y_0.$$

The solution can be written

$$x(\lambda) = x_0 \cos(\lambda) - y_0 \sin(\lambda),\ y(\lambda) = x_0 \sin(\lambda) + y_0 \cos(\lambda)$$

or

$$\begin{bmatrix} t \\ x \\ y \\ z \end{bmatrix} = \begin{bmatrix} 1 & 0 & 0 & 0 \\ 0 & \cos(\lambda) & -\sin(\lambda) & 0 \\ 0 & \sin(\lambda) & \cos(\lambda) & 0 \\ 0 & 0 & 0 & 1 \end{bmatrix} \begin{bmatrix} t_0 \\ x_0 \\ y_0 \\ z_0 \end{bmatrix}$$

where we easily recognize the one-parameter matrix group of rotations $\exp(i\,\lambda\,Jz)$ about the z axis. Differentiating with respect to the group parameter $\lambda$ and setting it $\lambda=0$ in that result, we recover the standard matrix,

$$iJ_z = \begin{bmatrix} 0 & 0 & 0 & 0 \\ 0 & 0 & -1 & 0 \\ 0 & 1 & 0 & 0 \\ 0 & 0 & 0 & 0 \end{bmatrix},$$

which corresponds to the vector field we started with. This illustrates how to pass between matrix and vector field representations of elements of the Lie algebra.

Reversing the procedure in the previous section, we see that the Möbius transformations that correspond to our six generators arise from exponentiating respectively $\beta/2$ (for the three boosts) or $i\theta/2$ (for the three rotations) times the three Pauli matrices

$$\sigma_1 = \begin{bmatrix} 0 & 1 \\ 1 & 0 \end{bmatrix}, \quad \sigma_2 = \begin{bmatrix} 0 & -i \\ i & 0 \end{bmatrix}, \quad \sigma_3 = \begin{bmatrix} 1 & 0 \\ 0 & -1 \end{bmatrix}.$$

For our purposes, another generating set is more convenient. The following table lists the six generators, in which

- The first column gives a generator of the flow under the Möbius action (after stereographic projection from the Riemann sphere) as a *real* vector field on the Euclidean plane.

- The second column gives the corresponding one-parameter subgroup of Möbius transformations.

- The third column gives the corresponding one-parameter subgroup of Lorentz transformations (the image under our homomorphism of preceding one-parameter subgroup).

- The fourth column gives the corresponding generator of the flow under the Lorentz action as a real vector field on Minkowski spacetime.

Notice that the generators consist of

- Two parabolics (null rotations)

- One hyperbolic (boost in the $\partial z$ direction)

- Three elliptics (rotations about the $x, y, z$ axes, respectively)

Let's verify one line in this table. Start with

$$\sigma_2 = \begin{bmatrix} 0 & i \\ -i & 0 \end{bmatrix}.$$

Exponentiate:

$$\exp\left(\frac{i\theta}{2}\sigma_2\right) = \begin{bmatrix} \cos(\theta/2) & -\sin(\theta/2) \\ \sin(\theta/2) & \cos(\theta/2) \end{bmatrix}.$$

This element of SL(2,**C**) represents the one-parameter subgroup of (elliptic) Möbius transformations:

$$\xi \mapsto \frac{\cos(\theta/2)\,\xi - \sin(\theta/2)}{\sin(\theta/2)\,\xi + \cos(\theta/2)}.$$

Next,

$$\frac{d\xi}{d\theta}\Big|_{\theta=0} = -\frac{1+\xi^2}{2}.$$

The corresponding vector field on **C** (thought of as the image of $S^2$ under stereographic projection) is

$$-\frac{1+\xi^2}{2}\,\partial_\xi.$$

Writing $\xi = u + iv$, this becomes the vector field on **R**$^2$

$$-\frac{1+u^2-v^2}{2}\,\partial_u - uv\,\partial_v.$$

Returning to our element of SL(2,C), writing out the action $X \mapsto PXP^*$ and collecting terms, we find that the image under the spinor map is the element of SO$^+$(1,3)

$$\begin{bmatrix} 1 & 0 & 0 & 0 \\ 0 & \cos(\theta) & 0 & \sin(\theta) \\ 0 & 0 & 1 & 0 \\ 0 & -\sin(\theta) & 0 & \cos(\theta) \end{bmatrix}.$$

Differentiating with respect to $\theta$ at $\theta=0$, yields the corresponding vector field on **R**$^4$,

$$z\partial_x - x\partial_z.$$

This is evidently the generator of counterclockwise rotation about the y axis.

## 23.4 Subgroups of the Lorentz group

The subalgebras of the Lie algebra of the Lorentz group can be enumerated, up to conjugacy, from which we can list the closed subgroups of the restricted Lorentz group, up to conjugacy. (See the book by Hall cited below for the details.) We can readily express the result in terms of the generating set given in the table above.

The one-dimensional subalgebras of course correspond to the four conjugacy classes of elements of the Lorentz group:

- $X_1$ generates a one-parameter subalgebra of parabolics SO(0,1),

- $X_3$ generates a one-parameter subalgebra of boosts SO(1,1),

- $X_4$ generates a one-parameter of rotations SO(2),

- $X_3 + aX_4$ (for any $a \neq 0$) generates a one-parameter subalgebra of loxodromic transformations.

(Strictly speaking the last corresponds to infinitely many classes, since distinct $a$ give different classes.) The two-dimensional subalgebras are:

- $X_1, X_2$ generate an abelian subalgebra consisting entirely of parabolics,

- $X_1, X_3$ generate a nonabelian subalgebra isomorphic to the Lie algebra of the affine group A(1),

- $X_3, X_4$ generate an abelian subalgebra consisting of boosts, rotations, and loxodromics all sharing the same pair of fixed points.

The three-dimensional subalgebras are:

- $X_1, X_2, X_3$ generate a **Bianchi V** subalgebra, isomorphic to the Lie algebra of Hom(2), the group of *euclidean homotheties*,

- $X_1, X_2, X_4$ generate a **Bianchi VII_0** subalgebra, isomorphic to the Lie algebra of E(2), the euclidean group,

- $X_2, X_2, X_3 + aX_4$, where $a \neq 0$, generate a **Bianchi VII_a** subalgebra,

- $X_1, X_3, X_5$ generate a **Bianchi VIII** subalgebra, isomorphic to the Lie algebra of SL(2,**R**), the group of isometries of the hyperbolic plane,

- $X_4, X_5, X_6$ generate a **Bianchi IX** subalgebra, isomorphic to the Lie algebra of SO(3), the rotation group.

(Here, the Bianchi types refer to the classification of three-dimensional Lie algebras by the Italian mathematician Luigi Bianchi.) The four-dimensional subalgebras are all conjugate to

- $X_1, X_2, X_3, X_4$ generate a subalgebra isomorphic to the Lie algebra of Sim(2), the group of Euclidean similitudes.

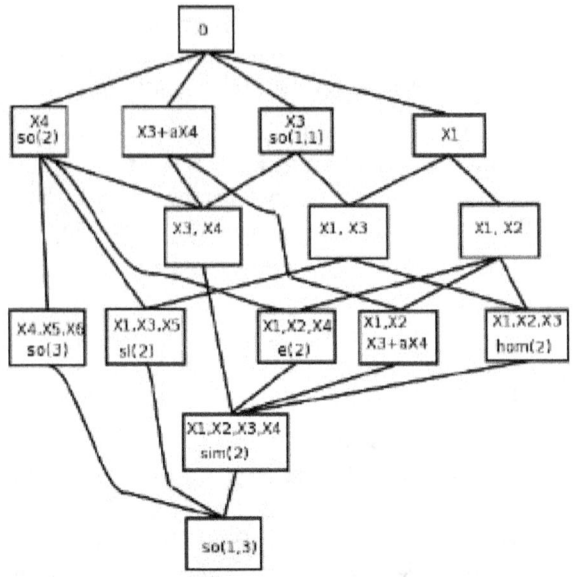

*The lattice of subalgebras of the Lie algebra SO(1,3), up to conjugacy.*

The subalgebras form a lattice (see the figure), and each subalgebra generates by exponentiation a closed subgroup of the restricted Lie group. From these, all subgroups of the Lorentz group can be constructed, up to conjugation, by multiplying by one of the elements of the Klein four-group.

As with any connected Lie group, the coset spaces of the closed subgroups of the restricted Lorentz group, or homogeneous spaces, have considerable mathematical interest. A few, brief descriptions:

- The group Sim(2) is the stabilizer of a *null line*, i.e., of a point on the Riemann sphere—so the homogeneous space $SO^+(1,3)/\mathrm{Sim}(2)$ is the Kleinian geometry that represents conformal geometry on the sphere S.[2]

- The (identity component of the) Euclidean group SE(2) is the stabilizer of a null vector, so the homogeneous space $SO^+(1,3)/\mathrm{SE}(2)$ is the momentum space of a massless particle; geometrically, this Kleinian geometry represents the *degenerate* geometry of the light cone in Minkowski spacetime.

- The rotation group SO(3) is the stabilizer of a timelike vector, so the homogeneous space $SO^+(1,3)/\mathrm{SO}(3)$ is the momentum space of a massive particle; geometrically, this space is none other than three-dimensional hyperbolic space $H^3$.

## 23.5 Covering groups

In a previous section, we constructed a homomorphism $SL(2, \mathbf{C}) \to SO^+(1, 3)$, which we called the spinor map. Since $SL(2,\mathbf{C})$ is simply connected, it is the covering group of the restricted Lorentz group $SO^+(1, 3)$. By restriction we obtain a homomorphism $SU(2) \to SO(3)$. Here, the special unitary group $SU(2)$, which is isomorphic to the group of unit norm quaternions, is also simply connected, so it is the covering group of the rotation group $SO(3)$. Each of these covering maps are twofold covers in the sense that precisely two elements of the covering group map to each element of the quotient. One often says that the restricted Lorentz group and the rotation group are **doubly connected**. This means that the fundamental group of the each group is isomorphic to the two-element cyclic group $Z_2$.

(In applications to quantum mechanics, the special linear group $SL(2, \mathbf{C})$ is sometimes called the Lorentz group.)

Twofold coverings are characteristic of spin groups. Indeed, in addition to the double coverings

$$Spin^+(1, 3) = SL(2, \mathbf{C}) \to SO^+(1, 3)$$
$$Spin(3) = SU(2) \to SO(3)$$

we have the double coverings

$$Pin(1, 3) \to O(1, 3)$$
$$Spin(1, 3) \to SO(1, 3)$$
$$Spin^+(1, 2) = SU(1, 1) \to SO(1, 2)$$

These spinorial double coverings are all closely related to Clifford algebras.

## 23.6 Topology

The left and right groups in the double covering

$$SU(2) \to SO(3)$$

are deformation retracts of the left and right groups, respectively, in the double covering

$$SL(2,\mathbf{C}) \to SO^+(1,3).$$

But the homogeneous space $SO^+(1,3)/SO(3)$ is homeomorphic to hyperbolic 3-space $H^3$, so we have exhibited the restricted Lorentz group as a principal fiber bundle with fibers $SO(3)$ and base $H^3$. Since the latter

is homeomorphic to $\mathbf{R}^3$, while $SO(3)$ is homeomorphic to three-dimensional real projective space $\mathbf{RP}^3$, we see that the restricted Lorentz group is *locally* homeomorphic to the product of $\mathbf{RP}^3$ with $\mathbf{R}^3$. Since the base space is contractible, this can be extended to a global homeomorphism.

## 23.7 Generalization to higher dimensions

The concept of the Lorentz group has a natural generalization to spacetime of any number of dimensions. Mathematically, the Lorentz group of $n+1$-dimensional Minkowski space is the group $O(n,1)$ (or $O(1,n)$) of linear transformations of $\mathbf{R}^{n+1}$ that preserves the quadratic form

$$(x_1, x_2, \ldots, x_n, x_{n+1}) \mapsto x_1^2 + x_2^2 + \cdots + x_n^2 - x_{n+1}^2.$$

Many of the properties of the Lorentz group in four dimensions (where $n = 3$) generalize straightforwardly to arbitrary $n$. For instance, the Lorentz group $O(n,1)$ has four connected components, and it acts by conformal transformations on the celestial $(n-1)$-sphere in $n+1$-dimensional Minkowski space. The identity component $SO^+(n,1)$ is an $SO(n)$-bundle over hyperbolic $n$-space $H^n$.

The low-dimensional cases $n = 1$ and $n = 2$ are often useful as "toy models" for the physical case $n = 3$, while higher-dimensional Lorentz groups are used in physical theories such as string theory that posit the existence of hidden dimensions. The Lorentz group $O(n,1)$ is also the isometry group of $n$-dimensional de Sitter space $dSn$, which may be realized as the homogeneous space $O(n,1)/O(n-1,1)$. In particular $O(4,1)$ is the isometry group of the de Sitter universe $dS_4$, a cosmological model.

## 23.8 Notes

[1] Gelfand, Minlos & Shapiro 1963

[2] Wigner 1939

## 23.9 See also

## 23.10 References

• Artin, Emil (1957). *Geometric Algebra*. New York: Wiley. ISBN 0-471-60839-4. *See Chapter III* for the orthogonal groups O(p,q).

- Carmeli, Moshe (1977). *Group Theory and General Relativity, Representations of the Lorentz Group and Their Applications to the Gravitational Field.* McGraw-Hill, New York. ISBN 0-07-009986-3. A canonical reference; *see chapters 1–6* for representations of the Lorentz group.

- Frankel, Theodore (2004). *The Geometry of Physics (2nd Ed.).* Cambridge: Cambridge University Press. ISBN 0-521-53927-7. An excellent resource for Lie theory, fiber bundles, spinorial coverings, and many other topics.

- Fulton, William; Harris, Joe (1991), *Representation theory. A first course*, Graduate Texts in Mathematics, Readings in Mathematics **129**, New York: Springer-Verlag, ISBN 978-0-387-97495-8, MR 1153249, ISBN 978-0-387-97527-6 *See Lecture 11* for the irreducible representations of SL(2,**C**).

- Gelfand, I.M.; Minlos, R.A.; Shapiro, Z.Ya. (1963), *Representations of the Rotation and Lorentz Groups and their Applications*, New York: Pergamon Press

- Hall, G. S. (2004). *Symmetries and Curvature Structure in General Relativity.* Singapore: World Scientific. ISBN 981-02-1051-5. *See Chapter 6* for the subalgebras of the Lie algebra of the Lorentz group.

- Hatcher, Allen (2002). *Algebraic topology.* Cambridge: Cambridge University Press. ISBN 0-521-79540-0. *See also* the "online version". Retrieved July 3, 2005. *See Section 1.3* for a beautifully illustrated discussion of covering spaces. *See Section 3D* for the topology of rotation groups.

- Naber, Gregory (1992). *The Geometry of Minkowski Spacetime.* New York: Springer-Verlag. ISBN 0486432351. (Dover reprint edition.) An excellent reference on Minkowski spacetime and the Lorentz group.

- Needham, Tristan (1997). *Visual Complex Analysis.* Oxford: Oxford University Press. ISBN 0-19-853446-9. *See Chapter 3* for a superbly illustrated discussion of Möbius transformations.

- Wigner, E. P. (1939), "On unitary representations of the inhomogeneous Lorentz group", *Annals of Mathematics* **40** (1): 149–204, Bibcode:1939AnMat..40..922E, doi:10.2307/1968551, MR 1503456.

# Chapter 24

# Lorentz transformation

In physics, the **Lorentz transformation** (or **transformations**) are coordinate transformations between two coordinate frames that move at constant velocity relative to each other.

Frames of reference can be divided into two groups, inertial (relative motion with constant velocity) and non-inertial (accelerating in curved paths, rotational motion with constant angular velocity, etc.). The term "Lorentz transformations" only refers to transformations between *inertial* frames, usually in the context of special relativity.

In each reference frame, an observer can use a local coordinate system (most exclusively Cartesian coordinates in this context) to measure lengths, and a clock to measure time intervals. An observer is a real or imaginary entity that can take measurements, say humans, or any other living organism—or even robots and computers. An event is something that happens at a point in space at an instant of time, or more formally a point in spacetime. The transformations connect the space and time coordinates of an event as measured by an observer in each frame.[nb 1]

They supersede the Galilean transformation of Newtonian physics, which assumes an absolute space and time (see Galilean relativity). The Galilean transformation is a good approximation only at relative speeds much smaller than the speed of light. Lorentz transformations have a number of unintuitive features that do not appear in Galilean transformations. For example, they reflect the fact that observers moving at different velocities may measure different distances, elapsed times, and even different orderings of events, but always such that the speed of light is the same in all inertial reference frames. The invariance of light speed is one of the postulates of special relativity.

Historically, the transformations were the result of attempts by Lorentz and others to explain how the speed of light was observed to be independent of the reference frame, and to understand the symmetries of the laws of electromagnetism. The Lorentz transformation is in accordance with special relativity, but was derived before special relativity. The transformations are named after the Dutch physicist Hendrik Lorentz.

The Lorentz transformation is a linear transformation. It may include a rotation of space; a rotation-free Lorentz transformation is called a **Lorentz boost**. In Minkowski space, the mathematical model of spacetime in special relativity, the Lorentz transformations preserve the spacetime interval between any two events. This property is the defining property of a Lorentz transformation. They describe only the transformations in which the spacetime event at the origin is left fixed. They can be considered as a hyperbolic rotation of Minkowski space. The more general set of transformations that also includes translations is known as the Poincaré group.

## 24.1  History

Main article: History of Lorentz transformations

Many physicists—including Woldemar Voigt, George FitzGerald, Joseph Larmor, and Hendrik Lorentz[1] himself—had been discussing the physics implied by these equations since 1887.[2] Early in 1889, Oliver Heaviside had shown from Maxwell's equations that the electric field surrounding a spherical distribution of charge should cease to have spherical symmetry once the charge is in motion relative to the ether. FitzGerald then conjectured that Heaviside's distortion result might be applied to a theory of intermolecular forces. Some months later, FitzGerald published the conjecture that bodies in motion are being contracted, in order to explain the baffling outcome of the 1887 ether-wind experiment of Michelson and Morley. In 1892, Lorentz independently presented the same idea in a more detailed manner, which was subsequently called FitzGerald–Lorentz contraction hypothesis.[3] Their explanation was widely known before 1905.[4]

Lorentz (1892–1904) and Larmor (1897–1900), who believed the luminiferous ether hypothesis, also looked for the transformation under which Maxwell's equations are invariant when transformed from the ether to a moving frame. They extended the FitzGerald–Lorentz contraction hypothesis and found out that the time coordinate has to be modified as well ("local time"). Henri Poincaré gave a physical interpretation to local time (to first order in v/c) as the consequence of clock synchronization, under the assumption that the speed of light is constant in moving frames.[5] Larmor is credited to have been the first to understand the crucial time dilation property inherent in his equations.[6]

In 1905, Poincaré was the first to recognize that the transformation has the properties of a mathematical group, and named it after Lorentz.[7] Later in the same year Albert Einstein published what is now called special relativity, by deriving the Lorentz transformation under the assumptions of the principle of relativity and the constancy of the speed of light in any inertial reference frame, and by abandoning the mechanical aether.[8]

## 24.2    Derivation

An *event* is something that happens at a certain point in spacetime, or more generally, the point in spacetime itself. In any inertial frame an event is specified by a time coordinate $t$ and a set of Cartesian coordinates $x$, $y$, $z$ to specify position in space in that frame. Subscripts label individual events.

From Einstein's second postulate of relativity follows immediately

$$c^2(t_2 - t_1)^2 - (x_2 - x_1)^2 - (y_2 - y_1)^2 - (z_2 - z_1)^2 = 0$$

in all inertial frames for events connected by *light signals*. The quantity on the left is called the *spacetime interval* between events $(t_1, x_1, y_1, z_1)$ and $(t_2, x_2, y_2, z_2)$. The interval between *any two* events, not necessarily separated by light signals, is in fact invariant, i.e., independent of the state of relative motion of observers in different inertial frames, as is shown here (where one can also find several more explicit derivations than presently given) using homogeneity and isotropy of space. The transformation sought after thus must possess the property that

$$c^2(t_2 - t_1)^2 - (x_2 - x_1)^2 - (y_2 - y_1)^2 - (z_2 - z_1)^2$$

$$= c^2(t_2' - t_1')^2 - (x_2' - x_1')^2 - (y_2' - y_1')^2 - (z_2' - z_1')^2.$$

where $t$, $x$, $y$, $z$ are the spacetime coordinates used to define events in one frame, and $t'$, $x'$, $y'$, $z'$ are the coordinates in another frame. Now one observes that a *linear* solution to the simpler problem

$$c^2 t^2 - x^2 - y^2 - z^2 = c^2 t'^2 - x'^2 - y'^2 - z'^2$$

solves the general problem too. Finding the solution to the simpler problem is just a matter of look-up in the theory of classical groups that preserve bilinear forms of various signature.[nb 2] The Lorentz transformation is thus an element of the group O(3, 1) or, for those that prefer the other metric signature, O(1, 3).[nb 3]

## 24.3    Generalities

The relations between the primed and unprimed spacetime coordinates are the **Lorentz transformations**, each coordinate in one frame is a linear function of all the coordinates in the other frame, and the inverse functions are the inverse transformation. Depending on how the frames move relative to each other, and how they are oriented in space relative to each other, other parameters that describe direction, speed, and orientation enter the transformation equations.

Transformations describing relative motion with constant (uniform) velocity and without rotation of the space coordinate axes are called a *boosts*, and the relative velocity between the frames is the parameter of the transformation. The other basic type of Lorentz transformations is rotations in the spatial coordinates only, these are also inertial frames since there is no relative motion, the frames are simply tilted (and not continuously rotating), and in this case quantities defining the rotation are the parameters of the transformation (e.g., axis–angle representation, or Euler angles, etc.). A combination of a rotation and boost is a *homogenous transformation*, which transforms the origin back to the origin.

The full Lorentz group O(3, 1) also contains special transformations that are neither rotations nor boosts, but rather reflections in a plane through the origin. Two of these can be singled out; spatial inversion in which the spatial coordinates of all events are reversed in sign and temporal inversion in which the time coordinate for each event gets its sign reversed.

Boosts should not be conflated with mere displacements in spacetime; in this case, the coordinate systems are simply shifted and there is no relative motion. However, these also count as symmetries forced by special relativity since they leave the spacetime interval invariant. A combination of a rotation with a boost, followed by a shift in spacetime, is an *inhomogenous Lorentz transformation*, an element of the Poincaré group, which is also called the inhomogeneous Lorentz group.

# 24.4 Physical formulation of Lorentz boosts

## 24.4.1 Coordinate transformation

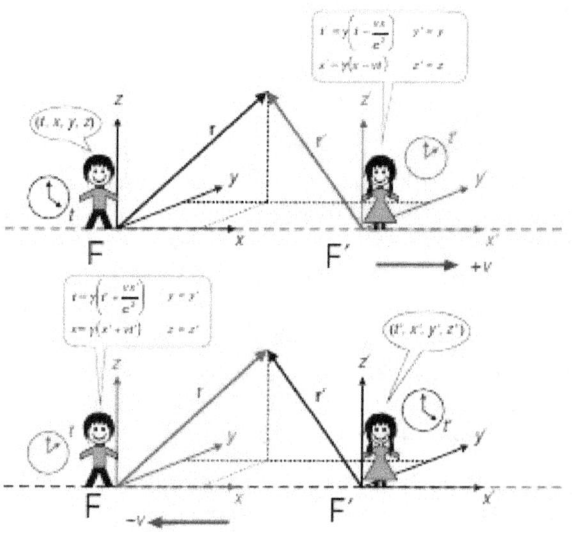

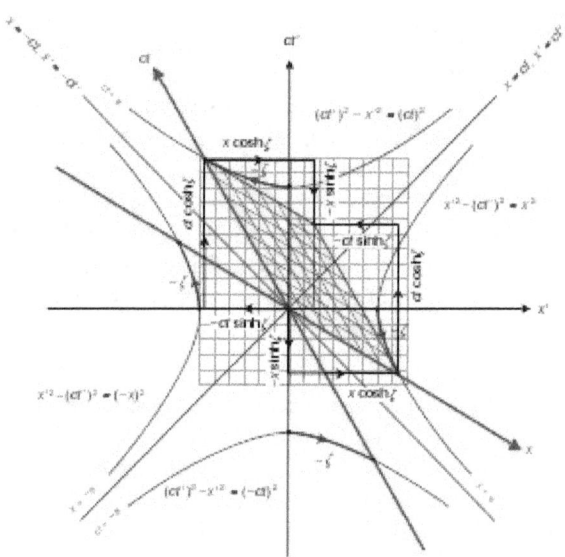

*This diagram actually shows the inverse configuration of F' "stationary" while F is boosted away along the negative x' direction, although it correctly gives the original transformation since the coordinates* ct, x *of* F *are projected onto the coordinates* ct', x' *of* F'. *The event* (ct, x) = (8, 6) *in* F *corresponds to approximately* (ct', x') ≈ (5.55, 1.67) *in* F', *with rapidity* ζ ≈ −0.66. *Notice the difference in length and time scales, such that the speed of light is invariant.*

*The spacetime coordinates of an event, as measured by each observer in their inertial reference frame (in standard configuration) are shown in the speech bubbles.*
**Top:** *frame* F' *moves at velocity* v *along the x-axis of frame* F.
**Bottom:** *frame* F *moves at velocity* −v *along the x'-axis of frame* F'.[9]

A "stationary" observer in frame $F$ defines events with coordinates $t$, $x$, $y$, $z$. Another frame $F'$ moves with velocity $v$ relative to $F$, and an observer in this "moving" frame $F'$ defines events using the coordinates $t'$, $x'$, $y'$, $z'$.

The coordinate axes in each frame are parallel (the $x$ and $x'$ axes are parallel, the $y$ and $y'$ axes are parallel, and the $z$ and $z'$ axes are parallel), remain mutually perpendicular, and relative motion is along the coincident $xx'$ axes. At $t = t' = 0$, the origins of both coordinate systems are the same, $(x, y, z) = (x', y', z') = (0, 0, 0)$. In other words, the times and positions are coincident at this event. If all these hold, then the coordinate systems are said to be in **standard configuration**, or **synchronized**.

If an observer in $F$ records an event $t$, $x$, $y$, $z$, then an observer in $F'$ records the *same* event with coordinates[10]

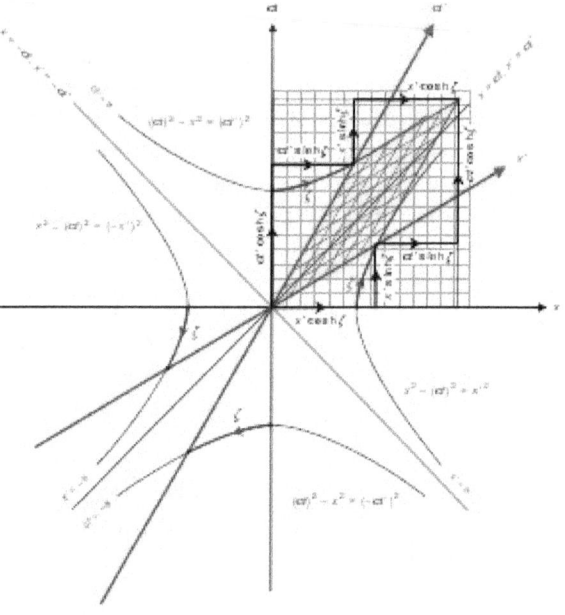

*This diagram shows the original configuration of* F *"stationary" while* F' *is boosted away along the positive* x *direction, although it correctly gives the inverse transformation since the coordinates* ct', x' *of* F' *are projected onto the coordinates* ct, x *of* F. *The event* (ct', x') = (8, 6) *in* F' *corresponds to approximately* (ct, x) ≈ (14.3, 13.28) *in* F, *with rapidity* ζ ≈ +0.66. *Again, the difference in length and time scales is such that the speed of light is invariant.*

where $v$ is the relative velocity between frames in the $x$-direction, $c$ is the speed of light, and

$$\gamma = \frac{1}{\sqrt{1 - \frac{v^2}{c^2}}}$$

(lowercase gamma) is the Lorentz factor.

Here, $v$ is the *parameter* of the transformation, for a given boost it is a constant number, but can take a continuous range of values. In the setup used here, positive relative velocity $v > 0$ is motion along the positive directions of the $xx'$ axes, zero relative velocity $v = 0$ is no relative motion, while negative relative velocity $v < 0$ is relative motion along the negative directions of the $xx'$ axes. The magnitude of relative velocity $v$ cannot equal or exceed $c$, so only subluminal speeds $-c < v < c$ are allowed. The corresponding range of $\gamma$ is $1 \leq \gamma < \infty$.

The transformations are not defined if $v$ is outside these limits. At the speed of light ($v = c$) $\gamma$ is infinite, and faster than light ($v > c$) $\gamma$ is a complex number, each of which make the transformations unphysical. The space and time coordinates are measurable quantities and numerically must be real numbers, not complex.

As an active transformation, an observer in F′ notices the coordinates of the event to be "boosted" in the negative directions of the $xx'$ axes, because of the $-v$ in the transformations. This has the equivalent effect of the *coordinate system* F′ boosted in the positive directions of the $xx'$ axes, while the event does not change and is simply represented in another coordinate system, a passive transformation.

The inverse relations ($t$, $x$, $y$, $z$ in terms of $t'$, $x'$, $y'$, $z'$) can be found by algebraically solving the original set of equations. A more efficient way is to use physical principles. Here F′ is the "stationary" frame while F is the "moving" frame. According to the principle of relativity, there is no privileged frame of reference, so the transformations from F′ to F must take exactly the same form as the transformations from F to F′. The only difference is F′ moves with velocity $-v$ relative to F (i.e., the relative velocity has the same magnitude but is oppositely directed). Thus if an observer in F′ notes an event $t'$, $x'$, $y'$, $z'$, then an observer in F notes the *same* event with coordinates

and the value of $\gamma$ remains unchanged. This "trick" of simply reversing the direction of relative velocity while preserving its magnitude, and exchanging primed and unprimed variables, always applies to finding the inverse transformation of every boost in any direction.

Sometimes it is more convenient to use $\beta = v/c$ (lowercase beta) instead of $v$, so that

$$ct' = \gamma (ct - \beta x) \ ,$$
$$x' = \gamma (x - \beta ct) \ ,$$

which shows clearer the symmetry in the transformation. From the allowed ranges of $v$ and the definition of $\beta$, it fol-

lows $-1 < \beta < 1$. The use of $\beta$ and $\gamma$ is standard throughout the literature.

The Lorentz transformations can also be derived in a way that resembles circular rotations in 3d space using the hyperbolic functions. For the boost in the $x$ direction, the results are

where $\zeta$ (lowercase zeta) is a parameter called *rapidity* (many other symbols are used, including $\theta$, $\phi$, $\varphi$, $\eta$, $\psi$, $\xi$). Given the strong resemblance to rotations of spatial coordinates in 3d space in the Cartesian xy, yz, and zx planes, a Lorentz boost can be thought of as a hyperbolic rotation of spacetime coordinates in the xt, yt, and zt Cartesian-time planes of 4d Minkowski space. The parameter $\zeta$ is the hyperbolic angle of rotation, analogous to the ordinary angle for circular rotations. This transformation can be illustrated with a Minkowski diagram.

The hyperbolic functions arise from the *difference* between the squares of the time and spatial coordinates in the spacetime interval, rather than a sum. The geometric significance of the hyperbolic functions can be visualized by taking $x = 0$ or $ct = 0$ in the transformations. Squaring and subtracting the results, one can derive hyperbolic curves of constant coordinate values but varying $\zeta$, which parametrizes the curves according to the identity

$$\cosh^2 \zeta - \sinh^2 \zeta = 1 \ .$$

Conversely the $ct$ and $x$ axes can be constructed for varying coordinates but constant $\zeta$. The definition

$$\tanh \zeta = \frac{\sinh \zeta}{\cosh \zeta} \ ,$$

provides the link between a constant value of rapidity, and the slope of the $ct$ axis in spacetime. A consequence these two hyperbolic formulae is an identity that matches the Lorentz factor

$$\cosh \zeta = \frac{1}{\sqrt{1 - \tanh^2 \zeta}} \ .$$

Comparing the Lorentz transformations in terms of the relative velocity and rapidity, or using the above formulae, the connections between $\beta$, $\gamma$, and $\zeta$ are

$$\beta = \tanh \zeta \ ,$$

$\gamma = \cosh \zeta$ ,

$\beta\gamma = \sinh \zeta$ .

Taking the inverse hyperbolic tangent gives the rapidity

$\zeta = \tanh^{-1} \beta$ .

Since $-1 < \beta < 1$, it follows $-\infty < \zeta < \infty$. From the relation between $\zeta$ and $\beta$, positive rapidity $\zeta > 0$ is motion along the positive directions of the $xx'$ axes, zero rapidity $\zeta = 0$ is no relative motion, while negative rapidity $\zeta < 0$ is relative motion along the negative directions of the $xx'$ axes.

The inverse transformations are obtained by exchanging primed and unprimed quantities to switch the coordinate frames, and negating rapidity $\zeta \to -\zeta$ since this is equivalent to negating the relative velocity. Therefore,

The inverse transformations can be similarly visualized by considering the cases when $x' = 0$ and $ct' = 0$.

So far the Lorentz transformations have been applied to *one event*. If there are two events, there is a spatial separation and time interval between them. It follows from the linearity of the Lorentz transformations that two values of space and time coordinates can be chosen, the Lorentz transformations can be applied to each, then subtracted to get the Lorentz transformations of the differences;

$$\Delta t' = \gamma \left( \Delta t - \frac{v\Delta x}{c^2} \right) ,$$

$$\Delta x' = \gamma \left( \Delta x - v\Delta t \right) ,$$

with inverse relations

$$\Delta t = \gamma \left( \Delta t' + \frac{v\Delta x'}{c^2} \right) ,$$

$$\Delta x = \gamma \left( \Delta x' + v\Delta t' \right) .$$

where $\Delta$ (capital Delta) indicates a difference of quantities, e.g., $\Delta x = x_2 - x_1$ for two values of $x$ coordinates, and so on.

These transformations on *differences* rather than spatial points or instants of time are useful for a number of reasons:

- in calculations and experiments, it is lengths between two points or time intervals that are measured or of interest (e.g., the length of a moving vehicle, or time duration it takes to travel from one place to another),

- the transformations of velocity can be readily derived by making the difference infinitesimally small and dividing the equations, and the process repeated for the transformation of acceleration,

- if the coordinate systems are never coincident (i.e., not in standard configuration), and if both observers can agree on an event $t_0$, $x_0$, $y_0$, $z_0$ in $F$ and $t_0{}'$, $x_0{}'$, $y_0{}'$, $z_0{}'$ in $F'$, then they can use that event as the origin, and the spacetime coordinate differences are the differences between their coordinates and this origin, e.g., $\Delta x = x - x_0$, $\Delta x' = x' - x_0{}'$, etc.

### 24.4.2 Physical implications

A critical requirement of the Lorentz transformations is the invariance of the speed of light, a fact used in their derivation, and contained in the transformations themselves. If in $F$ the equation for a pulse of light along the $x$ direction is $x = ct$, then in $F'$ the Lorentz transformations give $x' = ct'$, and vice versa, for any $-c < v < c$.

For relative speeds much less than the speed of light, the Lorentz transformations reduce to the Galilean transformation

$$t' \approx t$$

$$x' \approx x - vt$$

in accordance with the correspondence principle. It is sometimes said that nonrelativistic physics is a physics of "instantaneous action at a distance".[11]

Three unintuitive, but correct, predictions of the transformations are:

- **Time dilation.** Suppose there is a clock at rest in $F$. If a time interval (say a "tick") is measured at the same point so that $\Delta x = 0$, then the transformations give this tick in $F'$ by $\Delta t' = \gamma \Delta t$. Conversely, suppose there is a clock at rest in $F'$. If a tick is measured at the same point so that $\Delta x' = 0$, then the transformations give this tick in F by $\Delta t = \gamma \Delta t'$. Either way, the boosted observer measures longer time intervals than the observer in the other frame.

- **Relativity of simultaneity.** Suppose two events occur simultaneously ($\Delta t = 0$) along the x axis, but separated by a nonzero displacement $\Delta x$. Then in $F'$, we find that $\Delta t' = \gamma \frac{-v\Delta x}{c^2}$ , so the events are no longer simultaneous according to a moving observer.

- **Length contraction.** Suppose there is a rod at rest in $F$ aligned along the x axis, with length $\Delta x$. In $F'$, the rod moves with velocity -$v$, so its length must be measured by taking two simultaneous ($\Delta t' = 0$) measurements at opposite ends. Under these conditions, the inverse Lorentz transform shows that $\Delta x = \gamma \Delta x'$. In $F$ the two measurements are no longer simultaneous, but this does not matter because the rod is at rest in $F$. We conclude that the boosted observer measures a shorter length, by a factor of $\gamma$, than the observer in the rest frame of the rod. Length contraction affects any geometric quantity related to lengths, so from the perspective of a moving observer, areas and volumes will also appear to shrink along the direction of motion.

### 24.4.3   Vector transformations

Further information: Euclidean vector and vector projection

The use of vectors allows positions and velocities to be ex-

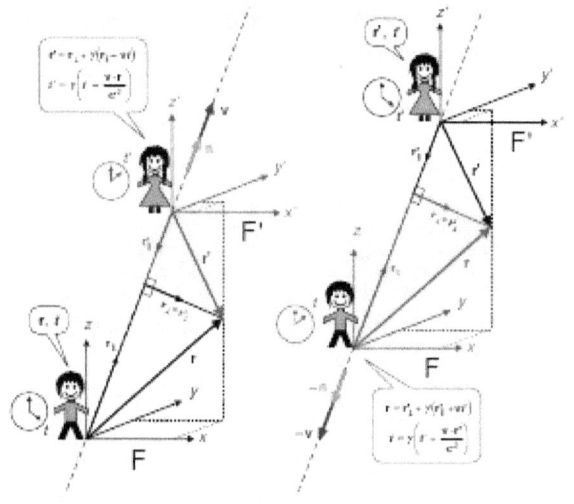

*An observer in frame* F *observes* F′ *to move with velocity* **v**, *while* F′ *observes* F *to move with velocity* −**v**. *The coordinate axes of each frame are still parallel and orthogonal. The position vector as measured in each frame is split into components parallel and perpendicular to the relative velocity vector* **v**. **Left:** *Standard configuration.* **Right:** *Inverse configuration.*

pressed in arbitrary directions compactly. A single boost in any direction depends on the full relative velocity vector **v** with a magnitude | **v** | = $v$ that cannot equal or exceed $c$, so that $0 \leq v < c$.

Only time and the coordinates parallel to the direction of relative motion change, while those coordinates perpendicular do not. With this in mind, split the spatial position vector **r** as measured in $F$, and **r**′ as measured in $F'$, each into components perpendicular ($\perp$) and parallel ($\parallel$) to **v**,

$$\mathbf{r} = \mathbf{r}_\perp + \mathbf{r}_\parallel, \quad \mathbf{r}' = \mathbf{r}'_\perp + \mathbf{r}'_\parallel,$$

then the transformations are

$$t' = \gamma \left( t - \frac{\mathbf{r}_\parallel \cdot \mathbf{v}}{c^2} \right)$$

$$\mathbf{r}'_\parallel = \gamma(\mathbf{r}_\parallel - \mathbf{v}t)$$

$$\mathbf{r}'_\perp = \mathbf{r}_\perp$$

where · is the dot product. The Lorentz factor $\gamma$ retains its definition for a boost in any direction, since it depends only on the magnitude of the relative velocity. The definition $\boldsymbol{\beta} = \mathbf{v}/c$ with magnitude $0 \leq \beta < 1$ is also used by some authors.

Introducing a unit vector $\mathbf{n} = \mathbf{v}/v = \boldsymbol{\beta}/\beta$ in the direction of relative motion, the relative velocity is $\mathbf{v} = v\mathbf{n}$ with magnitude $v$ and direction $\mathbf{n}$, and vector projection and rejection give respectively

$$\mathbf{r}_\parallel = (\mathbf{r} \cdot \mathbf{n})\mathbf{n}, \quad \mathbf{r}_\perp = \mathbf{r} - (\mathbf{r} \cdot \mathbf{n})\mathbf{n}$$

Accumulating the results gives the full transformations,

The projection and rejection also applies to **r**′. For the inverse transformations, exchange **r** and **r**′ to switch observed coordinates, and negate the relative velocity $\mathbf{v} \rightarrow -\mathbf{v}$ (or simply the unit vector $\mathbf{n} \rightarrow -\mathbf{n}$ since the magnitude $v$ is always positive) to obtain

The unit vector has the advantage of simplifying equations for a single boost, allows either **v** or $\boldsymbol{\beta}$ to be reinstated when convenient, and the rapidity parametrization is immediately obtained by replacing $\beta$ and $\beta\gamma$. It is not convenient for multiple boosts.

The vectorial relation between relative velocity and rapidity is[12]

$$\boldsymbol{\beta} = \beta\mathbf{n} = \mathbf{n}\tanh\zeta,$$

and the "rapidity vector" can be defined as

$$\boldsymbol{\zeta} = \zeta\mathbf{n} = \mathbf{n}\tanh^{-1}\beta,$$

each of which serves as a useful abbreviation in some contexts. The magnitude of $\boldsymbol{\zeta}$ is the absolute value of the rapidity scalar confined to $0 \leq \zeta < \infty$, which agrees with the range $0 \leq \beta < 1$.

### 24.4.4 Transformation of velocities

Further information: differential of a function and velocity addition formula

Defining the coordinate velocities and Lorentz factor by

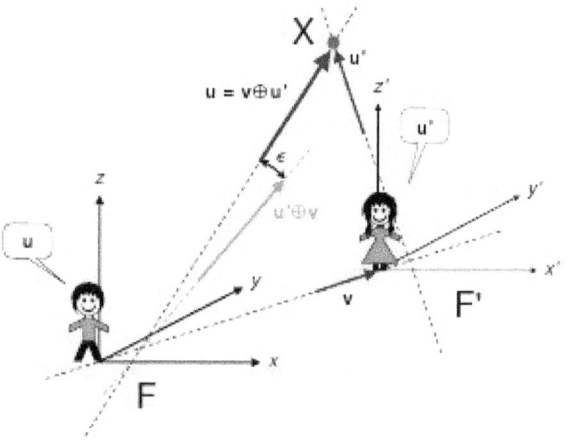

*The transformation of velocities provides the definition relativistic velocity addition ⊕, the ordering of vectors is chosen to reflect the ordering of the addition of velocities; first v (the velocity of F' relative to F) then u' (the velocity of X relative to F') to obtain u = v ⊕ u' (the velocity of X relative to F).*

$$\mathbf{u} = \frac{d\mathbf{r}}{dt}, \quad \mathbf{u}' = \frac{d\mathbf{r}'}{dt'}, \quad \gamma_v = \frac{1}{\sqrt{1 - \dfrac{\mathbf{v} \cdot \mathbf{v}}{c^2}}}$$

taking the differentials in the coordinates and time of the vector transformations, then dividing equations, leads to

$$\mathbf{u}' = \frac{1}{1 - \frac{\mathbf{v} \cdot \mathbf{u}}{c^2}} \left[ \frac{\mathbf{u}}{\gamma_v} - \mathbf{v} + \frac{1}{c^2} \frac{\gamma_v}{\gamma_v + 1} (\mathbf{u} \cdot \mathbf{v}) \mathbf{v} \right]$$

The velocities $\mathbf{u}$ and $\mathbf{u}'$ are the velocity of some massive object. They can also be for a third inertial frame (say F''), in which case they must be *constant*. Denote either entity by X. Then X moves with velocity $\mathbf{u}$ relative to F, or equivalently with velocity $\mathbf{u}'$ relative to F', in turn F' moves with velocity $\mathbf{v}$ relative to F. The inverse transformations can be obtained in a similar way, or as with position coordinates exchange $\mathbf{u}$ and $\mathbf{u}'$, and change $\mathbf{v}$ to $-\mathbf{v}$.

The transformation of velocity is useful in stellar aberration, the Fizeau experiment, and the relativistic Doppler effect.

The Lorentz transformations of acceleration can be similarly obtained by taking differentials in the velocity vectors, and dividing these by the time differential.

### 24.4.5 Transformation of coordinate derivatives

Numerous equations in physics are partial differential equations involving space and time coordinates. Since the space and time coordinates change under Lorentz transformations, the derivatives must also. Using the chain rule one finds the transformation of the coordinate time and space derivatives to be

$$\frac{\partial}{\partial t'} = \gamma \left( \frac{\partial}{\partial t} + v\mathbf{n} \cdot \nabla \right)$$

$$\nabla' = \nabla + (\gamma - 1)\mathbf{n}(\mathbf{n} \cdot \nabla) + \gamma \frac{v\mathbf{n}}{c^2} \frac{\partial}{\partial t}$$

with inverses

$$\frac{\partial}{\partial t} = \gamma \left( \frac{\partial}{\partial t'} - v\mathbf{n} \cdot \nabla \right)$$

$$\nabla = \nabla' + (\gamma - 1)\mathbf{n}(\mathbf{n} \cdot \nabla') - \gamma \frac{v\mathbf{n}}{c^2} \frac{\partial}{\partial t'}$$

These are not quite the same as the transformations of coordinates. It turns out many physical quantities transform either like the coordinates, or like the derivatives.

### 24.4.6 Transformation of other quantities

In general, given four quantities $A$ and $\mathbf{Z} = (Z_x, Z_y, Z_z)$ and their Lorentz-boosted counterparts $A'$ and $\mathbf{Z}' = (Z_x', Z_y', Z_z')$, a relation of the form

$$A^2 - \mathbf{Z} \cdot \mathbf{Z} = A'^2 - \mathbf{Z}' \cdot \mathbf{Z}'$$

implies the quantities transform under Lorentz transformations similar to the transformation of spacetime coordinates;

$$A' = \gamma \left( A - \frac{v\mathbf{n} \cdot \mathbf{Z}}{c} \right),$$

$$\mathbf{Z}' = \mathbf{Z} + (\gamma - 1)(\mathbf{Z} \cdot \mathbf{n})\mathbf{n} - \frac{\gamma A v\mathbf{n}}{c}.$$

The decomposition of $\mathbf{Z}$ (and $\mathbf{Z}'$) into components perpendicular and parallel to $\mathbf{v}$ is exactly the same as for the position vector, as is the process of obtaining the inverse transformations (exchange $(A, \mathbf{Z})$ and $(A', \mathbf{Z}')$ to switch observed quantities, and reverse the direction of relative motion by $\mathbf{n} \rightarrow -\mathbf{n}$);

$$A = \gamma \left( A' + \frac{Z' \cdot v\mathbf{n}}{c} \right) ,$$

$$\mathbf{Z} = \mathbf{Z}' + (\gamma - 1)(\mathbf{Z}' \cdot \mathbf{n})\mathbf{n} + \frac{\gamma A' v\mathbf{n}}{c} .$$

The quantities $(A, \mathbf{Z})$ collectively make up a *four vector*, where $A$ is the "timelike component", and $\mathbf{Z}$ the "spacelike component". Examples of $A$ and $\mathbf{Z}$ are the following:

For a given object (e.g. particle, fluid, field, material), if $A$ or $\mathbf{Z}$ correspond to properties specific to the object like its charge density, mass density, spin, etc., its properties can be fixed in the rest frame of that object. Then the Lorentz transformations give the corresponding properties in a frame moving relative to the object with constant velocity. This breaks some notions taken for granted in non-relativistic physics. For example, the energy $E$ of an object is a scalar in non-relativistic mechanics, but not in relativistic mechanics because energy changes under Lorentz transformations; its value is different for various inertial frames. In the rest frame of an object, it has a rest energy and zero momentum. In a boosted frame its energy is different and it appears to have a momentum. Similarly, in non-relativistic quantum mechanics the spin of a particle is a constant vector, but in relativistic quantum mechanics spin s depends on relative motion. In the rest frame of the particle, the spin pseudovector can be fixed to be its ordinary non-relativistic spin with a zero timelike quantity $st$, however a boosted observer will perceive a nonzero timelike component and an altered spin.[13]

Not all quantities are invariant in the form as shown above, for example orbital angular momentum $\mathbf{L}$ does not have a timelike quantity, and neither does the electric field $\mathbf{E}$ nor the magnetic field $\mathbf{B}$. The definition of angular momentum is $\mathbf{L} = \mathbf{r} \times \mathbf{p}$, and in a boosted frame the altered angular momentum is $\mathbf{L}' = \mathbf{r}' \times \mathbf{p}'$. Applying this definition using the transformations of coordinates and momentum leads to the transformation of angular momentum. It turns out $\mathbf{L}$ transforms with another vector quantity $\mathbf{N} = (E/c^2)\mathbf{r} - t\mathbf{p}$ related to boosts, see relativistic angular momentum for details. For the case of the $\mathbf{E}$ and $\mathbf{B}$ fields, the transformations cannot be obtained as directly using vector algebra. A method of deriving the EM field transformations in an efficient way which also illustrates the unit of the electromagnetic field uses tensor algebra, given below.

# 24.5  Mathematical formulation

Throughout, italic non-bold capital letters are 4×4 matrices, while non-italic bold letters are 3×3 matrices.

## 24.5.1  Boost matrix

The separate algebraic equations are often used in practical calculations, but for theoretical purposes it is useful to arrange the coordinates in column vectors and the quantities defining the transformation into a transformation matrix thus

$$X' = \begin{bmatrix} ct' \\ x' \\ y' \\ z' \end{bmatrix},$$

$$B(\mathbf{v}) = \begin{bmatrix} \gamma & -\gamma\beta n_x & -\gamma\beta n_y & -\gamma\beta n_z \\ -\gamma\beta n_x & 1+(\gamma-1)n_x^2 & (\gamma-1)n_x n_y & (\gamma-1)n_x n_z \\ -\gamma\beta n_y & (\gamma-1)n_y n_x & 1+(\gamma-1)n_y^2 & (\gamma-1)n_y n_z \\ -\gamma\beta n_z & (\gamma-1)n_z n_x & (\gamma-1)n_z n_y & 1+(\gamma-1)n_z^2 \end{bmatrix}, \quad X = \begin{bmatrix} ct \\ x \\ y \\ z \end{bmatrix}$$

and all the separate equations compress into one matrix equation;

$$X' = B(\mathbf{v})X$$

The boost matrix $\mathbf{B}$ is a symmetric matrix, it equals its transpose. In the inverse transformations the transformation matrix is the matrix inverse of the original transformation. Instead of explicitly calculating the inverse matrix by brute force, the simple change $\mathbf{v} \to -\mathbf{v}$ suffices, and the inverse transformation is

$$B(\mathbf{v})^{-1} = B(-\mathbf{v}) \quad \Rightarrow \quad X = B(-\mathbf{v})X'$$

The boosts along the Cartesian directions can be readily obtained, for example the unit vector in the x direction has components $nx = 1$ and $ny = nz = 0$. Looking at the patterns in the boost matrices along the Cartesian directions, the general boost matrix can be systematically rewritten by introducing

$$K_x = \begin{bmatrix} 0 & 1 & 0 & 0 \\ 1 & 0 & 0 & 0 \\ 0 & 0 & 0 & 0 \\ 0 & 0 & 0 & 0 \end{bmatrix}, \quad K_y = \begin{bmatrix} 0 & 0 & 1 & 0 \\ 0 & 0 & 0 & 0 \\ 1 & 0 & 0 & 0 \\ 0 & 0 & 0 & 0 \end{bmatrix}$$

$$, \quad K_z = \begin{bmatrix} 0 & 0 & 0 & 1 \\ 0 & 0 & 0 & 0 \\ 0 & 0 & 0 & 0 \\ 1 & 0 & 0 & 0 \end{bmatrix}$$

Collecting these into a vector of matrices $\mathbf{K} = (Kx, Ky, Kz)$, the matrix $\mathbf{n}\cdot\mathbf{K} = nxKx + nyKy + nzKz$ and its square allow the compact expression

$$B(\mathbf{v}) = I - \gamma\beta(\mathbf{n} \cdot \mathbf{K}) + (\gamma - 1)(\mathbf{n} \cdot \mathbf{K})^2$$

or in the rapidity parametrization

$$B(\zeta) = I - \sinh \zeta (\mathbf{n} \cdot \mathbf{K}) + (\cosh \zeta - 1)(\mathbf{n} \cdot \mathbf{K})^2$$

which resembles Rodrigues' rotation formula for spatial rotations.

The matrices make one or more successive transformations easier to handle, rather than rotely iterating the transformations to obtain the result of more than one transformation. For two boosts along the same direction, the result is another boost, and rapidity provides a natural way to handle this. If a frame $F'$ is boosted with rapidity $\zeta_1$ relative to frame $F$ in direction $\mathbf{n}$, and another frame $F''$ is boosted with rapidity $\zeta_2$ relative to $F'$ along the same direction, the separate boosts are

$$X'' = B(\zeta_2 \mathbf{n})X', \quad X' = B(\zeta_1 \mathbf{n})X$$

then $\zeta_1 + \zeta_2$ is the rapidity of the overall boost of $F''$ relative to $F$ in the same direction as $\mathbf{n}$,

$$X'' = B[(\zeta_1 + \zeta_2)\mathbf{n}]X$$

Moreover, the relative velocities are related to the rapidities by

$$\beta = \tanh(\zeta_1 + \zeta_2), \quad \beta_1 = \tanh \zeta_1, \quad \beta_2 = \tanh \zeta_2.$$

and the hyperbolic identity

$$\tanh(\zeta_1 + \zeta_2) = \frac{\tanh \zeta_1 + \tanh \zeta_2}{1 + \tanh \zeta_1 \tanh \zeta_2}$$

coincides with the resultant relative velocity of the two relative velocities along the same direction. Thus rapidities add if the boosts are collinear as they are here, while the relative velocities do not. The relative velocities can be in the same or opposite directions, but must be collinear.

For two or more consecutive boosts that are not collinear but in different directions, the result is still a Lorentz transformation, but not a single boost. Also, Lorentz boosts along different directions do not commute, changing their order changes the resultant transformation. The non-commutativity of Lorentz boosts is another unintuitive feature of special relativity that is unlike Galilean relativity. In Newtonian mechanics, any pair of Galilean boosts can be performed in either order, and both results are the same Galilean transformation.

The most general proper Lorentz transformation also contains a rotation of the three axes, because the composition of two boosts is not a pure boost but is a boost followed or preceded by a rotation. The rotation is the Wigner rotation, and gives rise to the Thomas precession. The boost is given by a symmetric matrix, but the general Lorentz transformation matrix need not be symmetric. Explicit formulae for the composite transformation matrices are given in the linked article.

### 24.5.2 Rotation matrix

A rotation on the spatial coordinates only, leaving the time coordinate alone, leaves the spacetime interval invariant. Therefore, ordinary spatial rotations are also Lorentz transformations. The 4d matrix is simply

$$R(\boldsymbol{\theta}) = \begin{bmatrix} 1 & 0 \\ 0 & \mathbf{R}(\boldsymbol{\theta}) \end{bmatrix}$$

where $\mathbf{R}$ is a 3d rotation matrix. For the purposes of this article the axis-angle representation will be used here, and the "axis-angle vector" $\boldsymbol{\theta} = \theta \mathbf{e}$ is a useful definition; the angle $\theta$ multiplied by a unit vector $\mathbf{e}$ parallel to the axis. The inverse of $\mathbf{R}$ corresponds to rotations using the same axis and angle, but in the opposite sense. The rotation matrix is orthogonal, so the transpose equals the inverse,

$$R(\boldsymbol{\theta})^{-1} = R(\boldsymbol{\theta})^{\mathsf{T}} = R(-\boldsymbol{\theta}).$$

Looking at the patterns in the rotation matrices about the Cartesian axes, it is useful to introduce the matrices

$$J_x = \begin{bmatrix} 0 & 0 & 0 & 0 \\ 0 & 0 & 0 & 0 \\ 0 & 0 & 0 & -1 \\ 0 & 0 & 1 & 0 \end{bmatrix}, \quad J_y = \begin{bmatrix} 0 & 0 & 0 & 0 \\ 0 & 0 & 0 & 1 \\ 0 & 0 & 0 & 0 \\ 0 & -1 & 0 & 0 \end{bmatrix}$$

$$, \quad J_z = \begin{bmatrix} 0 & 0 & 0 & 0 \\ 0 & 0 & -1 & 0 \\ 0 & 1 & 0 & 0 \\ 0 & 0 & 0 & 0 \end{bmatrix}$$

Collecting these into a vector $\mathbf{J} = (J_x, J_y, J_z)$, these matrices allow the 4d rotation matrix to be expressed in the Rodrigues quadratic,

$$R(\boldsymbol{\theta}) = I + \sin \theta (\mathbf{e} \cdot \mathbf{J}) + (1 - \cos \theta)(\mathbf{e} \cdot \mathbf{J})^2$$

In this article, the *right-handed* convention for the spatial coordinates is used (see orientation (vector space)), so that rotations are positive in the anticlockwise sense according to the right-hand rule, and negative in the clockwise sense. This matrix rotates any 3d vector about the axis $\mathbf{e}$ through angle $\theta$ anticlockwise (an active transformation), which has the equivalent effect of rotating the coordinate frame clockwise about the same axis through the same angle (a passive transformation).

## 24.6 Introduction to the Lorentz group

Main article: Lorentz group

It is a result of special relativity that the quantity

$$X \cdot X = X^{\mathrm{T}} \eta X = X'^{\mathrm{T}} \eta X'$$

is an invariant, where $\eta$ is the Minkowski metric as a square matrix

$$\eta = \begin{bmatrix} -1 & 0 & 0 & 0 \\ 0 & 1 & 0 & 0 \\ 0 & 0 & 1 & 0 \\ 0 & 0 & 0 & 1 \end{bmatrix}$$

and the coordinates change under a Lorentz transformation

$$X' = \Lambda X$$

where $\Lambda$ is a constant square matrix. Boosts and rotations themselves are Lorentz transformations since each operation leaves the spacetime interval invariant, and the composition of any two is also a Lorentz transformation. Specifically, two pure rotations (without boosts) is a rotation, but two pure boosts (without rotations) is generally a boost followed or preceded by a rotation.

The set of all Lorentz transformations $\Lambda$ is denoted $\mathcal{L}$. This set together with matrix multiplication forms a group, in this context known as the *Lorentz group*. Also, the above expression $X \cdot X$ is a quadratic form of signature (3,1) on spacetime, and the group of transformations which leaves this quadratic form invariant is the indefinite orthogonal group O(3,1), a Lie group. In other words, the Lorentz group is O(3,1). As presented in this article, any Lie groups mentioned are matrix Lie groups. In this context the operation of composition amounts to matrix multiplication.

For the specific cases that $\Lambda$ is a boost, rotation, or both, there is an additional detail; the determinant of any boost or rotation matrix is +1. The group of Lorentz transformations consisting only of boosts and rotations is called the "restricted Lorentz group", and is the special indefinite orthogonal group SO(3,1).

However, $\Lambda$ is not limited to boosts and rotations. Other Lorentz transformations may have a determinant of opposite sign and other properties, for example any boosts and/or rotation, combined with parity inversion and/or time reversal, will also leave the above quadratic form invariant. The other transformations are outlined later.

In fact, the above transformation does not include all the symmetries in spacetime. For the spacetime interval to be invariant, it can be shown[14] that it is necessary and sufficient for the coordinate transformation to be of the form

$$X' = \Lambda X + C$$

where $C$ is a constant column containing translations in time and space. If $C \neq 0$, this is an **inhomogeneous Lorentz transformation** or **Poincaré transformation**.[15][16] If $C = 0$, this is a **homogeneous Lorentz transformation**.

The set of Poincaré transformations also satisfy the properties of a group and is called the Poincaré group or inhomogeneous Lorentz group. The extra translations mean the Poincaré group is not O(3,1), details are given in the linked article. Under the Erlangen program, Minkowski space can be viewed as the geometry defined by the Poincaré group, which combines Lorentz transformations with translations. This is the full symmetry of special relativity.

### 24.6.1 Generators and parameters of the homogeneous Lorentz group

The axis-angle vector $\boldsymbol{\theta}$ and rapidity vector $\boldsymbol{\zeta}$ are altogether six continuous variables which make up the group parameters (in this particular representation), and $\mathbf{J}$ and $\mathbf{K}$ are the corresponding six generators of the group.[nb 4]

Physically, the generators of the Lorentz group are operators that correspond to important symmetries in spacetime: $\mathbf{J}$ are the *rotation generators* which correspond to angular momentum, and $\mathbf{K}$ are the *boost generators* which correspond to the motion of the system in spacetime.

Lorentz generators can be added together, or multiplied by real numbers, to get more Lorentz generators. For example,

$$\boldsymbol{\zeta} \cdot \mathbf{K} + \boldsymbol{\theta} \cdot \mathbf{J} = \zeta_x K_x + \zeta_y K_y + \zeta_z K_z + \theta_x J_x + \theta_y J_y + \theta_z J_z =$$

$$\begin{bmatrix} 0 & \zeta_x & \zeta_y & \zeta_z \\ \zeta_x & 0 & -\theta_z & \theta_y \\ \zeta_y & \theta_z & 0 & -\theta_x \\ \zeta_z & -\theta_y & \theta_x & 0 \end{bmatrix}$$

is a generator. Therefore, the set of all Lorentz generators

$$V = \{ \boldsymbol{\zeta} \cdot \mathbf{K} + \boldsymbol{\theta} \cdot \mathbf{J} \}$$

together with the operations of ordinary matrix addition and multiplication of a matrix by a number, forms a vector space over the real numbers.[nb 5] The generators $Jx$, $Jy$, $Jz$, $Kx$, $Ky$, $Kz$ form a basis set of $V$, and the components of the axis-angle and rapidity vectors, $\theta x$, $\theta y$, $\theta z$, $\zeta x$, $\zeta y$, $\zeta z$, are the coordinates of a Lorentz generator with respect to this basis.[nb 6]

Three of the commutation relations of the Lorentz generators are

$$[J_x, J_y] = J_z$$

$$[K_x, K_y] = -J_z$$

$$[J_x, K_y] = K_z$$

where the bracket $[A, B] = AB - BA$ is a binary operation known as the *commutator*, and the other relations can be found by taking cyclic permutations of x, y, z components (i.e. change x to y, y to z, and z to x, repeat).

These commutation relations, and the vector space of generators, fulfill the definition of the Lie algebra $so(3, 1)$. In summary, a Lie algebra is defined as a vector space $V$ over a field of numbers, and with a binary operation [ , ] (called a Lie bracket in this context) on the elements of the vector space, satisfying the axioms of bilinearity, alternatization, and the Jacobi identity. Here the operation [ , ] is the commutator which satisfies all of these axioms, the vector space is the set of Lorentz generators $V$ as given previously, and the field is the set of real numbers.

The exponential map (Lie theory) from the Lie algebra to the Lie group,

$$\exp : so(3, 1) \rightarrow SO(3, 1),$$

provides a one-to-one correspondence between small enough neighborhoods of the origin of the Lie algebra and neighborhoods of the identity element of the Lie group. It the case of the Lorentz group, the exponential map is just the matrix exponential. Globally, the exponential map is not one-to-one, but in the case of the Lorentz group, it is surjective (onto). Hence any group element can be expressed as an exponential of an element of the Lie algebra.

To see the exponential mapping heuristically, consider the infinitesimal Lorentz boost in the x direction for simplicity (the generalization to any direction follows an almost identical procedure). The infinitesimal transformation a small boost away from the identity, obtained by the Taylor expansion of the boost matrix to first order about $\zeta = 0$,

$$B(\zeta e_x) = I + \zeta \left. \frac{\partial B(\zeta e_x)}{\partial \zeta} \right|_{\zeta=0} + \cdots$$

where the higher order terms not shown are negligible because $\zeta$ is small. The derivative of the matrix is the matrix of the entries differentiated with respect to the same variable (see matrix calculus), and it is understood the derivatives are found first then evaluated at $\zeta = 0$, which turn out to give

$$\left. \frac{\partial B(\zeta e_x)}{\partial \zeta} \right|_{\zeta=0} = -K_x$$

The derivative of any smooth curve $A(t)$ with $A(0) = I$ in the group depending on some group parameter $t$ with respect to that group parameter, evaluated at $t = 0$, serves as a definition of a corresponding group generator $X$, and this reflects an infinitesimal transformation away from the identity. The smooth curve can always be taken as an exponential as the exponential will always map $X$ smoothly back into the group via $t \rightarrow \exp(tX)$ for all $t$; this curve will yield $X$ again when differentiated at $t = 0$. In other words, linking terminology used in mathematics and physics: A group generator is any element of the Lie algebra. A group parameter is a component of a coordinate vector representing an arbitrary element of the Lie algebra with respect to some basis. A basis, then, is a set of generators being a basis of the Lie algebra in the usual vector space sense.

In the limit of an infinite number of infinitely small steps, the finite boost transformation in the form of a matrix exponential is obtained

$$B(\zeta e_x) = \lim_{N \to \infty} \left( I - \frac{\zeta K_x}{N} \right)^N = e^{-\zeta K_x}$$

where the limit definition of the exponential has been used (see also characterizations of the exponential function).

Almost identical results appear for the other Cartesian directions, and the general boost matrix is

$$B(\zeta) = e^{-\zeta \cdot \mathbf{K}}$$

similarly the general rotation matrix is

$$R(\theta) = e^{\theta \cdot \mathbf{J}}$$

and the general Lorentz transformation is

$$\Lambda(\zeta, \theta) = e^{\zeta \cdot \mathbf{K} + \theta \cdot \mathbf{J}}.$$

This is in general a product of a rotation and a boost, but the decomposition of a general Lorentz transformation into such factors is nontrivial. In particular,

$$e^{\boldsymbol{\zeta}\cdot\mathbf{K}+\boldsymbol{\theta}\cdot\mathbf{J}} \neq e^{\boldsymbol{\zeta}\cdot\mathbf{K}}e^{\boldsymbol{\theta}\cdot\mathbf{J}},$$

because the generators do not commute. For a description of how to find the factors of a general Lorentz transformation in terms of a boost and a rotation *in principle* (this usually does not yield an intelligible expression in terms of generators $\mathbf{J}$ and $\mathbf{K}$), see Wigner rotation. If, on the other hand, *the decomposition is given* in terms of the generators, and one wants to find the product in terms of the generators, then the Baker–Campbell–Hausdorff formula applies.

### 24.6.2   Generators and parameters of the inhomogeneous Lorentz group

For inhomogenous Lorentz transformations, the additional generators are the components of the four-momentum: energy is the generator of time translation, and the 3d momentum components are the generators of spatial translations in those directions. The extra parameters corresponding to these generators are displacements in space and time. The commutation relations are enlarged to include the momenta with the boost and rotation generators.

### 24.6.3   Classification of the homogeneous Lorentz group

From the invariance of the spacetime interval it follows immediately

$$\eta = \Lambda^{\mathsf{T}}\eta\Lambda$$

and this matrix equation contains the general conditions on the Lorentz transformation to ensure invariance of the spacetime interval. Taking the determinant of the equation using the product rule[nb 7] gives immediately

$$[\det(\Lambda)]^2 = 1 \quad \Rightarrow \quad \det(\Lambda) = \pm 1$$

Writing the Minkowski metric as a block matrix, and the Lorentz transformation in the most general form,

$$\eta = \begin{bmatrix} -1 & 0 \\ 0 & \mathbf{I} \end{bmatrix}, \quad \Lambda = \begin{bmatrix} \Gamma & -\mathbf{a}^{\mathsf{T}} \\ -\mathbf{b} & \mathbf{M} \end{bmatrix},$$

carrying out the block matrix multiplications obtains general conditions on $\Gamma$, $\mathbf{a}$, $\mathbf{b}$, $\mathbf{M}$ to ensure relativistic invariance. Not much information can be directly extracted from all the conditions, however one of the results

$$\Gamma^2 = 1 + \mathbf{b}^{\mathsf{T}}\mathbf{b}$$

is useful; $\mathbf{b}^{\mathsf{T}}\mathbf{b} \geq 0$ always so it follows that

$$\Gamma^2 \geq 1 \quad \Rightarrow \quad \Gamma \leq -1, \quad \Gamma \geq 1$$

The negative inequality may be unexpected, because $\Gamma$ multiplies the time coordinate and this has an effect on time symmetry. If the positive equality holds, then $\Gamma$ is the Lorentz factor.

The determinant and inequality provide four ways to classify Lorentz transformations (herein LTs for brevity). However, any particular LT has only one determinant sign *and* only one inequality. There are four sets which include every possible pair given by the intersections ("n"-shaped symbol meaning "and") of these classifying sets. In set notation the four sets and their intersections are:

where "+" and "−" indicate the determinant sign, while "↑" for $\geq$ and "↓" for $\leq$ denote the inequalities.

The full Lorentz group splits into the union ("u"-shaped symbol meaning "or") of four disjoint sets

$$\mathcal{L} = \mathcal{L}_+^\uparrow \cup \mathcal{L}_-^\uparrow \cup \mathcal{L}_+^\downarrow \cup \mathcal{L}_-^\downarrow$$

A subgroup of a group must be closed under the same operation of the group (here matrix multiplication). In other words, for two Lorentz transformations $\Lambda$ and $L$ from a particular set, the composite Lorentz transformations $\Lambda L$ and $L\Lambda$ must return to the same set $\Lambda$ and $L$ came from. This will not always be the case; it can be shown that the composition of *any* two Lorentz transformations always has the positive determinant and positive inequality, a proper orthochronous transformation.

The orthochronous, proper, proper orthochronous sets of LTs are all subgroups. Another subgroup is the union of proper orthochronous and improper antichronous sets, $\mathcal{L}_0 = \mathcal{L}_+^\uparrow \cup \mathcal{L}_-^\downarrow$ . Rotations and boosts are elements of the proper orthochronous Lorentz group.

The other sets involving the improper and/or antichronous properties do not form subgroups, because the composite transformation always has a positive determinant or inequality, whereas the original separate transformations will have negative determinants and/or inequalities. However, the elements of these sets can be expressed in terms of proper orthochronous transformations with appropriate parity inversion $P$ and/or time reversal $T$. These are in matrix form

$$P = \begin{bmatrix} 1 & 0 \\ 0 & -\mathbf{I} \end{bmatrix}, \quad T = \begin{bmatrix} -1 & 0 \\ 0 & \mathbf{I} \end{bmatrix}$$

so if $\Lambda$ is proper orthochronous, then $T\Lambda$ is improper antichronous, $P\Lambda$ is improper orthochronous, and $TP\Lambda = PT\Lambda$ is proper antichronous.

## 24.7 Tensor formulation

Main article: Representation theory of the Lorentz group
For the notation used, see Ricci calculus.

### 24.7.1 Contravariant vectors

Writing the general matrix transformation of coordinates as the matrix equation

$$\begin{bmatrix} x'^0 \\ x'^1 \\ x'^2 \\ x'^3 \end{bmatrix} = \begin{bmatrix} \Lambda^0{}_0 & \Lambda^0{}_1 & \Lambda^0{}_2 & \Lambda^0{}_3 \\ \Lambda^1{}_0 & \Lambda^1{}_1 & \Lambda^1{}_2 & \Lambda^1{}_3 \\ \Lambda^2{}_0 & \Lambda^2{}_1 & \Lambda^2{}_2 & \Lambda^2{}_3 \\ \Lambda^3{}_0 & \Lambda^3{}_1 & \Lambda^3{}_2 & \Lambda^3{}_3 \end{bmatrix} \begin{bmatrix} x^0 \\ x^1 \\ x^2 \\ x^3 \end{bmatrix}$$

allows the transformation of other physical quantities that cannot be expressed as four-vectors, e.g., tensors or spinors of any order in 4d spacetime, to be defined. In the corresponding tensor index notation, the above matrix expression is

$$x'^\nu = \Lambda^\nu{}_\mu x^\mu,$$

where upper and lower indices label covariant and contravariant components respectively, and the summation convention is applied. It is a standard convention to use Greek indices that take the value 0 for time components, and 1, 2, 3 for space components, while Latin indices simply take the values 1, 2, 3, for spatial components. Note that the first index (reading left to right) corresponds in the matrix notation to a *row index*. The second index corresponds to the column index.

The transformation matrix is universal for all four-vectors, not just 4-dimensional spacetime coordinates. If $A$ is any four-vector, then in tensor index notation

$$A'^\nu = \Lambda^\nu{}_\mu A^\mu.$$

Alternatively, one writes

$$A^{\nu'} = \Lambda^{\nu'}{}_\mu A^\mu.$$

in which the primed indices denote the indices of A in the primed frame. This notation cuts risk of exhausting the Greek alphabet roughly in half.

For a general $n$-component object one may write

$$X'^\alpha = \Pi(\Lambda)^\alpha{}_\beta X^\beta,$$

where $\Pi$ is the appropriate representation of the Lorentz group, an $n{\times}n$ matrix for every $\Lambda$. In this case, the indices should *not* be thought of as spacetime indices (sometimes called Lorentz indices), and they run from 1 to $n$. E.g. if X is a bispinor, then the indices are called *Dirac indices*.

### 24.7.2 Covariant vectors

There are also vector quantities with covariant indices. They are generally obtained from their corresponding objects with contravariant indices by the operation of *lowering an index*, e.g.

$$x_\nu = \eta_{\mu\nu} x^\mu,$$

where $\eta$ is the metric tensor. (The linked article also provides more information about what the operation of raising and lowering indices really is mathematically.) The inverse of this transformation is given by

$$x^\mu = \eta^{\nu\mu} x_\nu,$$

where, when viewed as matrices, $\eta^{\mu\nu}$ is the inverse of $\eta\mu\nu$. As it happens, $\eta^{\mu\nu} = \eta\mu\nu$. This is referred to as *raising an index*. To transform a covariant vector $A\mu$, first raise its index, then transform it according to the same rule as for contravariant 4-vectors, then finally lower the index;

$$A'_\nu = \eta_{\rho\nu} \Lambda^\rho{}_\sigma \eta^{\mu\sigma} A_\mu.$$

But

$$\eta_{\rho\nu} \Lambda^\rho{}_\sigma \eta^{\mu\sigma} = (\Lambda^{-1})^\mu{}_\nu,$$

i. e. it is the $(\mu, \nu)$-component of the *inverse* Lorentz transformation. One defines (as a matter of notation),

$$\Lambda_\nu{}^\mu \equiv (\Lambda^{-1})^\mu{}_\nu,$$

and may in this notation write

$$A'_{\,\nu} = \Lambda_\nu{}^\mu A_\mu.$$

Now for a subtlety. The implied summation on the right hand side of

$$A'_{\,\nu} = \Lambda_\nu{}^\mu A_\mu = (\Lambda^{-1})^\mu{}_\nu A_\mu$$

is running over *a row index* of the matrix representing $\Lambda^{-1}$. Thus, in terms of matrices, this transformation should be thought of as the *inverse transpose* of $\Lambda$ acting on the column vector $A\mu$. That is, in pure matrix notation,

$$A' = (\Lambda^{-1})^{\mathsf{T}} A.$$

This means exactly that covariant vectors (thought of as column matrices) transform according to the dual representation of the standard representation of the Lorentz group. This notion generalizes to general representations, simply replace $\Lambda$ with $\Pi(\Lambda)$.

### 24.7.3 Tensors

If A and B are linear operators on vector spaces U and V, then a linear operator $A \otimes B$ may be defined on the tensor product of U and V, denoted $U \otimes V$ according to[17]

From this it is immediately clear that if u and v are a four-vectors in V, then $u \otimes v \in T_2 V \equiv V \otimes V$ transforms as

The second step uses the bilinearity of the tensor product and the last step defines a 2-tensor on component form, or rather, it just renames the tensor $u \otimes v$.

These observations generalize in an obvious way to more factors, and using the fact that a general tensor on a vector space $V$ can be written as a sum of a coefficient (component!) times tensor products of basis vectors and basis convectors, one arrives at the transformation law for any tensor quantity $T$. It is given by[18]

where $\Lambda\chi^{\nu}$ is defined above. This form can generally be reduced to the form for general $n$-component objects given above with a single matrix $(\Pi(\Lambda))$ operating on column vectors. This latter form is sometimes preferred, e. g. for the electromagnetic field tensor.

### Transformation of the electromagnetic field

Main article: Electromagnetic tensor
Further information: classical electromagnetism and special relativity

Lorentz transformations can also be used to illustrate that

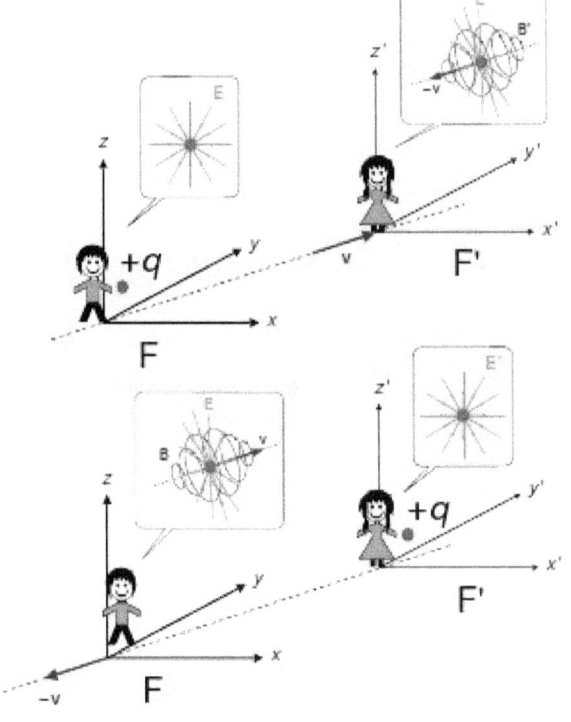

*Lorentz boost of an electric charge, the charge is at rest in one frame or the other.*

the magnetic field **B** and electric field **E** are simply different aspects of the same force — the electromagnetic force, as a consequence of relative motion between electric charges and observers.[19] The fact that the electromagnetic field shows relativistic effects becomes clear by carrying out a simple thought experiment.[20]

- An observer measures a charge at rest in frame F. The observer will detect a static electric field. As the charge is stationary in this frame, there is no electric current, so the observer does not observe any magnetic field.

- The other observer in frame F′ moves at velocity **v** relative to F and the charge. *This* observer sees a different electric field because the charge moves at velocity −**v** in their rest frame. The motion of the charge corresponds to an electric current, and thus the observer in frame F′ also sees a magnetic field.

The electric and magnetic fields transform differently from space and time, but exactly the same way as relativistic angular momentum and the boost vector.

The electromagnetic field strength tensor is given by

$$F^{\mu\nu} = \begin{bmatrix} 0 & -E_x/c & -E_y/c & -E_z/c \\ E_x/c & 0 & -B_z & B_y \\ E_y/c & B_z & 0 & -B_x \\ E_z/c & -B_y & B_x & 0 \end{bmatrix}$$

signature units, (SI(+, −, −, −)).

in SI units. In relativity, the Gaussian system of units is often preferred over SI units, even in texts whose main choice of units is SI units, because in it the electric field **E** and the magnetic induction **B** have the same units making the appearance of the electromagnetic field tensor more natural.[21] Consider a Lorentz boost in the x-direction. It is given by[22]

$$\Lambda^{\mu}{}_{\nu} = \begin{bmatrix} \gamma & -\gamma\beta & 0 & 0 \\ -\gamma\beta & \gamma & 0 & 0 \\ 0 & 0 & 1 & 0 \\ 0 & 0 & 0 & 1 \end{bmatrix}$$

$$, \qquad F^{\mu\nu} = \begin{bmatrix} 0 & E_x & E_y & E_z \\ -E_x & 0 & B_z & -B_y \\ -E_y & -B_z & 0 & B_x \\ -E_z & B_y & -B_x & 0 \end{bmatrix}$$

signature units, (Gaussian( −, +, +, +)),

where the field tensor is displayed side by side for easiest possible reference in the manipulations below.
The general transformation law (**T3**) becomes

$$F^{\mu'\nu'} = \Lambda^{\mu'}{}_{\mu}\Lambda^{\nu'}{}_{\nu}F^{\mu\nu}.$$

For the magnetic field one obtains

$$B_{x'} = F^{2'3'} = \Lambda^2{}_{\mu}\Lambda^3{}_{\nu}F^{\mu\nu} = \Lambda^2{}_2\Lambda^3{}_3F^{23} = 1 \times 1 \times B_x$$
$$= B_x,$$

$$B_{y'} = F^{3'1'} = \Lambda^3{}_{\mu}\Lambda^1{}_{\nu}F^{\mu\nu} = \Lambda^3{}_3\Lambda^1{}_{\nu}F^{3\nu} = \Lambda^3{}_3\Lambda^1{}_0F^{30}$$
$$+ \Lambda^3{}_3\Lambda^1{}_1F^{13} = 1 \times (-\beta\gamma)(-E_z) + 1 \times \gamma B_y = \gamma B_y$$
$$+ \beta\gamma E_z = \gamma(\mathbf{B} - \beta \times \mathbf{E})_y$$

$$B_{z'} = F^{1'2'} = \Lambda^1{}_{\mu}\Lambda^2{}_{\nu}F^{\mu\nu} = \Lambda^1{}_{\mu}\Lambda^2{}_2F^{\mu2} = \Lambda^1{}_0\Lambda^2{}_2F^{02}$$
$$+ \Lambda^1{}_1\Lambda^2{}_2F^{12} = (-\gamma\beta)\times 1 \times E_y + \gamma \times 1 \times B_z = \gamma B$$
$$z- \beta\gamma E_y = \gamma(\mathbf{B} - \beta \times \mathbf{E})_z$$

For the electric field results

$$E_{x'} = F^{0'1'} = \Lambda^0{}_{\mu}\Lambda^1{}_{\nu}F^{\mu\nu} = \Lambda^0{}_1\Lambda^1{}_0F^{10} + \Lambda^0{}_0\Lambda^1{}_1F^{01} = (-\gamma\beta)(-\gamma\beta)(-E_x) + \gamma\gamma E_x = -\gamma^2\beta^2(E_x) + \gamma^2 E_x=$$
$$= E_x, \qquad\qquad\qquad\qquad\qquad\qquad\qquad\qquad\qquad\qquad\qquad\qquad\qquad\qquad E_x(1 - \beta^2)\gamma^2$$

$$E_{y'} = F^{0'2'} = \Lambda^0{}_{\mu}\Lambda^2{}_{\nu}F^{\mu\nu} = \Lambda^0{}_{\mu}\Lambda^2{}_2F^{\mu2} = \Lambda^0{}_0\Lambda^2{}_2F^{02} + \Lambda^0{}_1\Lambda^2{}_2F^{12} = \gamma \times 1 \times E_y + (-\beta\gamma) \times 1 \times B_z = \gamma E_y - \beta\gamma B_z$$
$$= \gamma(\mathbf{E} + \beta \times \mathbf{B})_y$$

$$E_{z'} = F^{0'3'} = \Lambda^0{}_{\mu}\Lambda^3{}_{\nu}F^{\mu\nu} = \Lambda^0{}_{\mu}\Lambda^3{}_3F^{\mu3} = \Lambda^0{}_0\Lambda^3{}_3F^{03} + \Lambda^0{}_1\Lambda^3{}_3F^{13} = \gamma \times 1 \times E_z - \beta\gamma \times 1 \times (-B_y) = \gamma E_z + \beta\gamma B_y$$
$$= \gamma(\mathbf{E} + \beta \times \mathbf{B})_z.$$

Here, $\beta = (\beta, 0, 0)$ is used. These results can be summarized by

$$\mathbf{E}_{\parallel'} = \mathbf{E}_{\parallel}$$

$$\mathbf{B}_{\parallel'} = \mathbf{B}_{\parallel}$$

$$\mathbf{E}_{\perp'} = \gamma(\mathbf{E}_{\perp} + \beta \times \mathbf{B}_{\perp}) = \gamma(\mathbf{E} + \beta \times \mathbf{B})_{\perp},$$

$$\mathbf{B}_{\perp'} = \gamma(\mathbf{B}_{\perp} - \beta \times \mathbf{E}_{\perp}) = \gamma(\mathbf{B} - \beta \times \mathbf{E})_{\perp},$$

and are independent of the metric signature. For SI units, substitute $E \to {}^E\!/c$. Misner, Thorne & Wheeler (1973) refer to this last form as the 3 + 1 view as opposed to the *geometric view* represented by the tensor expression

$$F^{\mu'\nu'} = \Lambda^{\mu'}{}_{\mu}\Lambda^{\nu'}{}_{\nu}F^{\mu\nu},$$

and make a strong point of the ease with which results that are difficult to achieve using the 3 + 1 view can be obtained and understood. Only objects that have well defined Lorentz transformation properties (in fact under *any* smooth coordinate transformation) are geometric objects. In the geometric view, the electromagnetic field is a six-dimensional geometric object in *spacetime* as opposed to two interdependent, but separate, 3-vector fields in *space* and *time*. The fields **E** (alone) and **B** (alone) do not have well defined Lorentz transformation properties. The mathematical underpinnings are equations (**T1**) and (**T2**) that immediately yield (**T3**). One should note that the primed and unprimed tensors refer to the *same event in spacetime*. Thus the complete equation with spacetime dependence is

$$F^{\mu'\nu'}(x') = \Lambda^{\mu'}{}_{\mu}\Lambda^{\nu'}{}_{\nu}F^{\mu\nu}(\Lambda^{-1}x') = \Lambda^{\mu'}{}_{\mu}\Lambda^{\nu'}{}_{\nu}F^{\mu\nu}(x).$$

Length contraction has an effect on charge density $\varrho$ and current density **J**, and time dilation has an effect on the rate of flow of charge (current), so charge and current distributions must transform in a related way under a boost. It turns out they transform exactly like the space-time and energy-momentum four-vectors,

$$\mathbf{j}' = \mathbf{j} - \gamma\rho v\mathbf{n} + (\gamma - 1)(\mathbf{j} \cdot \mathbf{n})\mathbf{n},$$

$$\rho' = \gamma(\rho - \mathbf{j} \cdot v\mathbf{n}/c^2)$$

or, in the simpler geometric view,

$$j^{\mu'} = \Lambda^{\mu'}{}_{\mu}j^{\mu}.$$

One says that charge density transforms as the time component of a four-vector. It is a rotational scalar. The current density is a 3-vector.

The Maxwell equations are invariant under Lorentz transformations.

### 24.7.4   Spinors

Equation (**T1**) hold unmodified for any representation of the Lorentz group, including the bispinor representation. In (**T2**) one simply replaces all occurrences of $\Lambda$ by the bispinor representation $\Pi(\Lambda)$,

The above equation could, for instance, be the transformation of a state in Fock space describing two free electrons.

### Transformation of general fields

A general *noninteracting* multi-particle state (Fock space state) in quantum field theory transforms according to the rule[23]

where $W(\Lambda, p)$ is the Wigner rotation and $D^{(j)}$ is the $(2j + 1)$-dimensional representation of SO(3).

## 24.8   See also

- Ricci calculus

- Electromagnetic field

- Galilean transformation

- Hyperbolic rotation

- Invariance mechanics

- Lorentz group

- Representation theory of the Lorentz group

- Principle of relativity

- Velocity-addition formula

- Algebra of physical space

- Relativistic aberration

- Prandtl–Glauert transformation

- Split-complex number

- Gyrovector space

## 24.9   Footnotes

[1] One can imagine that in each inertial frame there are observers positioned throughout space, each endowed with a synchronized clock and at rest in the particular inertial frame. These observers then report to a central office, where a report is collected. When one speaks of a *particular* observer, one refers to someone having, at least in principle, a copy of this report. See, e.g., Sard (1970).

[2] It should be noted that the separate requirements of the three equations lead to three different groups. The second equation is satisfied for spacetime translations in addition to Lorentz transformations leading to the Poincare group or the *inhomogeneous Lorentz group*. The first equation (or the second restricted to lightlike separation) leads to a yet larger group, the conformal group of spacetime.

[3] The groups O(3, 1) and O(1, 3) are isomorphic. It is widely believed that the choice between the two metric signatures has no physical relevance, even though some objects related to O(3, 1) and O(1, 3) respectively, e.g., the Clifford algebras corresponding to the different signatures of the bilinear form associated to the two groups, are non-isomorphic.

[4] In quantum mechanics, relativistic quantum mechanics, and quantum field theory, a different convention is used for these matrices; the right hand sides are all multiplied by a factor of the imaginary unit $i = \sqrt{-1}$.

[5] Until now the term "vector" has exclusively referred to "Euclidean vector", examples are position $\mathbf{r}$, velocity $\mathbf{v}$, etc. The term "vector" applies much more broadly than Euclidean vectors, row or column vectors, etc., see linear algebra and vector space for details. The generators of a Lie group also form a vector space over a field of numbers (e.g. real numbers, complex numbers), since a linear combination of the generators is also a generator. They just live in a different space to the position vectors in ordinary 3d space.

[6] In ordinary 3d position space, the position vector $\mathbf{r} = x\mathbf{e}x + y\mathbf{e}y + z\mathbf{e}z$ is expressed as a linear combination of the Cartesian unit vectors $\mathbf{e}x$, $\mathbf{e}y$, $\mathbf{e}z$ which form a basis, and the Cartesian coordinates $x$, $y$, $z$ are coordinates with respect to this basis.

[7] For two square matrices $A$ and $B$, $\det(AB) = \det(A)\det(B)$

## 24.10   Notes

[1] Lorentz, Hendrik Antoon (1904), "Electromagnetic phenomena in a system moving with any velocity smaller than that of light", *Proceedings of the Royal Netherlands Academy of Arts and Sciences* **6**: 809–831

[2] John & O'Connor 1996

[3] Brown 2003

[4] Rothman 2006, pp. 112f.

[5] Darrigol 2005, pp. 1–22

[6] Macrossan 1986, pp. 232–34

[7] The reference is within the following paper:Poincaré 1905, pp. 1504–1508

[8] Einstein 1905, pp. 891–921

[9] Young & Freedman 2008

[10] Forshaw & Smith 2009

[11] Einstein 1916

[12] Barut 1964, p. 18–19

[13] Chaichian & Hagedorn 1997, p. 239

[14] Weinberg 1972

[15] Weinberg 2005, pp. 55–58

[16] Ohlsson 2011, p. 3–9

[17] Hall 2003, Chapter 4

[18] Carroll 2004, p. 22

[19] Grant & Phillips 2008

[20] Griffiths 2007

[21] Jackson 1999

[22] Misner, Thorne & Wheeler 1973

[23] Weinberg 2002, Chapter 3

## 24.11 References

### 24.11.1 Websites

- O'Connor, John J.; Robertson, Edmund F. (1996), *A History of Special Relativity*

- Brown, Harvey R. (2003), *Michelson, FitzGerald and Lorentz: the Origins of Relativity Revisited*

### 24.11.2 Papers

- Cushing, J. T. (1967). "Vector Lorentz transformations". *American Journal of Physics* **35**: 858–862. Bibcode:1967AmJPh..35..858C. doi:10.1119/1.1974267.

- Macfarlane, A. J. (1962). "On the Restricted Lorentz Group and Groups Homomorphically Related to It". *Journal of Mathematical Physics* **3** (6): 1116–1129. Bibcode:1962JMP.....3.1116M. doi:10.1063/1.1703854.

- Rothman, Tony (2006), "Lost in Einstein's Shadow" (PDF), *American Scientist* **94** (2): 112f.

- Darrigol, Olivier (2005), "The Genesis of the theory of relativity" (PDF), *Séminaire Poincaré* **1**: 1–22, doi:10.1007/3-7643-7436-5_1

- Macrossan, Michael N. (1986), "A Note on Relativity Before Einstein", *Brit. Journal Philos. Science* **37**: 232–34, doi:10.1093/bjps/37.2.232

- Poincaré, Henri (1905), "On the Dynamics of the Electron", *Comptes rendus hebdomadaires des séances de l'Académie des sciences* **140**: 1504–1508

- Einstein, Albert (1905), "Zur Elektrodynamik bewegter Körper" (PDF), *Annalen der Physik* **322** (10): 891–921, Bibcode:1905AnP...322..891E, doi:10.1002/andp.19053221004. See also: English translation.

- Einstein, A. (1916). "Relativity: The Special and General Theory" (PDF). Retrieved 2012-01-23.

- Ungar, A. A. (1988). "Thomas rotation and the parameterization of the Lorentz transformation group". *Foundations of Physics Letters* (Kluwer Academic Publishers-Plenum Publishers) **1** (1): 55–89. Bibcode:1988FoPhL...1...57U. doi:10.1007/BF00661317. ISSN 0894-9875. (subscription required (help)). eqn (55).

- Ungar, A. A. (1989). "The relativistic velocity composition paradox and the Thomas rotation". *Foundations of Physics* **19**: 1385–1396. Bibcode:1989FoPh...19.1385U. doi:10.1007/BF00732759.

- Ungar, A. A. (2000). "The relativistic composite-velocity reciprocity principle". *Foundations of Physics* (Springer) **30** (2): 331–342. CiteSeerX: 10.1.1.35.1131.

- Mocanu, C. I. (1986). "Some difficulties within the framework of relativistic electrodynamics". *Archiv für Elektrotechnik* (Springer) **69**: 97–110. doi:10.1007/bf01574845.

- Mocanu, C. I. (1992). "On the relativistic velocity composition paradox and the Thomas rotation". *Foundations of Physics* (Plenum) **5**: 443–456. Bibcode:1992FoPhL...5..443M. doi:10.1007/bf00690425.

- Weinberg, S. (2002). *The Quantum Theory of Fields, vol I*. Cambridge University Press. ISBN 0-521-55001-7.

### 24.11.3 Books

- Young, H. D.; Freedman, R. A. (2008). *University Physics – With Modern Physics* (12th ed.). ISBN 0-321-50130-6.

- Halpern, A. (1988). *3000 Solved Problems in Physics*. Schaum Series. Mc Graw Hill. p. 688. ISBN 978-0-07-025734-4.

- Forshaw, J. R.; Smith, A. G. (2009). *Dynamics and Relativity*. Manchester Physics Series. John Wiley & Sons Ltd. pp. 124–126. ISBN 978-0-470-01460-8.

- Wheeler, J. A.; Taylor, E. F (1971). *Spacetime Physics*. Freeman. ISBN 0-7167-0336-X.

- Wheeler, J. A.; Thorne, K. S.; Misner, C. W. (1973). *Gravitation*. Freeman. ISBN 0-7167-0344-0.

- Carroll, S. M. (2004). *Spacetime and Geometry: An Introduction to General Relativity* (illustrated ed.). Addison Wesley. p. 22. ISBN 0-8053-8732-3.

- Grant, I. S.; Phillips, W. R. (2008). "14". *Electromagnetism*. Manchester Physics (2nd ed.). John Wiley & Sons. ISBN 0-471-92712-0.

- Griffiths, D. J. (2007). *Introduction to Electrodynamics* (3rd ed.). Pearson Education, Dorling Kindersley,. ISBN 81-7758-293-3.

- Hall, Brian C. (2003). *Lie Groups, Lie Algebras, and Representations An Elementary Introduction*. Springer Publishing. ISBN 0-387-40122-9.

- Weinberg, S. (2008), *Cosmology*, Wiley, ISBN 978-0-19-852682-7

- Weinberg, S. (2005), *The quantum theory of fields (3 vol.)* **1**, Cambridge University Press, ISBN 978-0-521-67053-1

- Ohlsson, T. (2011), *Relativistic Quantum Physics*, Cambridge University Press, ISBN 978-0-521-76726-2

- Goldstein, H. (1980) [1950]. *Classical Mechanics* (2nd ed.). Reading MA: Addison-Wesley. ISBN 0-201-02918-9.

- Jackson, J. D. (1975) [1962]. "Chapter 11". *Classical Electrodynamics* (2nd ed.). John Wiley & Sons. pp. 542–545. ISBN 0-471-43132-X.

- Landau, L. D.; Lifshitz, E. M. (2002) [1939]. *The Classical Theory of Fields*. Course of Theoretical Physics 2 (4th ed.). Butterworth–Heinemann. pp. 9–12. ISBN 0 7506 2768 9.

- Feynman, R. P.; Leighton, R. B.; Sands, M. (1977) [1963]. "15". *The Feynman Lectures on Physics* **1**. Addison Wesley. ISBN 0-201-02117-X.

- Feynman, R. P.; Leighton, R. B.; Sands, M. (1977) [1964]. "13". *The Feynman Lectures on Physics* **2**. Addison Wesley. ISBN 0-201-02117-X.

- Misner, Charles W.; Thorne, Kip S.; Wheeler, John Archibald (1973). *Gravitation*. San Francisco: W. H. Freeman. ISBN 978-0-7167-0344-0.

- Rindler, W. (2006) [2001]. "Chapter 9". *Relativity Special, General and Cosmological* (2nd ed.). Dallas: Oxford University Press. ISBN 978-0-19-856732-5.

- Ryder, L. H. (1996) [1985]. *Quantum Field Theory* (2nd ed.). Cambridge: Cambridge University Press. ISBN 978-0521478144.

- Sard, R. D. (1970). *Relativistic Mechanics - Special Relativity and Classical Particle Dynamics*. New York: W. A. Benjamin. ISBN 978-0805384918.

- R. U. Sexl, H. K. Urbantke (2001) [1992]. *Relativity, Groups Particles. Special Relativity and Relativistic Symmetry in Field and Particle Physics*. Springer. ISBN 978-3211834435.

- Gourgoulhon, Eric (2013). *Special Relativity in General Frames: From Particles to Astrophysics*. Springer. p. 213. ISBN 978-3-642-37276-6.

- Chaichian, Masud; Hagedorn, Rolf (1997). *Symmetry in quantum mechanics:From angular momentum to supersymmetry*. IoP. p. 239. ISBN 0-7503-0408-1.

## 24.12 Further reading

- Einstein, Albert (1961), *Relativity: The Special and the General Theory*, New York: Three Rivers Press (published 1995), ISBN 0-517-88441-0

- Ernst, A.; Hsu, J.-P. (2001), "First proposal of the universal speed of light by Voigt 1887" (PDF), *Chinese Journal of Physics* **39** (3): 211–230, Bibcode:2001ChJPh..39..211E

- Thornton, Stephen T.; Marion, Jerry B. (2004), *Classical dynamics of particles and systems* (5th ed.), Belmont, [CA.]: Brooks/Cole, pp. 546–579, ISBN 0-534-40896-6

- Voigt, Woldemar (1887), "Über das Doppler'sche princip", *Nachrichten von der Königlicher Gesellschaft den Wissenschaft zu Göttingen* **2**: 41–51

## 24.13   External links

- Derivation of the Lorentz transformations. This web page contains a more detailed derivation of the Lorentz transformation with special emphasis on group properties.

- The Paradox of Special Relativity. This webpage poses a problem, the solution of which is the Lorentz transformation, which is presented graphically in its next page.

- Relativity – a chapter from an online textbook

- Warp Special Relativity Simulator. A computer program demonstrating the Lorentz transformations on everyday objects.

- Animation clip on YouTube visualizing the Lorentz transformation.

- Lorentz Frames Animated *from John de Pillis*. Online Flash animations of Galilean and Lorentz frames, various paradoxes, EM wave phenomena, *etc.*

# Chapter 25

# Derivations of the Lorentz transformations

There are many ways to derive the Lorentz transformations utilizing a variety of mathematical tools, spanning from elementary algebra and hyperbolic functions, to linear algebra and group theory.

This article provides a few of the easier ones to follow in the context of special relativity, for the simplest case of a Lorentz boost in standard configuration, i.e. two inertial frames moving relative to each other at constant (uniform) relative velocity less than the speed of light, and using Cartesian coordinates so that the $x$ and $x'$ axes are collinear.

## 25.1  Lorentz transformation

Main article: Lorentz transformation

In the fundamental branches of modern physics, namely general relativity and its widely applicable subset special relativity, as well as relativistic quantum mechanics and relativistic quantum field theory, the Lorentz transformation is the transformation rule under which all four-vectors and tensors containing physical quantities transform.

The prime examples of such four vectors are the four position and four momentum of a particle, and for fields the electromagnetic tensor and stress–energy tensor. The fact that these objects transform according to the Lorentz transformation is what mathematically *defines* them as vectors and tensors, see tensor.

Given the components of the four vectors or tensors in some frame, the "transformation rule" allows one to determine the altered components of the same four vectors or tensors in another frame, which could be boosted or accelerated, relative to the original frame. A "boost" should not be conflated with spatial translation, rather it's characterized by the relative velocity between frames. The transformation rule itself depends on the relative motion of the frames. In the simplest case of two inertial frames the relative velocity between enters the transformation rule. For rotating reference frames or general non-inertial reference frames,

more parameters are needed, including the relative velocity (magnitude and direction), the rotation axis and angle turned through.

## 25.2  Historical background

The usual treatment (e.g., Einstein's original work) is based on the invariance of the speed of light. However, this is not necessarily the starting point: indeed (as is exposed, for example, in the second volume of the *Course of Theoretical Physics* by Landau and Lifshitz), what is really at stake is the *locality* of interactions: one supposes that the influence that one particle, say, exerts on another can not be transmitted instantaneously. Hence, there exists a theoretical maximal speed of information transmission which must be invariant, and it turns out that this speed coincides with the speed of light in vacuum. Newton had himself called the idea of action at a distance philosophically "absurd", and held that gravity had to be transmitted by some agent according to certain laws.[1]

Michelson and Morley in 1887 designed an experiment, employing an interferometer and a half-silvered mirror, that was accurate enough to detect aether flow. The mirror system reflected the light back into the interferometer. If there were an aether drift, it would produce a phase shift and a change in the interference that would be detected. However, no phase shift was ever found. The negative outcome of the Michelson–Morley experiment left the concept of aether (or its drift) undermined. There was consequent perplexity as to why light evidently behaves like a wave, without any detectable medium through which wave activity might propagate.

In a 1964 paper,[2] Erik Christopher Zeeman showed that the causality preserving property, a condition that is weaker in a mathematical sense than the invariance of the speed of light, is enough to assure that the coordinate transformations are the Lorentz transformations. Norman Goldstein's paper shows a similar result using *inertiality* (the preservation of time-like lines) rather than *causality*.[3]

## 25.3  Physical principles

Assume the second postulate of special relativity stating the constancy of the speed of light, independent of reference frame, and consider a collection of reference systems moving with respect to each other with constant velocity, i.e. inertial systems, each endowed with its own set of cartesian coordinates labeling the points, i.e. events of spacetime. To express the invariance of the speed of light in mathematical form, fix two events in spacetime, to be recorded in each reference frame. Let the first event be the emission of a light signal, and the second event be it being absorbed.

Pick any reference frame in the collection. In its coordinates, the first event will be assigned coordinates $x_1, y_1, z_1, ct_1$, and the second $x_2, y_2, z_2, ct_2$. The spatial distance between emission and absorption is $\sqrt{(x_2-x_1)^2 + (y_2-y_1)^2 + (z_2-z_1)^2}$, but this is also the distance $c(t_2-t_1)$ traveled by the signal. One may therefore set up the equation

$$c^2(t_2 - t_1)^2 - (x_2 - x_1)^2 - (y_2 - y_1)^2 - (z_2 - z_1)^2 = 0.$$

Every other coordinate system will record, in its own coordinates, the same equation. This is the immediate mathematical consequence of the invariance of the speed of light. The quantity on the left is called the *interval*. The interval is, for events separated by light signals, the same (zero) in all reference frames, and is therefore called *invariant*.

### 25.3.1  Invariance of interval

For the Lorentz transformation to have physical significance, it is crucial that the interval is an invariant measure for *any* two events, not just for those separated by light signals. To establish this, one considers an *infinitesimal* interval.[4]

$$ds^2 = c^2 dt^2 - dx^2 - dy^2 - dz^2,$$

as recorded in a system K. Let $K'$ be another system assigning the interval $ds'^2$ to the same two infinitesimally separated events. Since if $ds^2 = 0$, then the interval will be null in any other system (second postulate), and since $ds^2$ and $ds'^2$ are infinitesimals of the same order, they must be proportional to each other,

$$ds^2 = a\,ds'^2.$$

On what may a depend? It may not depend on the two events position in spacetime. That would violate the postulated *homogeneity of spacetime*. It might depend on the relative velocity $V'$ between $K$ and $K'$ but only on the speed, not the direction, because that would violate the *isotropy of space*.

Now bring in systems $K_1$ and $K_2$,

$$ds^2 = a(V_1)ds_1^2, \quad ds^2 = a(V_2)ds_2^2, \quad ds_1^2 = a(V_{12})ds_2^2.$$

From these it follows,

$$\frac{a(V_2)}{a(V_1)} = a(V_{12}).$$

Now one observes that on the right, $V_{12}$ does not only depend on $V_1$ and $V_2$, but also on the angle between the *vectors* $\mathbf{V}_1$ and $\mathbf{V}_2$. The left hand side does not depend on this angle, and the only way for the equation to hold true is if the function $a(V)$ is a constant, and by the same equation this constant is unity. Thus,

$$ds^2 = ds'^2$$

for all systems $K'$, and since this holds for all infinitesimal intervals, it holds for *all* intervals. Most, if not all, derivations of the Lorentz transformations take this for granted, and use the constancy of the speed of light (invariance of lightlike separated events) only. This result ensures that the Lorentz transformation is the correct transformation.

## 25.4  Standard configuration

The invariant interval can be seen as non-positive definite distance function on spacetime. The set of transformations sought must leave this distance invariant. Due to the reference frames coordinate systems cartesian nature, one concludes that, as in the Euclidean case, the possible transformations are made up of translations and rotations, where a slightly broader meaning should be allowed for the term rotation.

The interval is quite trivially invariant under translation. For rotations, there are four coordinates. Hence there are six planes of rotation. Three of those are rotations in spatial planes. The interval is invariant under ordinary rotations too.[4]

It remains to find a "rotation" in the three remaining coordinate planes that leaves the interval invariant. Equivalently, to find a way to assign coordinates so that they coincide with the coordinates corresponding to a moving frame.

The general problem is to find a transformation such that

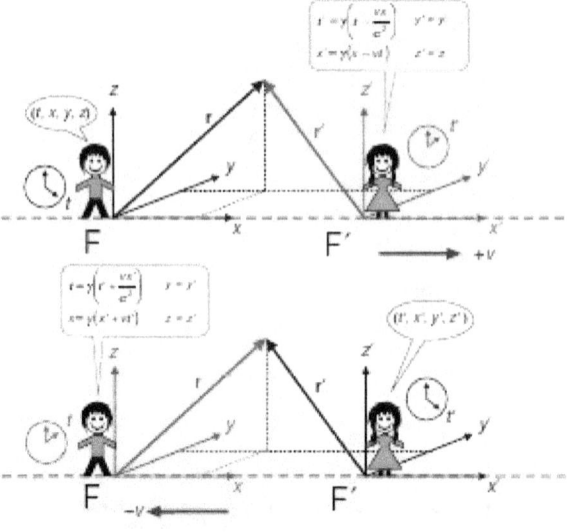

*The spacetime coordinates of an event, as measured by each observer in their inertial reference frame (in standard configuration) are shown in the speech bubbles.*
*Top: frame F′ moves at velocity v along the x-axis of frame F.*
*Bottom: frame F moves at velocity −v along the x′-axis of frame F′.[5]*

## 25.5  The solutions

As mentioned, the general problem is solved by translations in spacetime. These do not appear as a solution to the simpler problem posed, while the boosts do (and sometimes rotations depending on angle of attack). Even more solutions exist if one *only* insist on invariance of the interval for lightlike separated events. These are nonlinear conformal ("angle preserving") transformations. One has

> Lorentz transformations ⊂ Poincare transformations ⊂ conformal group transformations.

Some equations of physics are conformal invariant, e.g. the Maxwell's equations in source-free space,[6] but not all. The relevance of the conformal transformations in spacetime is not known at present, but the conformal group in two dimensions is highly relevant in conformal field theory and statistical mechanics.[7] It is thus the Poincare group that is singled out by the postulates of special relativity. It is the presence of Lorentz boosts (for which velocity addition is different from mere vector addition that would allow for speeds greater than the speed of light) as opposed to ordinary boosts that separates it from the Galilean group of Galilean relativity. Spatial rotations, spatial and temporal inversions and translations are present in both groups and have the same consequences in both theories (conservation laws of momentum, energy, and angular momentum). Not all accepted theories respect symmetry under the inversions.

$$c^2(t_2 - t_1)^2 - (x_2 - x_1)^2 - (y_2 - y_1)^2 - (z_2 - z_1)^2$$
$$= c^2(t_2' - t_1')^2 - (x_2' - x_1')^2 - (y_2' - y_1')^2 - (z_2' - z_1')^2.$$

To solve the general problem, one may use the knowledge about invariance of the interval of translations and ordinary rotations to assume, without loss of generality,[4] that the frames F and F′ are aligned in such a way that their coordinate axes all meet at $t = t' = 0$ and that the x and x′ axes are permanently aligned and system F′ has speed V along the positive x-axis. Call this the *standard configuration*. It reduces the general problem to finding a transformation such that

$$c^2(t_2 - t_1)^2 - (x_2 - x_1)^2 = c^2(t_2' - t_1')^2 - (x_2' - x_1')^2.$$

The standard configuration is used in most examples below. It should be noted that a *linear* solution of the simpler problem

$$c^2 t^2 - x^2 = c^2 t'^2 - x'^2$$

solves the more general problem since coordinate *differences* then transform the same way. Linearity is often assumed or argued somehow in the literature when this simpler problem is considered. If the solution to the simpler problem is *not* linear, then it doesn't solve the original problem because of the cross terms appearing when expanding the squares.

## 25.6  Landau & Lifshitz solution

Three useful hyperbolic function formulae (H1–H3).

$$1.\ \cosh^2 \Psi - \sinh^2 \Psi = 1, \quad 2.\ \sinh \Psi = \frac{\tanh \Psi}{\sqrt{1 - \tanh^2 \Psi}}$$
$$, \quad 3.\ \cosh \Psi = \frac{1}{\sqrt{1 - \tanh^2 \Psi}},$$

The problem posed in standard configuration for a boost in the x-direction, where the primed coordinates refer to the *moving* system is solved by finding a *linear* solution to the simpler problem

$$c^2 t^2 - x^2 = c^2 t'^2 - x'^2.$$

The most general solution, the detailed derivation of which can be found in the section hyperbolic rotation, is, as can be verified by direct substitution using (H1),[4]

To find the role of $\Psi$ in the physical setting, record the origin's of F′ progression, i.e. $x' = 0$, $x = vt$. The equations become (using first $x' = 0$),

$$x = ct' \sinh \Psi, \quad ct = ct' \cosh \Psi.$$

Now divide,

$$\frac{x}{ct} = \tanh \Psi = \frac{v}{c} \Rightarrow \quad \sinh \Psi = \frac{\frac{v}{c}}{\sqrt{1 - \frac{v^2}{c^2}}}$$

$$, \quad \cosh \Psi = \frac{1}{\sqrt{1 - \frac{v^2}{c^2}}} \quad ,$$

where $x = vt$ was used in the first step, (H2) and (H3) in the second, which, when plugged back in (1), gives

$$x = \frac{x' + vt'}{\sqrt{1 - \frac{v^2}{c^2}}}, \quad t = \frac{t' + \frac{v}{c^2}x'}{\sqrt{1 - \frac{v^2}{c^2}}},$$

or, with the usual abbreviations,

This calculation is repeated with more detail in section hyperbolic rotation.

## 25.7 From physical principles

The problem is usually restricted to two dimensions by using a velocity along the $x$ axis such that the $y$ and $z$ coordinates do not intervene, as described in standard configuration above. The following is similar to that of Einstein.[8][9] As in the Galilean transformation, the Lorentz transformation is linear since the relative velocity of the reference frames is constant as a vector; otherwise, inertial forces would appear. They are called inertial or Galilean reference frames. According to relativity no Galilean reference frame is privileged. Another condition is that the speed of light must be independent of the reference frame, in practice of the velocity of the light source.

### 25.7.1 Spherical wavefronts of light

Consider two inertial frames of reference $O$ and $O'$, assuming $O$ to be at rest while $O'$ is moving with a velocity $v$ with respect to $O$ in the positive $x$-direction. The origins of $O$ and $O'$ initially coincide with each other. A light signal is emitted from the common origin and travels as a spherical wave front. Consider a point $P$ on a spherical wavefront at a distance $r$ and $r'$ from the origins of $O$ and $O'$ respectively. According to the second postulate of the special theory of relativity the speed of light is the same in both frames, so for the point $P$:

$$r = ct$$
$$r' = ct'.$$

The equation of a sphere in frame $O$ is given by

$$x^2 + y^2 + z^2 = r^2.$$

For the spherical wavefront that becomes

$$x^2 + y^2 + z^2 = c^2 t^2.$$

Similarly, the equation of a sphere in frame $O'$ is given by

$$x'^2 + y'^2 + z'^2 = r'^2,$$

so the spherical wavefront satisfies

$$x'^2 + y'^2 + z'^2 = c^2 t'^2.$$

The origin $O'$ is moving along $x$-axis. Therefore,

$$y' = y$$
$$z' = z.$$

$x'$ must vary linearly with $x$ and $t$. Therefore, the transformation has the form

$$x' = \gamma x + \sigma t.$$

For the origin of $O'$ $x'$ and $x$ are given by

$$x' = 0$$
$$x = vt,$$

so, for all $t$,

$$0 = \gamma vt + \sigma t$$

and thus

$$\sigma = -\gamma v.$$

This simplifies the transformation to

$$x' = \gamma (x - vt)$$

where $\gamma$ is to be determined. At this point $\gamma$ is not necessarily a constant, but is required to reduce to 1 for $v \ll c$.

The inverse transformation is the same except that the sign of $v$ is reversed:

$$x = \gamma \left( x' + vt' \right).$$

The above two equations give the relation between $t$ and $t'$ as:

$$x = \gamma \left[ \gamma \left( x - vt \right) + vt' \right]$$

or

$$t' = \gamma t + \frac{\left( 1 - \gamma^2 \right) x}{\gamma v}.$$

Replacing $x'$, $y'$, $z'$ and $t'$ in the spherical wavefront equation in the $O'$ frame,

$$x'^2 + y'^2 + z'^2 = c^2 t'^2,$$

with their expressions in terms of $x$, $y$, $z$ and $t$ produces:

$$\gamma^2 \left( x - vt \right)^2 + y^2 + z^2 = c^2 \left[ \gamma t + \frac{\left( 1 - \gamma^2 \right) x}{\gamma v} \right]^2$$

and therefore,

$$\gamma^2 x^2 + \gamma^2 v^2 t^2 - 2\gamma^2 vtx + y^2 + z^2 = c^2 \gamma^2 t^2 + \frac{\left( 1 - \gamma^2 \right)^2 c^2 x}{\gamma^2 v^2}$$

$$+ 2 \frac{\left( 1 - \gamma^2 \right) txc^2}{v}$$

which implies,

$$\left[ \gamma^2 - \frac{\left( 1 - \gamma^2 \right)^2 c^2}{\gamma^2 v^2} \right] x^2 - 2\gamma^2 vtx + y^2 + z^2 = \left( c^2 \gamma^2 \right)$$

$$- v^2 \gamma^2 \right) t^2 + 2 \frac{\left[ 1 - \gamma^2 \right] txc^2}{v}$$

or

$$\left[ \gamma^2 - \frac{\left( 1 - \gamma^2 \right)^2 c^2}{\gamma^2 v^2} \right] x^2 - \left[ 2\gamma^2 v + 2 \frac{\left( 1 - \gamma^2 \right) c^2}{v} \right] tx$$

$$+ y2 + z2 = \left[ c^2 \gamma^2 - v^2 \gamma^2 \right] t^2$$

Comparing the coefficient of $t^2$ in the above equation with the coefficient of $t^2$ in the spherical wavefront equation for frame $O$ produces:

$$c^2 \gamma^2 - v^2 \gamma^2 = c^2$$

Equivalent expressions for $\gamma$ can be obtained by matching the $x^2$ coefficients or setting the $tx$ coefficient to zero. Rearranging:

$$\gamma^2 = \frac{1}{1 - \frac{v^2}{c^2}}$$

or, choosing the positive root to ensure that the x and x' axes and the time axes point in the same direction,

$$\gamma = \frac{1}{\sqrt{1 - \frac{v^2}{c^2}}}$$

which is called the Lorentz factor. This produces the Lorentz transformation from the above expression. It is given by

$$x' = \gamma \left( x - vt \right)$$
$$t' = \gamma \left( t - \frac{vx}{c^2} \right)$$
$$y' = y$$
$$z' = z$$

The Lorentz transformation is not the only transformation leaving invariant the shape of spherical waves, as there is a wider set of spherical wave transformations in the context of conformal geometry, leaving invariant the expression $\lambda \left( \delta x^2 + \delta y^2 + \delta z^2 - c^2 \delta t^2 \right)$ . However, scale changing conformal transformations cannot be used to symmetrically describe all laws of nature including mechanics, whereas the Lorentz transformations (the only one implying $\lambda = 1$) represent a symmetry of all laws of nature and reduce to Galilean transformations at $v \ll c$ .

## 25.7.2    Galilean and Einstein's relativity

### Galilean reference frames

In classical kinematics, the total displacement $x$ in the R frame is the sum of the relative displacement $x'$ in frame R$'$ and of the distance between the two origins $x - x'$. If $v$ is the relative velocity of R$'$ relative to R, the transformation is: $x = x' + vt$, or $x' = x - vt$. This relationship is linear for a constant $v$, that is when $R$ and $R'$ are Galilean frames of reference.

In Einstein's relativity, the main difference from Galilean relativity is that space and time coordinates are intertwined, and in different inertial frames $t \neq t'$.

Since space is assumed to be homogeneous, the transformation must be linear. The most general linear relationship is obtained with four constant coefficients, $A$, $B$, $\gamma$, and $b$:

$$x' = \gamma x + bt$$

$$t' = Ax + Bt.$$

The Lorentz transformation becomes the Galilean transformation when $\gamma = B = 1$, $b = -v$ and $A = 0$.

An object at rest in the $R'$ frame at position $x' = 0$ moves with constant velocity $v$ in the R frame. Hence the transformation must yield $x' = 0$ if $x = vt$. Therefore, $b = -\gamma v$ and the first equation is written as

$$x' = \gamma(x - vt).$$

**Principle of relativity**

According to the principle of relativity, there is no privileged Galilean frame of reference: therefore the inverse transformation for the position from frame $R'$ to frame $R$ should have the same form as the original but with the velocity in the opposite direction, i.o.w. replacing $v$ with $-v$:

$$x = \gamma(x' - (-v)t'),$$

and thus

$$x = \gamma(x' + vt').$$

**The speed of light is constant**

Since the speed of light is the same in all frames of reference, for the case of a light signal, the transformation must guarantee that $t = x/c$ when $t' = x'/c$.

Substituting for $t$ and $t'$ in the preceding equations gives:

$$x' = \gamma(1 - v/c)x,$$

$$x = \gamma(1 + v/c)x'.$$

Multiplying these two equations together gives,

$$xx' = \gamma^2 \left(1 - v^2/c^2\right)xx'.$$

At any time after $t = t' = 0$, $xx'$ is not zero, so dividing both sides of the equation by $xx'$ results in

$$\gamma = \frac{1}{\sqrt{1 - \frac{v^2}{c^2}}},$$

which is called the "Lorentz factor".

When the transformation equations are required to satisfy the light signal equations in the form $x = ct$ and $x' = ct'$, by substituting the x and x'-values, the same technique produces the same expression for the Lorentz factor.[10][11]

**Transformation of time**

The transformation equation for time can be easily obtained by considering the special case of a light signal, satisfying

$$\begin{cases} x' = ct' \\ x = ct. \end{cases}$$

Substituting term by term into the earlier obtained equation for the spatial coordinate

$$x' = \gamma(x - vt),$$

gives

$$ct' = \gamma \left(ct - \frac{v}{c}x\right),$$

so that

$$t' = \gamma \left(t - \frac{v}{c^2}x\right),$$

which determines the transformation coefficients $A$ and $B$ as

$$A = -\gamma v/c^2,$$

$$B = \gamma.$$

So $A$ and $B$ are the unique coefficients necessary to preserve the constancy of the speed of light in the primed system of coordinates.

### 25.7.3 Einstein's popular derivation

In his popular book[12] Einstein derived the Lorentz transformation by arguing that there must be two non-zero coupling constants $\lambda$ and $\mu$ such that

$$\begin{cases} x' - ct' = \lambda(x - ct) \\ x' + ct' = \mu(x + ct) \end{cases}$$

that correspond to light traveling along the positive and negative x-axis, respectively. For light $x = ct$ if and only if $x' = ct'$. Adding and subtracting the two equations and defining

$$\begin{cases} \gamma = (\lambda + \mu)/2 \\ b = (\lambda - \mu)/2, \end{cases}$$

gives

$$\begin{cases} x' = \gamma x - bct \\ ct' = \gamma ct - bx. \end{cases}$$

Substituting $x' = 0$ corresponding to $x = vt$ and noting that the relative velocity is $v = bc/\gamma$, this gives

$$\begin{cases} x' = \gamma(x - vt) \\ t' = \gamma\left(t - \frac{v}{c^2}x\right) \end{cases}$$

The constant $\gamma$ can be evaluated by demanding $c^2 t^2 - x^2 = c^2 t'^2 - x'^2$ as per standard configuration.

### 25.7.4  Hyperbolic rotation

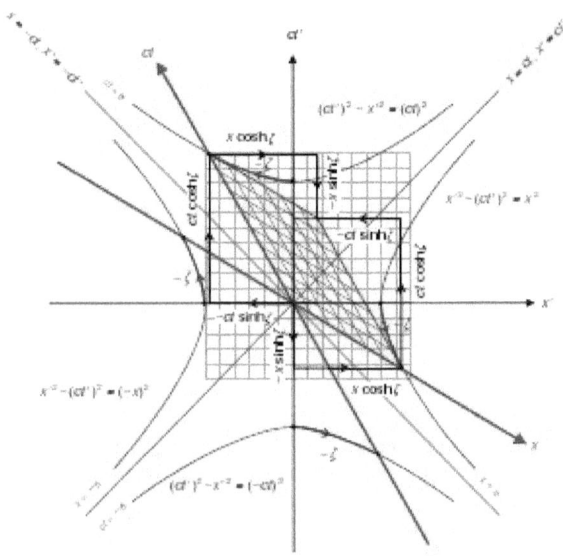

This diagram actually shows the inverse configuration of F′ "stationary" while F is boosted away along the negative x′ direction, although it correctly gives the original transformation since the coordinates ct, x of F are projected onto the coordinates ct′, x′ of F′. Notice the difference in length and time scales, such that the speed of light is invariant.

The Lorentz transformations can also be derived by simple application of the special relativity postulates and using hyperbolic identities.[13]

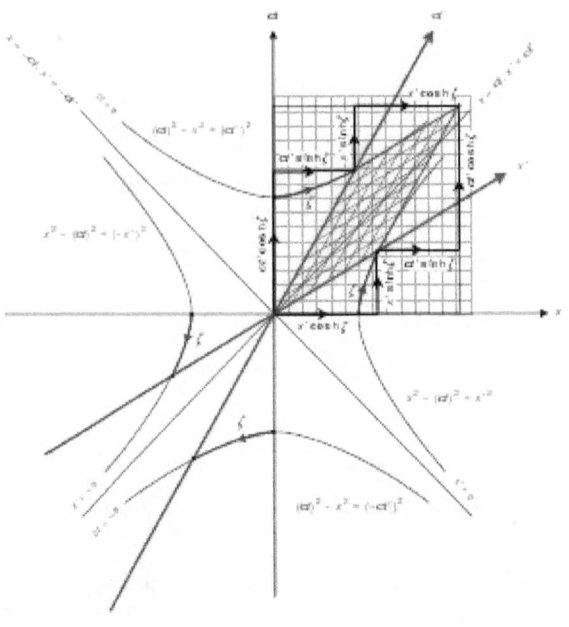

This diagram shows the original configuration of F "stationary" while F′ is boosted away along the positive x direction, although it correctly gives the inverse transformation since the coordinates ct′, x′ of F′ are projected onto the coordinates ct, x of F. Again, the difference in length and time scales is such that the speed of light is invariant.

**Relativity postulates**

Start from the equations of the spherical wave front of a light pulse, centred at the origin:

$$(ct)^2 - (x^2 + y^2 + z^2) = (ct')^2 - (x'^2 + y'^2 + z'^2) = 0$$

which take the same form in both frames because of the special relativity postulates. Next, consider relative motion along the x-axes of each frame, in standard configuration above, so that $y = y'$, $z = z'$, which simplifies to

$$(ct)^2 - x^2 = (ct')^2 - x'^2$$

**Linearity**

Now assume that the transformations take the linear form:

$$x' = Ax + Bct$$
$$ct' = Cx + Dct$$

where $A$, $B$, $C$, $D$ are to be found. If they were non-linear, they would not take the same form for all observers, since fictitious forces (hence accelerations) would occur in one frame even if the velocity was constant in another, which is inconsistent with inertial frame transformations.[14]

Substituting into the previous result:

$$(ct)^2 - x^2 = [(Cx)^2 + (Dct)^2 + 2CDcxt] - [(Ax)^2 + (Bct)^2$$
$$+ 2ABcxt]$$

and comparing coefficients of $x^2$, $t^2$, $xt$:

$$-1 = C^2 - A^2 \Rightarrow \quad A^2 - C^2 = 1$$
$$c^2 = (Dc)^2 - (Bc)^2 \Rightarrow \quad D^2 - B^2 = 1$$
$$2CDc - 2ABc = 0 \Rightarrow \quad AB = CD$$

**Hyperbolic rotation**

The formulae resemble the hyperbolic identity

$$\cosh^2 \phi - \sinh^2 \phi = 1$$

Introducing the rapidity parameter $\phi$ as a parametric hyperbolic angle allows the self-consistent identifications

$$A = D = \cosh \phi, \quad C = B = -\sinh \phi$$

where the signs after the square roots are chosen so that $x$ and $t$ increase. The hyperbolic transformations have been solved for:

$$x' = \cosh \phi x - \sinh \phi ct$$
$$ct' = -\sinh \phi x + \cosh \phi ct$$

If the signs were chosen differently the position and time coordinates would need to be replaced by $-x$ and/or $-t$ so that $x$ and $t$ increase not decrease.

To find how $\phi$ relates to the relative velocity, from the standard configuration the origin of the primed frame $x' = 0$ is measured in the unprimed frame to be $x = vt$ (or the equivalent and opposite way round; the origin of the unprimed frame is $x = 0$ and in the primed frame it is at $x' = -vt$):

$$0 = \cosh \phi vt - \sinh \phi ct \Rightarrow \tanh \phi = \frac{v}{c} = \beta$$

and manipulation of hyperbolic identities leads to the relations between $\beta$, $\gamma$, and $\phi$,

$$\cosh \phi = \gamma, \quad \sinh \phi = \beta\gamma.$$

## 25.8  From group postulates

Following is a classical derivation (see, e.g., and references therein) based on group postulates and isotropy of the space.

**Coordinate transformations as a group**

The coordinate transformations between inertial frames form a group (called the proper Lorentz group) with the group operation being the composition of transformations (performing one transformation after another). Indeed, the four group axioms are satisfied:

1. Closure: the composition of two transformations is a transformation: consider a composition of transformations from the inertial frame $K$ to inertial frame $K'$, (denoted as $K \to K'$), and then from $K'$ to inertial frame $K''$, $[K' \to K'']$, there exists a transformation, $[K \to K']\,[K' \to K'']$, directly from an inertial frame $K$ to inertial frame $K''$.

2. Associativity: the transformations ( $[K \to K']\,[K' \to K'']$ ) $[K'' \to K''']$ and $[K \to K']$ ( $[K' \to K'']\,[K'' \to K''']$ ) are identical.

3. Identity element: there is an identity element, a transformation $K \to K$.

4. Inverse element: for any transformation $K \to K'$ there exists an inverse transformation $K' \to K$.

**Transformation matrices consistent with group axioms**

Let us consider two inertial frames, $K$ and $K'$, the latter moving with velocity $v$ with respect to the former. By rotations and shifts we can choose the $x$ and $x'$ axes along the relative velocity vector and also that the events $(t, x) = (0,0)$ and $(t', x') = (0,0)$ coincide. Since the velocity boost is along the $x$ (and $x'$) axes nothing happens to the perpendicular coordinates and we can just omit them for brevity. Now since the transformation we are looking after connects two inertial frames, it has to transform a linear motion in $(t, x)$ into a linear motion in $(t', x')$ coordinates. Therefore, it must be a linear transformation. The general form of a linear transformation is

$$\begin{bmatrix} t' \\ x' \end{bmatrix} = \begin{bmatrix} \gamma & \delta \\ \beta & \alpha \end{bmatrix} \begin{bmatrix} t \\ x \end{bmatrix},$$

where $\alpha$, $\beta$, $\gamma$ and $\delta$ are some yet unknown functions of the relative velocity $v$.

Let us now consider the motion of the origin of the frame $K'$. In the $K'$ frame it has coordinates $(t', x'=0)$, while in the $K$ frame it has coordinates $(t, x=vt)$. These two points are connected by the transformation

$$\begin{bmatrix} t' \\ 0 \end{bmatrix} = \begin{bmatrix} \gamma & \delta \\ \beta & \alpha \end{bmatrix} \begin{bmatrix} t \\ vt \end{bmatrix},$$

from which we get

$$\beta = -v\alpha$$

Analogously, considering the motion of the origin of the frame $K$, we get

$$\begin{bmatrix} t' \\ -vt' \end{bmatrix} = \begin{bmatrix} \gamma & \delta \\ \beta & \alpha \end{bmatrix} \begin{bmatrix} t \\ 0 \end{bmatrix},$$

from which we get

$$\beta = -v\gamma$$

Combining these two gives $\alpha = \gamma$ and the transformation matrix has simplified,

$$\begin{bmatrix} t' \\ x' \end{bmatrix} = \begin{bmatrix} \gamma & \delta \\ -v\gamma & \gamma \end{bmatrix} \begin{bmatrix} t \\ x \end{bmatrix}.$$

Now let us consider the group postulate *inverse element*. There are two ways we can go from the $K'$ coordinate system to the $K$ coordinate system. The first is to apply the inverse of the transform matrix to the $K'$ coordinates:

$$\begin{bmatrix} t \\ x \end{bmatrix} = \frac{1}{\gamma^2 + v\delta\gamma} \begin{bmatrix} \gamma & -\delta \\ v\gamma & \gamma \end{bmatrix} \begin{bmatrix} t' \\ x' \end{bmatrix}.$$

The second is, considering that the $K'$ coordinate system is moving at a velocity $v$ relative to the $K$ coordinate system, the $K$ coordinate system must be moving at a velocity $-v$ relative to the $K'$ coordinate system. Replacing $v$ with $-v$ in the transformation matrix gives:

$$\begin{bmatrix} t \\ x \end{bmatrix} = \begin{bmatrix} \gamma(-v) & \delta(-v) \\ v\gamma(-v) & \gamma(-v) \end{bmatrix} \begin{bmatrix} t' \\ x' \end{bmatrix},$$

Now the function $\gamma$ can not depend upon the direction of $v$ because it is apparently the factor which defines the relativistic contraction and time dilation. These two (in an isotropic world of ours) cannot depend upon the direction

of $v$. Thus, $\gamma(-v) = \gamma(v)$ and comparing the two matrices, we get

$$\gamma^2 + v\delta\gamma = 1.$$

According to the *closure* group postulate a composition of two coordinate transformations is also a coordinate transformation, thus the product of two of our matrices should also be a matrix of the same form. Transforming $K$ to $K'$ and from $K'$ to $K''$ gives the following transformation matrix to go from $K$ to $K''$:

$$\begin{bmatrix} t'' \\ x'' \end{bmatrix} = \begin{bmatrix} \gamma(v') & \delta(v') \\ -v'\gamma(v') & \gamma(v') \end{bmatrix} \begin{bmatrix} \gamma(v) & \delta(v) \\ -v\gamma(v) & \gamma(v) \end{bmatrix} \begin{bmatrix} t \\ x \end{bmatrix}$$

$$= \begin{bmatrix} \gamma(v')\gamma(v) - v\delta(v')\gamma(v) & \gamma(v')\delta(v) + \delta(v')\gamma(v) \\ -(v' + v)\gamma(v')\gamma(v) & -v'\gamma(v')\delta(v) + \gamma(v')\gamma(v) \end{bmatrix} \begin{bmatrix} t \\ x \end{bmatrix}.$$

In the original transform matrix, the main diagonal elements are both equal to $\gamma$, hence, for the combined transform matrix above to be of the same form as the original transform matrix, the main diagonal elements must also be equal. Equating these elements and rearranging gives:

$$\gamma(v')\gamma(v) - v\delta(v')\gamma(v) = -v'\gamma(v')\delta(v) + \gamma(v')\gamma(v)$$

$$v\delta(v')\gamma(v) = v'\gamma(v')\delta(v)$$

$$\frac{\delta(v)}{v\gamma(v)} = \frac{\delta(v')}{v'\gamma(v')}.$$

The denominator will be nonzero for nonzero $v$, because $\gamma(v)$ is always nonzero;

$$\gamma^2 + v\delta\gamma = 1$$

If $v=0$ we have the identity matrix which coincides with putting $v=0$ in the matrix we get at the end of this derivation for the other values of $v$, making the final matrix valid for all nonnegative $v$.

For the nonzero $v$, this combination of function must be a universal constant, one and the same for all inertial frames. Define this constant as $\delta(v)/v\ \gamma(v) = \kappa$, where $\kappa$ has the dimension of $1/v^2$. Solving

$$1 = \gamma^2 + v\delta\gamma = \gamma^2(1 + \kappa v^2)$$

we finally get

$$\gamma = 1/\sqrt{1 + \kappa v^2}$$

and thus the transformation matrix, consistent with the group axioms, is given by

$$\begin{bmatrix} t' \\ x' \end{bmatrix} = \frac{1}{\sqrt{1 + \kappa v^2}} \begin{bmatrix} 1 & \kappa v \\ -v & 1 \end{bmatrix} \begin{bmatrix} t \\ x \end{bmatrix}.$$

If $\kappa > 0$, then there would be transformations (with $\kappa v^2 \gg 1$) which transform time into a spatial coordinate and vice versa. We exclude this on physical grounds, because time can only run in the positive direction. Thus two types of transformation matrices are consistent with group postulates:

1. with the universal constant $\kappa = 0$, and

2. with $\kappa < 0$.

### Galilean transformations

If $\kappa = 0$ then we get the Galilean-Newtonian kinematics with the Galilean transformation,

$$\begin{bmatrix} t' \\ x' \end{bmatrix} = \begin{bmatrix} 1 & 0 \\ -v & 1 \end{bmatrix} \begin{bmatrix} t \\ x \end{bmatrix},$$

where time is absolute, $t'=t$, and the relative velocity $v$ of two inertial frames is not limited.

### Lorentz transformations

If $\kappa < 0$, then we set $c = \sqrt{-\kappa}$ which becomes the invariant speed, the speed of light in vacuum. This yields $\kappa = -1/c^2$ and thus we get special relativity with Lorentz transformation

$$\begin{bmatrix} t' \\ x' \end{bmatrix} = \frac{1}{\sqrt{1 - \frac{v^2}{c^2}}} \begin{bmatrix} 1 & \frac{-v}{c^2} \\ -v & 1 \end{bmatrix} \begin{bmatrix} t \\ x \end{bmatrix},$$

where the speed of light is a finite universal constant determining the highest possible relative velocity between inertial frames.

If $v \ll c$ the Galilean transformation is a good approximation to the Lorentz transformation.

Only experiment can answer the question which of the two possibilities, $\kappa = 0$ or $\kappa < 0$, is realised in our world. The experiments measuring the speed of light, first performed by a Danish physicist Ole Rømer, show that it is finite, and the Michelson–Morley experiment showed that it is an absolute speed, and thus that $\kappa < 0$.

## 25.9  Boost from generators

Using rapidity $\phi$ to parametrize the Lorentz transformation, the boost in the $x$ direction is

$$\begin{bmatrix} ct' \\ x' \\ y' \\ z' \end{bmatrix} = \begin{bmatrix} \cosh\phi & -\sinh\phi & 0 & 0 \\ -\sinh\phi & \cosh\phi & 0 & 0 \\ 0 & 0 & 1 & 0 \\ 0 & 0 & 0 & 1 \end{bmatrix} \begin{bmatrix} ct \\ x \\ y \\ z \end{bmatrix},$$

likewise for a boost in the y-direction

$$\begin{bmatrix} ct' \\ x' \\ y' \\ z' \end{bmatrix} = \begin{bmatrix} \cosh\phi & 0 & -\sinh\phi & 0 \\ 0 & 1 & 0 & 0 \\ -\sinh\phi & 0 & \cosh\phi & 0 \\ 0 & 0 & 0 & 1 \end{bmatrix} \begin{bmatrix} ct \\ x \\ y \\ z \end{bmatrix},$$

and the z-direction

$$\begin{bmatrix} ct' \\ x' \\ y' \\ z' \end{bmatrix} = \begin{bmatrix} \cosh\phi & 0 & 0 & -\sinh\phi \\ 0 & 1 & 0 & 0 \\ 0 & 0 & 1 & 0 \\ -\sinh\phi & 0 & 0 & \cosh\phi \end{bmatrix} \begin{bmatrix} ct \\ x \\ y \\ z \end{bmatrix}.$$

where $e_x$, $e_y$, $e_z$ are the Cartesian basis vectors, a set of mutually perpendicular unit vectors along their indicated directions. If one frame is boosted with velocity $\mathbf{v}$ relative to another, it is convenient to introduce a unit vector $\mathbf{n} = \mathbf{v}/v = \boldsymbol{\beta}/\beta$ in the direction of relative motion. The general boost is

$$\begin{bmatrix} ct' \\ x' \\ y' \\ z' \end{bmatrix} = \begin{bmatrix} \cosh\phi & -n_x\sinh\phi & -n_y\sinh\phi & -n_z\sinh\phi \\ -n_x\sinh\phi & 1+(\cosh\phi-1)n_x^2 & (\cosh\phi-1)n_xn_y & (\cosh\phi-1)n_xn_z \\ -n_y\sinh\phi & (\cosh\phi-1)n_yn_x & 1+(\cosh\phi-1)n_y^2 & (\cosh\phi-1)n_yn_z \\ -n_z\sinh\phi & (\cosh\phi-1)n_zn_x & (\cosh\phi-1)n_zn_y & 1+(\cosh\phi-1)n_z^2 \end{bmatrix} \begin{bmatrix} ct \\ x \\ y \\ z \end{bmatrix}.$$

Notice the matrix depends on the direction of the relative motion as well as the rapidity, in all three numbers (two for direction, one for rapidity).

We can cast each of the boost matrices in another form as follows. First consider the boost in the $x$ direction. The Taylor expansion of the boost matrix about $\phi = 0$ is

$$B(\mathbf{e}_x, \phi) = \sum_{n=0}^{\infty} \frac{\phi^n}{n!} \left. \frac{\partial^n B(\mathbf{e}_x, \phi)}{\partial\phi^n} \right|_{\phi=0}$$

where the derivatives of the matrix with respect to $\phi$ are given by differentiating each entry of the matrix separately, and the notation $|\phi = 0$ indicates $\phi$ is set to zero *after* the derivatives are evaluated. Expanding to first order gives the *infinitesimal* transformation

$$B(\mathbf{e}_x, \phi) = I + \phi \left. \frac{\partial B}{\partial \phi}\right|_{\phi=0} = \begin{bmatrix} 1 & 0 & 0 & 0 \\ 0 & 1 & 0 & 0 \\ 0 & 0 & 1 & 0 \\ 0 & 0 & 0 & 1 \end{bmatrix}$$

$$-\phi \begin{bmatrix} 0 & 1 & 0 & 0 \\ 1 & 0 & 0 & 0 \\ 0 & 0 & 0 & 0 \\ 0 & 0 & 0 & 0 \end{bmatrix}$$

which is valid if $\phi$ is small (hence $\phi^2$ and higher powers are negligible), and can be interpreted as no boost (the first term $I$ is the 4×4 identity matrix), followed by a small boost. The matrix

$$K_x = \begin{bmatrix} 0 & 1 & 0 & 0 \\ 1 & 0 & 0 & 0 \\ 0 & 0 & 0 & 0 \\ 0 & 0 & 0 & 0 \end{bmatrix}$$

is the *generator* of the boost in the $x$ direction, so the infinitesimal boost is

$$B(\mathbf{e}_x, \phi) = I - \phi K_x$$

Now, $\phi$ is small, so dividing by a positive integer $N$ gives an even smaller increment of rapidity $\phi/N$, and $N$ of these infinitesimal boosts will give the original infinitesimal boost with rapidity $\phi$,

$$B(\mathbf{e}_x, \phi) = \left( I - \frac{\phi K_x}{N} \right)^N$$

In the limit of an infinite number of infinitely small steps, we obtain the finite boost transformation

$$B(\mathbf{e}_x, \phi) = \lim_{N \to \infty} \left( I - \frac{\phi K_x}{N} \right)^N = e^{-\phi K_x}$$

which is the limit definition of the exponential due to Leonhard Euler, and is now true for any $\phi$.

Repeating the process for the boosts in the $y$ and $z$ directions obtains the other generators

$$K_y = \begin{bmatrix} 0 & 0 & 1 & 0 \\ 0 & 0 & 0 & 0 \\ 1 & 0 & 0 & 0 \\ 0 & 0 & 0 & 0 \end{bmatrix}, \quad K_z = \begin{bmatrix} 0 & 0 & 0 & 1 \\ 0 & 0 & 0 & 0 \\ 0 & 0 & 0 & 0 \\ 1 & 0 & 0 & 0 \end{bmatrix}$$

and the boosts are

$$B(\mathbf{e}_y, \phi) = e^{-\phi K_y}, \quad B(\mathbf{e}_z, \phi) = e^{-\phi K_z}.$$

For any direction, the infinitesimal transformation is (small $\phi$ and expansion to first order)

$$B(\mathbf{n}, \phi) = I + \phi \left. \frac{\partial B}{\partial \phi}\right|_{\phi=0} = \begin{bmatrix} 1 & 0 & 0 & 0 \\ 0 & 1 & 0 & 0 \\ 0 & 0 & 1 & 0 \\ 0 & 0 & 0 & 1 \end{bmatrix} - \phi \begin{bmatrix} 0 & n_x & n_y & n_z \\ n_x & 0 & 0 & 0 \\ n_y & 0 & 0 & 0 \\ n_z & 0 & 0 & 0 \end{bmatrix}$$

where

$$\begin{bmatrix} 0 & n_x & n_y & n_z \\ n_x & 0 & 0 & 0 \\ n_y & 0 & 0 & 0 \\ n_z & 0 & 0 & 0 \end{bmatrix} = n_x K_x + n_y K_y + n_z K_z = \mathbf{n} \cdot \mathbf{K}$$

is the generator of the boost in direction $\mathbf{n}$. It is the full boost generator, a vector of matrices $\mathbf{K} = (Kx, Ky, Kz)$, projected into the direction of the boost $\mathbf{n}$. The infinitesimal boost is

$$B(\mathbf{n}, \phi) = I - \phi(\mathbf{n} \cdot \mathbf{K})$$

Then in the limit of an infinite number of infinitely small steps, we obtain the finite boost transformation

$$B(\mathbf{n}, \phi) = \lim_{N \to \infty} \left( I - \frac{\phi(\mathbf{n} \cdot \mathbf{K})}{N} \right)^N = e^{-\phi(\mathbf{n} \cdot \mathbf{K})}$$

which is now true for any $\phi$. Expanding the matrix exponential of $-\phi(\mathbf{n} \cdot \mathbf{K})$ in its power series

$$e^{-\phi \mathbf{n} \cdot \mathbf{K}} = \sum_{n=0}^{\infty} \frac{1}{n!} (-\phi \mathbf{n} \cdot \mathbf{K})^n$$

we now need the powers of the generator. The square is

$$(\mathbf{n} \cdot \mathbf{K})^2 = \begin{bmatrix} 1 & 0 & 0 & 0 \\ 0 & n_x^2 & n_x n_y & n_x n_z \\ 0 & n_y n_x & n_y^2 & n_y n_z \\ 0 & n_z n_x & n_z n_y & n_z^2 \end{bmatrix}$$

but the cube $(\mathbf{n} \cdot \mathbf{K})^3$ returns to $(\mathbf{n} \cdot \mathbf{K})$, and as always the zeroth power is the 4×4 identity, $(\mathbf{n} \cdot \mathbf{K})^0 = I$. In general the odd powers $n = 1, 3, 5, \ldots$ are

$$(\mathbf{n} \cdot \mathbf{K})^n = (\mathbf{n} \cdot \mathbf{K})$$

while the even powers $n = 2, 4, 6, \ldots$ are

$$(\mathbf{n} \cdot \mathbf{K})^n = (\mathbf{n} \cdot \mathbf{K})^2$$

therefore the explicit form of the boost matrix depends only the generator and its square. Splitting the power series into an odd power series and an even power series, using the odd and even powers of the generator, and the Taylor series of $\sinh\phi$ and $\cosh\phi$ about $\phi = 0$ obtains a more compact but detailed form of the boost matrix

$$e^{-\phi\mathbf{n}\cdot\mathbf{K}} = -\sum_{n=1,3,5\ldots}^{\infty} \frac{1}{n!}\phi^n(\mathbf{n}\cdot\mathbf{K})^n + \sum_{n=0,2,4\ldots}^{\infty} \frac{1}{n!}\phi^n(\mathbf{n}\cdot\mathbf{K})^n$$

$$= -\left[\phi + \frac{\phi^3}{3!} + \frac{\phi^5}{5!} + \cdots\right]$$

$$(\mathbf{n}\cdot\mathbf{K}) + I + \left[-1 + 1 + \frac{1}{2!}\phi^2 + \frac{1}{4!}\phi^4 + \frac{1}{6!}\phi^6 + \cdots\right]$$

$$= -\sinh\phi(\mathbf{n}\cdot\mathbf{K}) + I + (-1 + \cosh\phi)(\mathbf{n}\cdot\mathbf{K})^2$$

where $0 = -1 + 1$ is introduced for the even power series to complete the Taylor series for $\cosh\phi$. The boost is similar to Rodrigues' rotation formula,

$$B(\mathbf{n}, \phi) = e^{-\phi\mathbf{n}\cdot\mathbf{K}} = I - \sinh\phi(\mathbf{n}\cdot\mathbf{K}) + (\cosh\phi - 1)(\mathbf{n}\cdot\mathbf{K})^2.$$

Negating the rapidity in the exponential gives the inverse transformation matrix,

$$B(\mathbf{n}, -\phi) = e^{\phi\mathbf{n}\cdot\mathbf{K}} = I + \sinh\phi(\mathbf{n}\cdot\mathbf{K}) + (\cosh\phi - 1)(\mathbf{n}\cdot\mathbf{K})^2.$$

In quantum mechanics, relativistic quantum mechanics, and quantum ▨d theory ▨▨ ▨rent convention ▨ u▨d for ▨ ▨ boost generators all ▨▨▨ boost generators are multiplied by a factor ▨▨▨ imaginary unit $i$ ▨▨$-1$.

## 25.< From experiments

Howard Percy Robertson and others showed that the Lorentz transformation can also be derived empirically.[15][16] In order to achieve this, it's necessary to write down coordinate transformations that include experimentally testable parameters. For instance, let there be given a single "preferred" inertial frame $X, Y, Z, T$ in which the speed of light is constant, isotropic, and independent of the velocity of the source. It is also assumed that Einstein synchronization and synchronization by slow clock transport are equivalent in this frame. Then

assume another frame $x, y, z, t$ in relative motion, in which clocks and rods have the same internal constitution as in the preferred frame. The following relations, however, are left undefined:

- $a(v)$ differences in time measurements,

- $b(v)$ differences in measured longitudinal lengths,

- $d(v)$ differences in measured transverse lengths,

- $\varepsilon(v)$ depends on the clock synchronization procedure in the moving frame,

then the transformation formulas (assumed to be linear) between those frames are given by: $(\mathbf{n}\cdot\mathbf{K})^2$

$$t = a(v)T + \varepsilon(v)x$$
$$x = b(v)(X - vT)$$
$$y = d(v)Y$$
$$z = d(v)Z$$

$\varepsilon(v)$ depends on the synchronization convention and is not determined experimentally, it obtains the value $-v/c^2$ by using Einstein synchronization in both frames. The ratio between $b(v)$ and $d(v)$ is determined by the Michelson–Morley experiment, the ratio between $a(v)$ and $b(v)$ is determined by the Kennedy–Thorndike experiment, and $a(v)$ alone is determined by the Ives–Stilwell experiment. In this way, they have been determined with great precision to $1/a(v) = b(v) = \gamma$ and $d(v) = 1$, which converts the above transformation into the Lorentz transformation.

## 25.11 See also

- Gyrovector space

- Lorentz group

- Noether's theorem

- Poincaré group

- Proper time

- Relativistic metric

- Spinor

## 25.12   Notes

[1] "Newton's Philosophy". *stanford.edu*.

[2] Zeeman, Erik Christopher (1964), "Causality implies the Lorentz group", *Journal of Mathematical Physics* **5** (4): 490–493, Bibcode:1964JMP.....5..490Z, doi:10.1063/1.1704140

[3] Goldstein, Norman (2007). "Inertiality Implies the Lorentz Group" (PDF). *Mathematical Physics Electronic Journal* **13**. ISSN 1086-6655. Retrieved 14 February 2016.

[4] (Landau & Lifshitz 2002)

[5] University Physics – With Modern Physics (12th Edition), H.D. Young, R.A. Freedman (Original edition), Addison-Wesley (Pearson International), 1st Edition: 1949, 12th Edition: 2008, ISBN 978-0-321-50130-1

[6] Greiner & Bromley 2000, Chapter 16

[7] Weinberg 2002, Footnote p. 56

[8] Einstein, Albert (1916). "Relativity: The Special and General Theory" (PDF). Retrieved 2012-01-23.

[9] Stauffer, Dietrich; Stanley, Harry Eugene (1995). *From Newton to Mandelbrot: A Primer in Theoretical Physics* (2nd enlarged ed.). Springer-Verlag. p. 80,81. ISBN 978-3-540-59191-7.

[10] Born, Max (2012). *Einstein's Theory of Relativity* (revised ed.). Courier Dover Publications. pp. 236–237. ISBN 0-486-14212-4. Extract of page 237

[11] Gupta, S. K. (2010). *Engineering Physics: Vol. 1* (18th ed.). Krishna Prakashan Media. pp. 12–13. ISBN 81-8283-098-2. Extract of page 12

[12] Einstein, Albert (1916). "Relativity: The Special and General Theory" (PDF). Retrieved 2012-01-23.

[13] Relativity DeMystified, D. McMahon, Mc Graw Hill (USA), 2006, ISBN 0-07-145545-0

[14] An Introduction to Mechanics, D. Kleppner, R.J. Kolenkow, Cambridge University Press, 2010, ISBN 978-0-521-19821-9

[15] Robertson, H. P. (1949). "Postulate versus Observation in the Special Theory of Relativity". *Reviews of Modern Physics* **21** (3): 378–382. Bibcode:1949RvMP...21..378R. doi:10.1103/RevModPhys.21.378.

[16] Mansouri R., Sexl R.U. (1977). "A test theory of special relativity. I: Simultaneity and clock synchronization". *Gen. Rel. Gravit.* **8** (7): 497–513. Bibcode:1977GReGr...8..497M. doi:10.1007/BF00762634.

## 25.13   References

• Greiner, W.; Bromley, D. A. (2000). *Relativistic Quantum Mechanics* (3rd ed.). springer. ISBN 9783540674573.

• Landau, L.D.; Lifshitz, E.M. (2002) [1939]. *The Classical Theory of Fields*. Course of Theoretical Physics **2** (4th ed.). Butterworth–Heinemann. ISBN 0 7506 2768 9.

• Weinberg, S. (2002), *The Quantum Theory of Fields* **1**, Cambridge University Press, ISBN 0-521-55001-7

# Chapter 26

# Causal structure

This article is about the possible causal relationships among points in a Lorentzian manifold. For classification of Lorentzian manifolds according to the types of causal structures they admit, see Causality conditions.

In mathematical physics, the **causal structure** of a Lorentzian manifold describes the causal relationships between points in the manifold.

## 26.1 Introduction

In modern physics (especially general relativity) spacetime is represented by a Lorentzian manifold. The causal relations between points in the manifold are interpreted as describing which events in spacetime can influence which other events.

Minkowski spacetime is a simple example of a Lorentzian manifold. The causal relationships between points in Minkowski spacetime take a particularly simple form since the space is flat. See Causal structure of Minkowski spacetime for more information.

The causal structure of an arbitrary (possibly curved) Lorentzian manifold is made more complicated by the presence of curvature. Discussions of the causal structure for such manifolds must be phrased in terms of smooth curves joining pairs of points. Conditions on the tangent vectors of the curves then define the causal relationships.

### 26.1.1 Tangent vectors

If $(M, g)$ is a Lorentzian manifold (for metric $g$ on manifold $M$) then the tangent vectors at each point in the manifold can be classed into three different types. A tangent vector $X$ is

- **timelike** if $g(X, X) > 0$
- **null** or **lightlike** if $g(X, X) = 0$

- **spacelike** if $g(X, X) < 0$

(Here we use the $(+, -, -, -, \cdots)$ metric signature). A tangent vector is called "non-spacelike" if it is null or timelike.

These names come from the simpler case of Minkowski spacetime (see Causal structure of Minkowski spacetime).

### 26.1.2 Time-orientability

At each point in $M$ the timelike tangent vectors in the point's tangent space can be divided into two classes. To do this we first define an equivalence relation on pairs of timelike tangent vectors.

If $X$ and $Y$ are two timelike tangent vectors at a point we say that $X$ and $Y$ are equivalent (written $X \sim Y$) if $g(X, Y) > 0$.

There are then two equivalence classes which between them contain all timelike tangent vectors at the point. We can (arbitrarily) call one of these equivalence classes "future-directed" and call the other "past-directed". Physically this designation of the two classes of future- and past-directed timelike vectors corresponds to a choice of an arrow of time at the point. The future- and past-directed designations can be extended to null vectors at a point by continuity.

A Lorentzian manifold is **time-orientable**[1] if a continuous designation of future-directed and past-directed for non-spacelike vectors can be made over the entire manifold.

### 26.1.3 Curves

A **path** in $M$ is a continuous map $\mu : \Sigma \to M$ where $\Sigma$ is a nondegenerate interval (i.e., a connected set containing more than one point) in $\mathbb{R}$. A **smooth** path has $\mu$ differentiable an appropriate number of times (typically $C^\infty$), and a **regular** path has nonvanishing derivative.

A **curve** in $M$ is the image of a path or, more properly, an equivalence class of path-images related by reparametrisation, i.e. homeomorphisms or diffeomorphisms of $\Sigma$. When $M$ is time-orientable, the curve is **oriented** if the parameter change is required to be monotonic.

Smooth regular curves (or paths) in $M$ can be classified depending on their tangent vectors. Such a curve is

- **chronological** (or **timelike**) if the tangent vector is timelike at all points in the curve.

- **null** if the tangent vector is null at all points in the curve.

- **spacelike** if the tangent vector is spacelike at all points in the curve.

- **causal** (or **non-spacelike**) if the tangent vector is timelike *or* null at all points in the curve.

The requirements of regularity and nondegeneracy of $\Sigma$ ensure that closed causal curves (such as those consisting of a single point) are not automatically admitted by all spacetimes.

If the manifold is time-orientable then the non-spacelike curves can further be classified depending on their orientation with respect to time.

A chronological, null or causal curve in $M$ is

- **future-directed** if, for every point in the curve, the tangent vector is future-directed.

- **past-directed** if, for every point in the curve, the tangent vector is past-directed.

These definitions only apply to causal (chronological or null) curves because only timelike or null tangent vectors can be assigned an orientation with respect to time.

- A **closed timelike curve** is a closed curve which is everywhere future-directed timelike (or everywhere past-directed timelike).

- A **closed null curve** is a closed curve which is everywhere future-directed null (or everywhere past-directed null).

- The holonomy of the ratio of the rate of change of the affine parameter around a closed null geodesic is the **redshift factor**.

## 26.1.4   Causal relations

There are two types of causal relations between points $x$ and $y$ in the manifold $M$.

- $x$ **chronologically precedes** $y$ (often denoted $x \ll y$) if there exists a future-directed chronological (timelike) curve from $x$ to $y$.

- $x$ **strictly causally precedes** $y$ (often denoted $x < y$) if there exists a future-directed causal (non-spacelike) curve from $x$ to $y$.

- $x$ **causally precedes** $y$ (often denoted $x \prec y$ or $x \leq y$) if $x$ strictly causally precedes $y$ or $x = y$.

- $x$ **horismos** $y$ [2] (often denoted $x \to y$ or $x \nearrow y$) if $x \prec y$ and $x \not\ll y$.

These relations are transitive:[3]

- $x \ll y$, $y \ll z$ implies $x \ll z$

- $x \prec y$, $y \prec z$ implies $x \prec z$

and satisfy[3]

- $x \ll y$ implies $x \prec y$ (this follows trivially from the definition)

- $x \ll y$, $y \prec z$ implies $x \ll z$

- $x \prec y$, $y \ll z$ implies $x \ll z$

For a point $x$ in the manifold $M$ we define[3]

- The **chronological future** of $x$, denoted $I^+(x)$, as the set of all points $y$ in $M$ such that $x$ chronologically precedes $y$:

$$I^+(x) = \{y \in M \,|\, x \ll y\}$$

- The **chronological past** of $x$, denoted $I^-(x)$, as the set of all points $y$ in $M$ such that $y$ chronologically precedes $x$:

$$I^-(x) = \{y \in M \,|\, y \ll x\}$$

We similarly define

- The **causal future** (also called the **absolute future**) of $x$, denoted $J^+(x)$, as the set of all points $y$ in $M$ such that $x$ causally precedes $y$:

$$J^+(x) = \{y \in M \,|\, x \prec y\}$$

- The **causal past** (also called the **absolute past**) of $x$, denoted $J^-(x)$, as the set of all points $y$ in $M$ such that $y$ causally precedes $x$:

$$J^-(x) = \{y \in M | y \prec x\}$$

Points contained in $I^+(x)$, for example, can be reached from $x$ by a future-directed timelike curve. The point $x$ can be reached, for example, from points contained in $J^-(x)$ by a future-directed non-spacelike curve.

As a simple example, in Minkowski spacetime the set $I^+(x)$ is the interior of the future light cone at $x$. The set $J^+(x)$ is the full future light cone at $x$, including the cone itself.

These sets $I^+(x), I^-(x), J^+(x), J^-(x)$ defined for all $x$ in $M$, are collectively called the **causal structure** of $M$.

For $S$ a subset of $M$ we define[3]

$$I^\pm(S) = \bigcup_{x \in S} I^\pm(x)$$

$$J^\pm(S) = \bigcup_{x \in S} J^\pm(x)$$

For $S, T$ two subsets of $M$ we define

- The **chronological future of** $S$ **relative to** $T$, $I^+(S; T)$, is the chronological future of $S$ considered as a submanifold of $T$. Note that this is quite a different concept from $I^+(S) \cap T$ which gives the set of points in $T$ which can be reached by future-directed timelike curves starting from $S$. In the first case the curves must lie in $T$ in the second case they do not. See Hawking and Ellis.

- The **causal future of** $S$ **relative to** $T$, $J^+(S; T)$, is the causal future of $S$ considered as a submanifold of $T$. Note that this is quite a different concept from $J^+(S) \cap T$ which gives the set of points in $T$ which can be reached by future-directed causal curves starting from $S$. In the first case the curves must lie in $T$ in the second case they do not. See Hawking and Ellis.

- A **future set** is a set closed under chronological future.

- A **past set** is a set closed under chronological past.

- An **indecomposable past set** is a past set which isn't the union of two different open past proper subsets.

- $I^-(x)$ is a **proper indecomposable past set** (PIP).

- A **terminal indecomposable past set** (TIP) is an IP which isn't a PIP.

- The future **Cauchy development** of $S$, $D^+(S)$ is the set of all points $x$ for which every past directed inextendible causal curve through $x$ intersects $S$ at least once. Similarly for the past Cauchy development. The

Cauchy development is the union of the future and past Cauchy developments. Cauchy developments are important for the study of determinism.

- A subset $S \subset M$ is **achronal** if there do not exist $q, r \in S$ such that $r \in I^+(q)$, or equivalently, if $S$ is disjoint from $I^+(S)$.

- A **Cauchy surface** is an closed achronal set whose Cauchy development is $M$.

- A metric is **globally hyperbolic** if it can be foliated by Cauchy surfaces.

- The **chronology violating set** is the set of points through which closed timelike curves pass.

- The **causality violating set** is the set of points through which closed causal curves pass.

- For a causal curve $\gamma$, the **causal diamond** is $J^+(\gamma) \cap J^-(\gamma)$ (here we are using the looser definition of 'curve' whereon it is just a set of points). In words: the causal diamond of a particle's world-line $\gamma$ is the set of all events that lie in both the past of some point in $\gamma$ and the future of some point in $\gamma$.

### 26.1.5 Properties

See Penrose, p13.

- A point $x$ is in $I^-(y)$ if and only if $y$ is in $I^+(x)$.

- $x \prec y \implies I^-(x) \subset I^-(y)$

- $x \prec y \implies I^+(y) \subset I^+(x)$

- $I^+[S] = I^+[I^+[S]] \subset J^+[S] = J^+[J^+[S]]$

- $I^-[S] = I^-[I^-[S]] \subset J^-[S] = J^-[J^-[S]]$

- The horismos is generated by null geodesic congruences.

Topological properties:

- $I^\pm(x)$ is open for all points $x$ in $M$.

- $I^\pm[S]$ is open for all subsets $S \subset M$.

- $I^\pm[S] = I^\pm[\overline{S}]$ for all subsets $S \subset M$. Here $\overline{S}$ is the closure of a subset $S$.

- $J^\pm[S] \subset \overline{I^\pm[S]}$

## 26.2   Conformal geometry

Two metrics $g$ and $\hat{g}$ are **conformally related**[4] if $\hat{g} = \Omega^2 g$ for some real function $\Omega$ called the **conformal factor**. (See conformal map).

Looking at the definitions of which tangent vectors are timelike, null and spacelike we see they remain unchanged if we use $g$ or $\hat{g}$. As an example suppose $X$ is a timelike tangent vector with respect to the $g$ metric. This means that $g(X, X) > 0$ . We then have that $\hat{g}(X, X) = \Omega^2 g(X, X) > 0$ so $X$ is a timelike tangent vector with respect to the $\hat{g}$ too.

It follows from this that the causal structure of a Lorentzian manifold is unaffected by a conformal transformation.

## 26.3   See also

- Causal dynamical triangulation (CDT)
- Causality conditions
- Cauchy surface
- Closed timelike curve
- Globally hyperbolic manifold
- Lorentzian manifold
- Penrose diagram
- Spacetime

## 26.4   Notes

[1] Hawking & Israel 1979, p. 255

[2] Penrose 1972, p. 15

[3] Penrose 1972, p. 12

[4] Hawking & Ellis 1973, p. 42

## 26.5   References

- Hawking, S.W.; Ellis, G.F.R. (1973), *The Large Scale Structure of Space-Time*, Cambridge: Cambridge University Press, ISBN 0-521-20016-4
- Hawking, S.W.; Israel, W. (1979), *General Relativity, an Einstein Centenary Survey*, Cambridge University Press, ISBN 0-521-22285-0
- Penrose, R. (1972), *Techniques of Differential Topology in Relativity*, SIAM, ISBN 0898710057

## 26.6   Further reading

- G. W. Gibbons, S. N. Solodukhin; *The Geometry of Small Causal Diamonds* arXiv:hep-th/0703098 (Causal intervals)
- S.W. Hawking, A.R. King, P.J. McCarthy; *A new topology for curved space–time which incorporates the causal, differential, and conformal structures*; J. Math. Phys. 17 2:174-181 (1976); (Geometry, Causal Structure)
- A.V. Levichev; *Prescribing the conformal geometry of a lorentz manifold by means of its causal structure*; Soviet Math. Dokl. 35:452-455, (1987); (Geometry, Causal Structure)
- D. Malament; *The class of continuous timelike curves determines the topology of spacetime*; J. Math. Phys. 18 7:1399-1404 (1977); (Geometry, Causal Structure)
- A.A. Robb ; *A theory of time and space*; Cambridge University Press, 1914; (Geometry, Causal Structure)
- A.A. Robb ; *The absolute relations of time and space*; Cambridge University Press, 1921; (Geometry, Causal Structure)
- A.A. Robb ; *Geometry of Time and Space*; Cambridge University Press, 1936; (Geometry, Causal Structure)
- R.D. Sorkin, E. Woolgar; *A Causal Order for Spacetimes with C^0 Lorentzian Metrics: Proof of Compactness of the Space of Causal Curves*; Classical & Quantum Gravity 13: 1971-1994 (1996); arXiv: gr-qc/9508018 (Causal Structure)

## 26.7   External links

- Turing Machine Causal Networks by Enrique Zeleny, the Wolfram Demonstrations Project
- Weisstein, Eric W., "Causal Network", *MathWorld*.

# Chapter 27

# Light cone

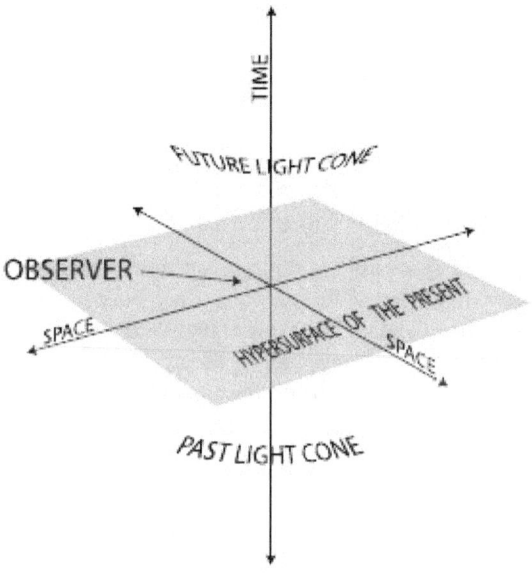

*Light cone in 2D space plus a time dimension.*

In special and general relativity, a **light cone** is the path that a flash of light, emanating from a single event (localized to a single point in space and a single moment in time) and traveling in all directions, would take through spacetime. If one imagines the light confined to a two-dimensional plane, the light from the flash spreads out in a circle after the event E occurs, and if we graph the growing circle with the vertical axis of the graph representing time, the result is a cone, known as the future light cone. The past light cone behaves like the future light cone in reverse, a circle which contracts in radius at the speed of light until it converges to a point at the exact position and time of the event E. In reality, there are three space dimensions, so the light would actually form an expanding or contracting sphere in three-dimensional (3D) space rather than a circle in 2D, and the light cone would actually be a four-dimensional version of a cone whose cross-sections form 3D spheres (analogous to a normal three-dimensional cone whose cross-sections form 2D circles), but the concept is easier to visualize with the

number of spatial dimensions reduced from three to two.

Because signals and other causal influences cannot travel faster than light (see special relativity and quantum entanglement), the light cone plays an essential role in defining the concept of causality: for a given event E, the set of events that lie on or inside the past light cone of E would also be the set of all events that could send a signal that would have time to reach E and influence it in some way. For example, at a time ten years before E, if we consider the set of all events in the past light cone of E which occur at that time, the result would be a sphere (2D: disk) with a radius of ten light-years centered on the future position E will occur. So, any point on or inside the sphere could send a signal moving at the speed of light or slower that would have time to influence the event E, while points outside the sphere at that moment would not be able to have any causal influence on E. Likewise, the set of events that lie on or inside the *future* light cone of E would also be the set of events that could receive a signal sent out from the position and time of E, so the future light cone contains all the events that could potentially be causally influenced by E. Events which lie neither in the past or future light cone of E cannot influence or be influenced by E in relativity.

## 27.1 Mathematical construction

In special relativity, a **light cone** (or **null cone**) is the surface describing the temporal evolution of a flash of light in Minkowski spacetime. This can be visualized in 3-space if the two horizontal axes are chosen to be spatial dimensions, while the vertical axis is time.[1]

The light cone is constructed as follows. Taking as event $p$ a flash of light (light pulse) at time $t_0$, all events that can be reached by this pulse from $p$ form the **future light cone** of $p$, while those events that can send a light pulse to $p$ form the **past light cone** of $p$.

Given an event $E$, the light cone classifies all events in space+time into 5 distinct categories:

- Events *on the future light cone* of $E$ .

- Events *on the past light cone* of $E$ .

- Events *inside the future light cone* of $E$ are those affected by a material particle emitted at $E$ .

- Events *inside the past light cone* of $E$ are those that can emit a material particle and affect what is happening at $E$ .

- All other events are in the *(absolute) elsewhere* of $E$ and are those that cannot affect or be affected by $E$ .

The above classifications hold true in any frame of reference; that is, an event judged to be in the light cone by one observer, will also be judged to be in the same light cone by all other observers, no matter their frame of reference. This is why the concept is so powerful.

The above refers to an event occurring at a specific location and at a specific time. To say that one event cannot affect another means that light cannot get from the location of one to the other *in a given amount of time*. Light from each event will ultimately make it to the *former* location of the other, but *after* those events have occurred.

As time progresses, the future light cone of a given event will eventually grow to encompass more and more locations (in other words, the 3D sphere that represents the cross-section of the 4D light cone at a particular moment in time becomes larger at later times). Likewise, if we imagine running time backwards from a given event, the event's past light cone would likewise encompass more and more locations at earlier and earlier times. The farther locations will of course be at later times: for example, if we are considering the past light cone of an event which takes place on Earth today, a star 10,000 light years away would only be inside the past light cone at times 10,000 years or more in the past. The past light cone of an event on present-day Earth, at its very edges, includes very distant objects (every object in the observable universe), but only as they looked long ago, when the Universe was young.

Two events at different locations, at the same time (according to a specific frame of reference), are always outside of each other's past and future light cones; light cannot travel instantaneously. Other observers, of course, might see the events happening at different times and at different locations, but one way or another, the two events will likewise be seen to be outside of each other's cones.

If using a system of units where the speed of light in vacuum is defined as exactly 1, for example if space is measured in light-seconds and time is measured in seconds, then the cone will have a slope of 45°, because light travels a distance of one light-second in vacuum during one second. Since special relativity requires the speed of light to be equal in

every inertial frame, all observers must arrive at the same angle of 45° for their light cones. Commonly a Minkowski diagram is used to illustrate this property of Lorentz transformations. Elsewhere, an integral part of light cones is the region of spacetime outside the light cone at a given event (a point in spacetime). Events that are elsewhere from each other are mutually unobservable, and cannot be causally connected.

(The 45° figure really only has meaning in space-space, as we try to understand space-time by making space-space drawings. Space-space tilt is measured by angles, and calculated with trig functions. Space-time tilt is measured by rapidity, and calculated with hyperbolic functions.)

## 27.2   Light-cones in general relativity

In flat spacetime, the future light cone of an event is the boundary of its causal future and its past light cone is the boundary of its causal past.

In a curved spacetime, assuming spacetime is globally hyperbolic, it is still true that the future light cone of an event includes the boundary of its causal future (and similarly for the past). However gravitational lensing can cause part of the light cone to fold in on itself, in such a way that part of the cone is strictly inside the causal future (or past), and not on the boundary.

Light cones also cannot all be tilted so that they are 'parallel'; this reflects the fact that the spacetime is curved and is essentially different from Minkowski space. In vacuum regions (those points of spacetime free of matter), this inability to tilt all the light cones so that they are all parallel is reflected in the non-vanishing of the Weyl tensor.

## 27.3   See also

- Absolute future

- Absolute past

- Hyperbolic partial differential equation

- Hypercone

- Light cone coordinates

- Method of characteristics

- Minkowski diagram

- Monge cone

- Wave equation

## 27.4 References

[1] Penrose, Roger (2005), *The Road to Reality*, London: Vintage Books, ISBN 0-09-944068-7

## 27.5 External links

- The Einstein-Minkowski Spacetime: Introducing the Light Cone

- The Paradox of Special Relativity

- RSS feed of stars in one's personal light cone

# Chapter 28

# Length contraction

**Length contraction** is the phenomenon of a decrease in length of an object as measured by an observer which is traveling at any non-zero velocity relative to the object. This contraction (more formally called **Lorentz contraction** or **Lorentz–FitzGerald contraction** after Hendrik Lorentz and George FitzGerald) is usually only noticeable at a substantial fraction of the speed of light. Length contraction is only in the direction parallel to the direction in which the observed body is travelling. This effect is negligible at everyday speeds, and can be ignored for all regular purposes. Only at greater speeds does it become relevant. At a speed of 13,400,000 m/s (30 million mph, $0.0447c$) contracted length is 99.9% of the length at rest; at a speed of 42,300,000 m/s (95 million mph, $0.141c$), the length is still 99%. As the magnitude of the velocity approaches the speed of light, the effect becomes dominant, as can be seen from the formula:

$$L = \frac{L_0}{\gamma(v)} = L_0\sqrt{1 - v^2/c^2}$$

where

$L_0$ is the proper length (the length of the object in its rest frame),

$L$ is the length observed by an observer in relative motion with respect to the object,

$v$ is the relative velocity between the observer and the moving object,

$c$ is the speed of light,

and the *Lorentz factor*, $\gamma(v)$, is defined as

$$\gamma(v) \equiv \frac{1}{\sqrt{1 - v^2/c^2}}$$

In this equation it is assumed that the object is parallel with its line of movement. For the observer in relative movement, the length of the object is measured by subtracting the simultaneously measured distances of both ends of the object. For more general conversions, see the Lorentz transformations. An observer at rest viewing an object travelling very close to the speed of light would observe the length of the object in the direction of motion as very near zero.

## 28.1  History

Main article: History of special relativity

Length contraction was postulated by George FitzGerald (1889) and Hendrik Antoon Lorentz (1892) to explain the negative outcome of the Michelson–Morley experiment and to rescue the hypothesis of the stationary aether (Lorentz–FitzGerald contraction hypothesis).[1][2] Although both FitzGerald and Lorentz alluded to the fact that electrostatic fields in motion were deformed ("Heaviside-Ellipsoid" after Oliver Heaviside, who derived this deformation from electromagnetic theory in 1888), it was considered an ad hoc hypothesis, because at this time there was no sufficient reason to assume that intermolecular forces behave the same way as electromagnetic ones. In 1897 Joseph Larmor developed a model in which all forces are considered to be of electromagnetic origin, and length contraction appeared to be a direct consequence of this model. Yet it was shown by Henri Poincaré (1905) that electromagnetic forces alone cannot explain the electron's stability. So he had to introduce another ad hoc hypothesis: non-electric binding forces (Poincaré stresses) that ensure the electron's stability, give a dynamical explanation for length contraction, and thus hide the motion of the stationary aether.[3]

Eventually, Albert Einstein (1905) was the first[3] to completely remove the ad hoc character from the contraction hypothesis, by demonstrating that this contraction did not require motion through a supposed aether, but could be explained using special relativity, which changed our notions of space, time, and simultaneity.[4] Einstein's view was further elaborated by Hermann Minkowski, who demonstrated the geometrical interpretation of all relativistic effects by introducing his concept of four-dimensional spacetime.[5]

## 28.2  Basis in relativity

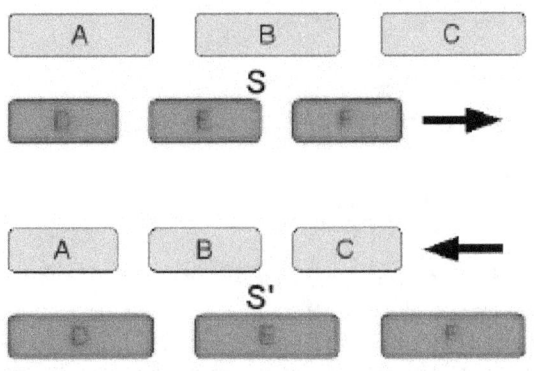

Length contraction: *Three blue rods are at rest in S, and three red rods in S'. At the instant when the left ends of A and D attain the same position on the axis of x, the lengths of the rods shall be compared. In S the simultaneous positions of the left side of A and the right side of C are more distant than those of D and F. While in S' the simultaneous positions of the left side of D and the right side of F are more distant than those of A and C.*

First it is necessary to carefully consider the methods for measuring the lengths of resting and moving objects.[6] Here, "object" simply means a distance with endpoints that are always mutually at rest, *i.e.*, that are at rest in the same inertial frame of reference. If the relative velocity between an observer (or his measuring instruments) and the observed object is zero, then the proper length $L_0$ of the object can simply be determined by directly superposing a measuring rod. However, if the relative velocity > 0, then one can proceed as follows:

The observer installs a row of clocks that either are synchronized a) by exchanging light signals according to the Poincaré-Einstein synchronization, or b) by "slow clock transport", that is, one clock is transported along the row of clocks in the limit of vanishing transport velocity. Now, when the synchronization process is finished, the object is moved along the clock row and every clock stores the exact time when the left or the right end of the object passes by. After that, the observer only has to look after the position of a clock A that stored the time when the left end of the object was passing by, and a clock B at which the right end of the object was passing by *at the same time*. It's clear that distance AB is equal to length $L$ of the moving object.[6] Using this method, the definition of simultaneity is crucial for measuring the length of moving objects.

Another method is to use a clock indicating its proper time

$T_0$, which is traveling from one endpoint of the rod to the other in time $T$ as measured by clocks in the rod's rest frame. The length of the rod can be computed by multiplying its travel time by its velocity, thus $L_0 = T \cdot v$ in the rod's rest frame or $L = T_0 \cdot v$ in the clock's rest frame.[7]

In Newtonian mechanics, simultaneity and time duration are absolute and therefore both methods lead to the equality of $L$ and $L_0$. Yet in relativity theory the constancy of light velocity in all inertial frames in connection with relativity of simultaneity and time dilation destroys this equality. In the first method an observer in one frame claims to have measured the object's endpoints simultaneously, but the observers in all other inertial frames will argue that the object's endpoints were *not* measured simultaneously. In the second method, times $T$ and $T_0$ are not equal due to time dilation, resulting in different lengths too.

The deviation between the measurements in all inertial frames is given by the formulas for Lorentz transformation and time dilation (see Derivation). It turns out, that the proper length remains unchanged and always denotes the greatest length of an object, yet the length of the same object as measured in another inertial frame is shorter than the proper length. This contraction only occurs in the line of motion, and can be represented by the following relation (where $v$ is the relative velocity and $c$ the speed of light)

$$L = L_0/\gamma.$$

## 28.3  Symmetry

Minkowski diagram

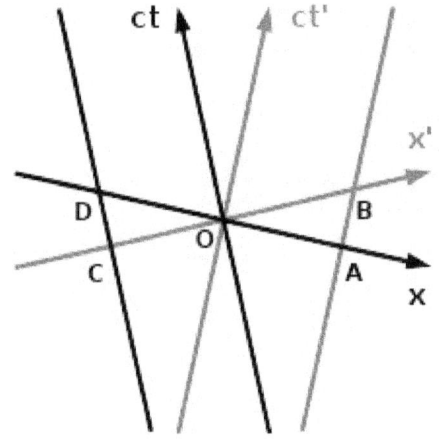

In S all events parallel to the axis of x are simultaneous, while in S' all events parallel to the axis of x' are simultaneous.

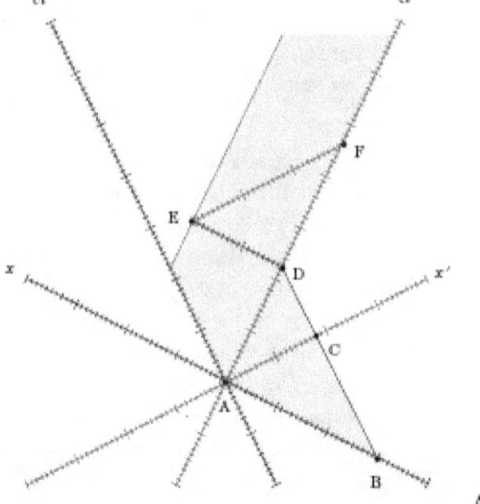

A rod is transported from S to S'

again according to the methods given above, and now the proper length $L_0' = $ EF $ = 30$ cm will be measured in S' (the rod has become larger in that system), while in S the rod is in motion and therefore its length is contracted (the rod has become smaller in that system):

$$L = \text{DE} = L_0'/\gamma = 18 \text{ cm.}$$

## 28.4 Experimental verifications

See also: Tests of special relativity

Any observer co-moving with the observed object cannot measure the object's contraction, because he can judge himself and the object as at rest in the same inertial frame in accordance with the principle of relativity (as it was demonstrated by the Trouton-Rankine experiment). So length contraction cannot be measured in the object's rest frame, but only in a frame in which the observed object is in motion. In addition, even in such a non-co-moving frame, *direct* experimental confirmations of length contraction are hard to achieve, because at the current state of technology, objects of considerable extension cannot be accelerated to relativistic speeds. And the only objects traveling with the speed required are atomic particles, yet whose spatial extensions are too small to allow a direct measurement of contraction.

However, there are *indirect* confirmations of this effect in a non-co-moving frame:

The principle of relativity (according to which the laws of nature must assume the same form in all inertial reference frames) requires that length contraction is symmetrical: If a rod rests in inertial frame S, it has its proper length in S and its length is contracted in S'. However, if a rod rests in S', it has its proper length in S' and its length is contracted in S. This can be vividly illustrated using symmetric Minkowski diagrams (or Loedel diagrams), because the Lorentz transformation geometrically corresponds to a rotation in four-dimensional spacetime.[8][9]

*First image*: If a rod at rest in S' is given, then its endpoints are located upon the ct' axis and the axis parallel to it. In this frame the simultaneous (parallel to the axis of x') positions of the endpoints are O and B, thus the *proper* length is given by OB. But in S the simultaneous (parallel to the axis of x) positions are O and A, thus the *contracted* length is given by OA.

On the other hand, if another rod is at rest in S, then its endpoints are located upon the ct axis and the axis parallel to it. In this frame the simultaneous (parallel to the axis of x) positions of the endpoints are O and D, thus the *proper* length is given by OD. But in S' the simultaneous (parallel to the axis of x') positions are O and C, thus the *contracted* length is given by OC.

*Second image*: A train at rest in S and a station at rest in S' with relative velocity of $v = 0.8c$ are given. In S a rod with proper length $L_0 = $ AB $ = 30$ cm is located, so its contracted length $L'$ in S' is given by:

$$L' = \text{AC} = L_0/\gamma = 18 \text{ cm.}$$

Then the rod will be thrown out of the train in S and will come to rest at the station in S'. Its length has to be measured

- It was the negative result of a famous experiment, that required the introduction of length contraction: the Michelson-Morley experiment (and later also the Kennedy–Thorndike experiment). In special relativity its explanation is as follows: In its rest frame the interferometer can be regarded as at rest in accordance with the relativity principle, so the propagation time of light is the same in all directions. Although in a frame in which the interferometer is in motion, the transverse beam must traverse a longer, diagonal path with respect to the non-moving frame thus making its travel time longer, the factor by which the longitudinal beam would be delayed by taking times L/(c-v) & L/(c+v) for the forward and reverse trips respectively is even longer. Therefore, in the longitudinal direction the interferometer is supposed to be contracted, in order to restore the equality of both travel times in accordance with the negative experimental result(s). Thus the two-way speed of light remains constant and the round trip propagation time along perpendicular arms of the interferometer is independent of its motion & orientation.

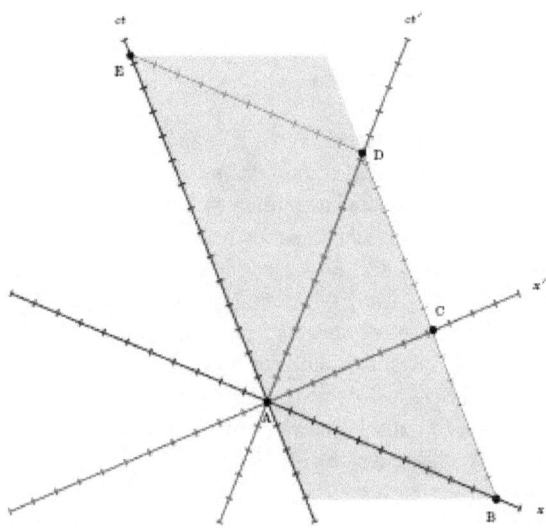

*Muon-atmosphere-scenario*

- The range of action of muons at high velocities is much higher than that of slower ones. The atmosphere has its proper length in the Earth frame, while the increased muon range is explained by their longer lifetimes due to time dilation (see Time dilation of moving particles). However, in the muon frame their lifetime is unchanged but the atmosphere is contracted so that even their small range is sufficient to reach the surface of earth.[10]

- Heavy ions that are spherical when at rest should assume the form of "pancakes" or flat disks when travelling nearly at the speed of light. And in fact, the results obtained from particle collisions can only be explained when the increased nucleon density due to length contraction is considered.[11][12][13]

- The ionization ability of electrically charged particles with large relative velocities is higher than expected. In pre-relativistic physics the ability should decrease at high velocities, because the time in which ionizing particles in motion can interact with the electrons of other atoms or molecules is diminished. Though in relativity, the higher-than-expected ionization ability can be explained by length contraction of the Coulomb field in frames in which the ionizing particles are moving, which increases their electrical field strength normal to the line of motion.[10][14]

- In free-electron lasers, relativistic electrons were injected into an undulator, so that synchrotron radiation

is generated. In the proper frame of the electrons, the undulator is contracted which leads to an increased radiation frequency. Additionally, to find out the frequency as measured in the laboratory frame, one has to apply the relativistic Doppler effect. So, only with the aid of length contraction and the relativistic Doppler effect, the extremely small wavelength of undulator radiation can be explained.[15][16]

## 28.5 Reality of length contraction

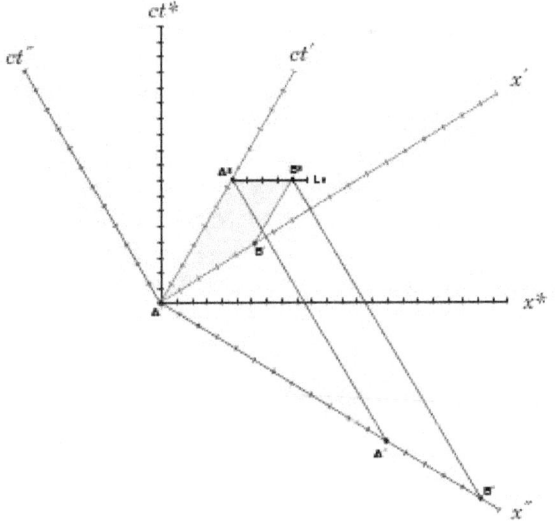

*Minkowski diagram of Einstein's 1911 thought experiment on length contraction. Two rods of rest length $A'B' = A''B'' = L_0$ are moving with $0.6c$ in opposite direction, resulting in $A^*B^* < L_0$ .*

In 1911 Vladimir Varićak asserted that length contraction is "real" according to Lorentz, while it is "apparent or subjective" according to Einstein.[17] Einstein replied:

> The author unjustifiably stated a difference of Lorentz's view and that of mine *concerning the physical facts*. The question as to whether length contraction *really* exists or not is misleading. It doesn't "really" exist, in so far as it doesn't exist for a comoving observer; though it "really" exists, *i.e.* in such a way that it could be demonstrated in principle by physical means by a non-comoving observer.[18]
> — Albert Einstein, 1911

Einstein also argued in that paper, that length contraction is not simply the product of *arbitrary* definitions concerning the way clock regulations and length measurements are performed. He presented the following thought experiment: Let A'B' and A"B" be the endpoints of two rods of the same proper length. Let them move in opposite directions at the same speed with respect to a resting coordinate x-axis. Endpoints A'A" meet at point A*, and B'B" meet at point B*, both points being marked on that axis. Einstein pointed out that length A*B* is shorter than A'B' or A"B", which can also be demonstrated by one of the rods when brought to rest with respect to that axis.[18]

## 28.6 Paradoxes

Due to superficial application of the contraction formula some paradoxes can occur. For examples see the Ladder paradox or Bell's spaceship paradox. However, those paradoxes can simply be solved by a correct application of relativity of simultaneity. Another famous paradox is the Ehrenfest paradox, which proves that the concept of rigid bodies is not compatible with relativity, reducing the applicability of Born rigidity, and showing that for a co-rotating observer the geometry is in fact non-euclidean.

## 28.7 Visual effects

Main article: Terrell rotation

Length contraction refers to measurements of position made at simultaneous times according to a coordinate system. This could suggest that if one could take a picture of a fast moving object, that the image would show the object contracted in the direction of motion. However, such visual effects are completely different measurements, as such a photograph is taken from a distance, while length contraction can only directly be measured at the exact location of the object's endpoints. It was shown by several authors such as Roger Penrose and James Terrell that moving objects generally do not appear length contracted on a photograph.[19] For instance, for a small angular diameter, a moving sphere remains circular and is rotated.[20] This kind of visual rotation effect is called Penrose-Terrell rotation.[21]

## 28.8 Derivation

### 28.8.1 Lorentz transformation

Length contraction can be derived from the Lorentz transformation in several ways:

$$x' = \gamma (x - vt),$$
$$t' = \gamma (t - vx/c^2).$$

**Moving length is known**

In an inertial reference frame S, $x_1$ and $x_2$ shall denote the endpoints of an object in motion in this frame. There, its length $L$ was measured according to the above convention by determining the simultaneous positions of its endpoints at $t_1 = t_2$. Now, the proper length of this object in S' shall be calculated by using the Lorentz transformation. Transforming the time coordinates from S into S' results in different times, but this is not problematic, as the object is at rest in S' where it does not matter when the endpoints are measured. Therefore, the transformation of the spatial coordinates suffices, which gives:[6]

$$x_1' = \gamma (x_1 - vt_1) \quad \text{and} \quad x_2' = \gamma (x_2 - vt_2).$$

Since $t_1 = t_2$, and by setting $L = x_2 - x_1$ and $L_0' = x_2' - x_1'$, the proper length in S' is given by

$$L_0' = L \cdot \gamma. \qquad (1),$$

with respect to which the measured length in S is contracted by

$$L = L_0'/\gamma. \qquad (2)$$

According to the relativity principle, objects that are at rest in S have to be contracted in S' as well. By exchanging the above signs and primes symmetrically, it follows:

$$L_0 = L' \cdot \gamma. \qquad (3)$$

Thus the contracted length as measured in S' is given by:

$$L' = L_0/\gamma. \qquad (4)$$

**Proper length is known**

Conversely, if the object rests in S and its proper length is known, the simultaneity of the measurements at the object's endpoints has to be considered in another frame S', as the object constantly changes its position there. Therefore, both spatial and temporal coordinates must be transformed:[22]

$$x'_1 = \gamma(x_1 - vt_1) \quad \text{and} \quad x'_2 = \gamma(x_2 - vt_2)$$
$$t'_1 = \gamma(t_1 - vx_1/c^2) \quad \text{and} \quad t'_2 = \gamma(t_2 - vx_2/c^2)$$

With $t_1 = t_2$ and $L_0 = x_2 - x_1$ this results in non-simultaneous differences:

$$\Delta x' = \gamma L_0$$
$$\Delta t' = \gamma v L_0 / c^2$$

In order to obtain the simultaneous positions of both endpoints, the distance traveled by the second endpoint with $v$ during $\Delta t'$ must be subtracted from $\Delta x'$ :

$$L' = \Delta x' - v\Delta t'$$
$$= \gamma L_0 - \gamma v^2 L_0 / c^2$$
$$= L_0 / \gamma$$

So the moving length in S' is contracted. Likewise, the preceding calculation gives a symmetric result for an object at rest in S':

$$L = L'_0 / \gamma$$

### 28.8.2   Time dilation

Length contraction can also be derived from time dilation,[23] according to which the rate of a single "moving" clock (indicating its proper time $T_0$ ) is lower with respect to two synchronized "resting" clocks (indicating $T$ ). Time dilation was experimentally confirmed multiple times, and is represented by the relation:

$$T = T_0 \cdot \gamma$$

Suppose a rod of proper length $L_0$ at rest in $S$ and a clock at rest in $S'$ are moving along each other. The respective travel times of the clock between the rod's endpoints are given by $T = L_0/v$ in $S$ and $T'_0 = L'/v$ in $S'$ , thus $L_0 = Tv$ and $L' = T'_0 v$ . By inserting the time dilation formula, the ratio between those lengths is:

$$\frac{L'}{L_0} = \frac{T'_0 v}{Tv} = 1/\gamma$$

Therefore, the length measured in $S'$ is given by

$$L' = L_0 / \gamma$$

So the effect that the moving clock indicates a lower travel time in $S$ due to time dilation, is interpreted in $S'$ as due to length contraction of the moving rod. Likewise, if the clock were at rest in $S$ and the rod in $S'$ , the above procedure would give

$$L = L'_0 / \gamma$$

### 28.8.3   Geometrical considerations

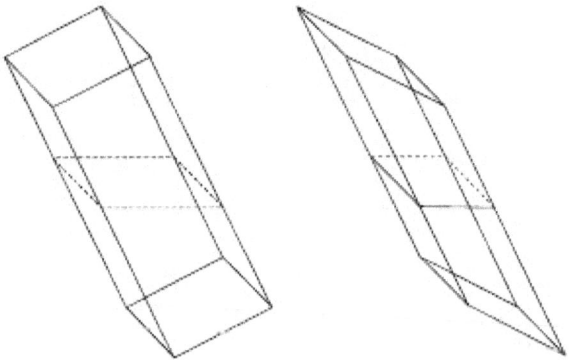

*Cuboids in Euclidean and Minkowski spacetime*

Additional geometrical considerations show, that length contraction can be regarded as a *trigonometric* phenomenon, with analogy to parallel slices through a cuboid before and after a *rotation* in $\mathbf{E}^3$ (see left half figure at the right). This is the Euclidean analog of *boosting* a cuboid in $\mathbf{E}^{1,2}$. In the latter case, however, we can interpret the boosted cuboid as the *world slab* of a moving plate.

*Image*: Left: a *rotated cuboid* in three-dimensional euclidean space $\mathbf{E}^3$. The cross section is *longer* in the direction of the rotation than it was before the rotation. Right: the *world slab* of a moving thin plate in Minkowski spacetime (with one spatial dimension suppressed) $\mathbf{E}^{1,2}$, which is a *boosted cuboid*. The cross section is *thinner* in the direction of the boost than it was before the boost. In both cases, the transverse directions are unaffected and the three planes meeting at each corner of the cuboids are *mutually orthogonal* (in the sense of $\mathbf{E}^{1,2}$ at right, and in the sense of $\mathbf{E}^3$ at left).

In special relativity, Poincaré transformations are a class of affine transformations which can be characterized as the transformations between alternative Cartesian coordinate charts on Minkowski spacetime corresponding to alternative states of inertial motion (and different choices of an origin). Lorentz transformations are Poincaré transformations which are linear transformations (preserve the origin). Lorentz transformations play the same role in Minkowski geometry (the Lorentz group forms the *isotropy group* of the self-isometries of the spacetime) which are played by rotations in euclidean geometry. Indeed, special relativity largely comes down to studying a kind of noneuclidean trigonometry in Minkowski spacetime, as suggested by the following table:

## 28.9   References

[1] FitzGerald, George Francis (1889), "The Ether and the Earth's Atmosphere", *Science* **13** (328): 390, Bibcode:1889Sci....13..390F, doi:10.1126/science.ns-13.328.390, PMID 17819387

[2] Lorentz, Hendrik Antoon (1892), "The Relative Motion of the Earth and the Aether", *Zittingsverlag Akad. V. Wet.* **1**: 74–79

[3] Pais, Abraham (1982), *Subtle is the Lord: The Science and the Life of Albert Einstein*, New York: Oxford University Press, ISBN 0-19-520438-7

[4] Einstein, Albert (1905a), "Zur Elektrodynamik bewegter Körper" (PDF), *Annalen der Physik* **322** (10): 891–921, Bibcode:1905AnP...322..891E, doi:10.1002/andp.19053221004. See also: English translation.

[5] Minkowski, Hermann (1909), "Raum und Zeit", *Physikalische Zeitschrift* **10**: 75–88

   • Various English translations on Wikisource: Space and Time

[6] Born, Max (1964), *Einstein's Theory of Relativity*, Dover Publications, ISBN 0-486-60769-0

[7] Edwin F. Taylor, John Archibald Wheeler (1992). *Spacetime Physics: Introduction to Special Relativity*. New York: W. H. Freeman. ISBN 0-7167-2327-1.

[8] Albert Shadowitz (1988). *Special relativity* (Reprint of 1968 ed.). Courier Dover Publications. pp. 20–22. ISBN 0-486-65743-4.

[9] Leo Sartori (1996). *Understanding Relativity: a simplified approach to Einstein's theories*. University of California Press. pp. 151ff. ISBN 0-520-20029-2.

[10] Sexl, Roman & Schmidt, Herbert K. (1979), *Raum-Zeit-Relativität*, Braunschweig: Vieweg, ISBN 3-528-17236-3

[11] Brookhaven National Laboratory. "The Physics of RHIC". Retrieved 2013.

[12] Manuel Calderon de la Barca Sanchez. "Relativistic heavy ion collisions". Retrieved 2013.

[13] Hands, Simon (2001). "The phase diagram of QCD". *Contemporary Physics* **42** (4): 209–225. arXiv:physics/0105022. Bibcode:2001ConPh..42..209H. doi:10.1080/00107510110063843.

[14] Williams, E. J. (1931), "The Loss of Energy by β -Particles and Its Distribution between Different Kinds of Collisions", *Proceedings of the Royal Society of London. Series A* **130** (813): 328–346, Bibcode:1931RSPSA.130..328W, doi:10.1098/rspa.1931.0008

[15] DESY photon science. "What is SR, how is it generated and what are its properties?". Retrieved 2013.

[16] DESY photon science. "FLASH The Free-Electron Laser in Hamburg (PDF 7,8 MB)" (PDF). Retrieved 2013.

[17] Miller, A.I. (1981), "Varičak and Einstein", *Albert Einstein's special theory of relativity. Emergence (1905) and early interpretation (1905–1911)*, Reading: Addison–Wesley, pp. 249–253, ISBN 0-201-04679-2

[18] Einstein, Albert (1911). "Zum Ehrenfestschen Paradoxon. Eine Bemerkung zu V. Varičaks Aufsatz". *Physikalische Zeitschrift* **12**: 509–510.; Original: Der Verfasser hat mit Unrecht einen Unterschied der *Lorentz*schen Auffassung von der meinigen *mit Bezug auf die physikalischen Tatsachen* statuiert. Die Frage, ob die *Lorentz*-Verkürzung *wirklich* besteht oder nicht, ist irreführend. Sie besteht nämlich nicht „wirklich", insofern sie für einen mitbewegten Beobachter nicht existiert; sie besteht aber „wirklich", d. h. in solcher Weise, daß sie prinzipiell durch physikalische Mittel nachgewiesen werden könnte, für einen nicht mitbewegten Beobachter.

[19] Kraus, U. (2000). "Brightness and color of rapidly moving objects: The visual appearance of a large sphere revisited" (PDF). *American Journal of Physics* **68** (1): 56–60. Bibcode:2000AmJPh..68...56K. doi:10.1119/1.19373.

[20] Penrose, Roger (2005). *The Road to Reality*. London: Vintage Books. pp. 430–431. ISBN 978-0-09-944068-0.

[21] Can You See the Lorentz-Fitzgerald Contraction? Or: Penrose-Terrell Rotation

[22] Bernard Schutz (2009). "Lorentz contraction". *A First Course in General Relativity*. Cambridge University Press. p. 18. ISBN 0521887054.

[23] David Halliday, Robert Resnick, Jearl Walker (2010), *Fundamentals of Physics, Chapters 33-37*, John Wiley & Son, pp. 1032f, ISBN 0470547944

## 28.10   External links

• Physics FAQ: Can You See the Lorentz–Fitzgerald Contraction? Or: Penrose-Terrell Rotation; The Barn and the Pole

# Chapter 29

# Time dilation

*Time dilation explains why two working clocks will report different times after different accelerations. For example, ISS astronauts return from missions having aged slightly less than they would have been if they had remained on Earth, and GPS satellites work because they adjust for similar bending of spacetime to coordinate with systems on Earth.[1]*

In the theory of relativity, **time dilation** is a difference of elapsed time between two events as measured by observers either moving relative to each other or differently situated from a gravitational mass or masses.

A clock at rest with respect to one observer may be measured to tick at a different rate when compared to a second observer's clock. This effect arises neither from technical aspects of the clocks nor from the propagation time of signals, but from the nature of spacetime.

## 29.1 Overview

Clocks on the Space Shuttle ran slightly slower than reference clocks on Earth, while clocks on GPS and Galileo satellites run slightly faster.[1] Such time dilation has been repeatedly demonstrated (see experimental confirmation below), for instance by small disparities in atomic clocks on Earth and in space, even though both clocks work perfectly (it is not a mechanical malfunction). The nature of spacetime is such that time measured along different trajec-

tories is affected by differences in either gravity or velocity – each of which affects time in different ways.[2][3]

In theory, and to make a clearer example, time dilation could affect planned meetings for astronauts with advanced technologies and greater travel speeds. The astronauts would have to set their clocks to count exactly 80 years, whereas mission control – back on Earth – might need to count 81 years. The astronauts would return to Earth, after their mission, having aged one year less than the people staying on Earth. What is more, the local experience of time passing never actually changes for anyone. In other words, the astronauts on the ship as well as the mission control crew on Earth each feel normal, despite the effects of time dilation (i.e. to the traveling party, those stationary are living "faster"; while to those who stood still, their counterparts in motion live "slower" at any given moment).

With technology limiting the velocities of astronauts, these differences are minuscule: after 6 months on the International Space Station (ISS), the astronaut crew has indeed aged less than those on Earth, but only by about 0.005 seconds (nowhere near the 1 year disparity from the theoretical example). The effects would be greater if the astronauts were traveling nearer to the speed of light (299,792,458 m/s), instead of their actual speed – which is the speed of the orbiting ISS, about 7,700 m/s.[4]

Time dilation is caused by differences in either gravity or relative velocity. In the case of ISS, time is slower due to the velocity in circular orbit; this effect is slightly reduced by the opposing effect of less gravitational potential.

### 29.1.1 Relative velocity time dilation

When two observers are in relative uniform motion and uninfluenced by any gravitational mass, the point of view of each will be that the other's (moving) clock is ticking at a *slower* rate than the local clock. The faster the relative velocity, the greater the magnitude of time dilation. This case is sometimes called special relativistic time dilation.

For instance, two rocket ships (A and B) speeding past one

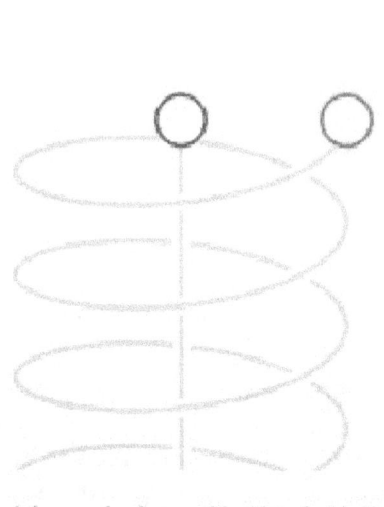

*From the local frame of reference (the blue clock), the relatively accelerated red clock moves slower*

*Time passes more quickly further from a center of gravity, as is witnessed with massive objects (like the Earth).*

another in space would experience time dilation. If they somehow had a clear view into each other's ships, each crew would see the others' clocks and movement as going more slowly. That is, inside the frame of reference of Ship A, everything is moving normally, but everything over on Ship B appears to be moving more slowly (and vice versa).

From a local perspective, time registered by clocks that are at rest with respect to the local frame of reference (and far from any gravitational mass) always appears to pass at the same rate. In other words, if a new ship, Ship C, travels alongside Ship A, it is "at rest" relative to Ship A. From the point of view of Ship A, new Ship C's time would appear normal too.[5]

A question arises: If Ship A and Ship B both think each other's time is moving slower, who will have aged more if they decided to meet up? With a more sophisticated understanding of relative velocity time dilation, this seeming twin paradox turns out not to be a paradox at all (the resolution of the paradox involves a jump in time, as a result of the accelerated observer turning around). Similarly, understanding the twin paradox would help explain why astronauts on the ISS age slower (e.g. 0.007 seconds behind for every six months) even though they are experiencing relative velocity time dilation.

## 29.1.2   Gravitational time dilation

Main article: Gravitational time dilation

Gravitational time dilation is at play for ISS astronauts too, and it has the opposite effect of the relative velocity time

dilation. With respect to ground observers the ISS astronaut's relative velocity slows down their time, whereas the reduced gravitational influence at their location speeds it up. The two opposing effects are not equally strong. At the ISS altitude the net effect is a slowing down of clocks, whereas in much higher orbits clocks run faster than on the ground.

The key is that both observers are differently situated in their distance from a significant gravitational mass. The general theory of relativity describes how, for both observers, the clock that is closer to the gravitational mass, i.e. deeper in its "gravity well", appears to go more slowly than the clock that is more distant from the mass. This effect is not restricted to astronauts in space; a climber's time is passing slightly faster at the top of a mountain (a high altitude, farther from the Earth's center of gravity) compared to people at sea level. As with all time dilation, the local experience of time is normal (nobody notices a difference within their own frame of reference). In the situations of velocity time dilation, both observers saw the other as moving slower (a reciprocal effect). Now, with gravitational time dilation, both observers – those at sea level, versus the climber – agree that the clock nearer the mass is slower in rate, and they agree on the ratio of the difference (time dilation from gravity is therefore not reciprocal). That is, the climber sees the sea level clocks as moving more slowly, and those living at sea level see the climber's clock as moving faster.

### 29.1.3 Time dilation: special vs. general theories of relativity

In Albert Einstein's theory of relativity, time dilation in these two circumstances can be summarized:

- In special relativity (or, hypothetically far from all gravitational mass), clocks that are moving with respect to an inertial system of observation are measured to be running more slowly. This effect is described precisely by the Lorentz transformation.

- In general relativity, clocks at a position with lower gravitational potential – such as in closer proximity to a planet – are found to be running more slowly. The articles on gravitational time dilation and gravitational redshift give a more detailed discussion.

Special and general relativistic effects can combine (as seen with ISS astronauts).

In special relativity, the time dilation effect is reciprocal: as observed from the point of view of either of two clocks which are in motion with respect to each other, it will be the other clock that is time dilated. (This presumes that the relative motion of both parties is uniform; that is, they do not accelerate with respect to one another during the course of the observations.) In contrast, gravitational time dilation (as treated in general relativity) is not reciprocal: an observer at the top of a tower will observe that clocks at ground level tick slower, and observers on the ground will agree about the direction and the magnitude of the difference. There is still some disagreement in a sense, because all the observers believe their own local clocks are correct, but the direction and ratio of gravitational time dilation is agreed by all observers, independent of their altitude.

### 29.1.4 Science fiction implications

Science fiction enthusiasts have noted the implications time dilation has on forward time travel, technically making it possible.[6] The Hafele and Keating experiment involved flying planes around the world with atomic clocks on board. Upon the trips' completion the clocks were compared to a static, ground based atomic clock. It was found that 273±7 nanoseconds had been gained on the planes' clocks.[7] The current human time travel record holder is Russian cosmonaut Sergei Krikalev,[8] who beat the previous record of about 20 milliseconds by cosmonaut Sergei Avdeyev.[9]

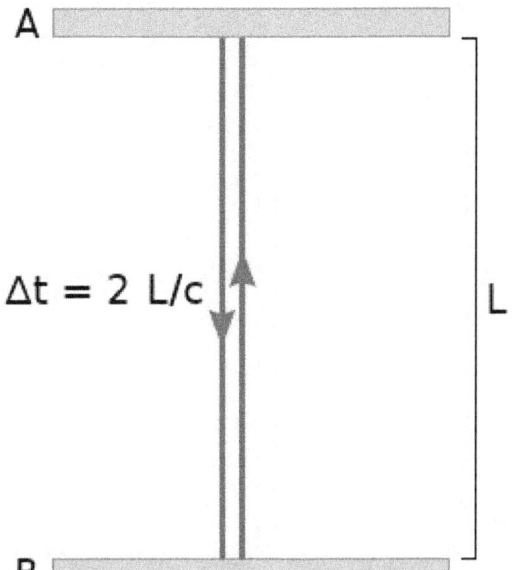

*Observer at rest measures time 2L/c.*

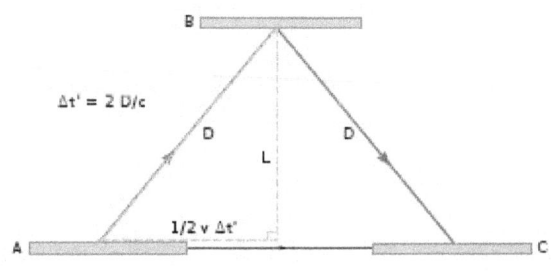

*Observer moving parallel relative to setup, measures longer path and thus, with same speed of light c, measures longer time 2D/c > 2L/c.*
*A: location of bottom mirror when signal is generated at time t'=0.*
*B: location of top mirror when signal gets reflected at time t'=D/c.*
*C: location of bottom mirror when signal hits bottom at time t'=2D/c.*

## 29.2 Simple inference of time dilation due to relative velocity

Time dilation can be inferred from the observed constancy of the speed of light in all reference frames.[10][11][12][13]

This constancy of the speed of light means, counter to intuition, that speeds of material objects and light are not additive. It is not possible to make the speed of light appear greater by approaching at speed towards the material source that is emitting light. It is not possible to make the speed of light appear less by receding from the source at speed. From one point of view, it is the implications of this unexpected constancy that take away from constancies expected elsewhere.

Consider a simple clock consisting of two mirrors A and B, between which a light pulse is bouncing. The separation of the mirrors is $L$ and the clock ticks once each time the light pulse hits a given mirror.

In the frame where the clock is at rest (diagram at right), the light pulse traces out a path of length $2L$ and the period of the clock is $2L$ divided by the speed of light

$$\Delta t = \frac{2L}{c}.$$

From the frame of reference of a moving observer traveling at the speed $v$ relative to the rest frame of the clock (diagram at lower right), the light pulse traces out a *longer*, angled path. The second postulate of special relativity states that the speed of light in free space is constant for all inertial observers, which implies a lengthening of the period of this clock from the moving observer's perspective. That is to say, in a frame moving relative to the clock, the clock appears to be running more slowly. Straightforward application of the Pythagorean theorem leads to the well-known prediction of special relativity:

The total time for the light pulse to trace its path is given by

$$\Delta t' = \frac{2D}{c}.$$

The length of the half path can be calculated as a function of known quantities as

$$D = \sqrt{\left(\frac{1}{2}v\Delta t'\right)^2 + L^2}.$$

Substituting $D$ from this equation into the previous and solving for $\Delta t'$ gives:

$$\Delta t' = \frac{\sqrt{(v\Delta t')^2 + (2L)^2}}{c}$$

$$(\Delta t')^2 = \frac{v^2}{c^2}(\Delta t')^2 + \left(\frac{2L}{c}\right)^2$$

$$(1 - \frac{v^2}{c^2})(\Delta t')^2 = \left(\frac{2L}{c}\right)^2$$

$$(\Delta t')^2 = \frac{\left(\frac{2L}{c}\right)^2}{\left(1 - \frac{v^2}{c^2}\right)}$$

$$\Delta t' = \frac{\frac{2L}{c}}{\sqrt{1 - \frac{v^2}{c^2}}}$$

and thus, with the definition of $\Delta t$:

$$\Delta t' = \frac{\Delta t}{\sqrt{1 - \frac{v^2}{c^2}}}$$

which expresses the fact that for the moving observer the period of the clock is longer than in the frame of the clock itself.

## 29.3   Due to relative velocity symmetric between observers

Minkowski diagram

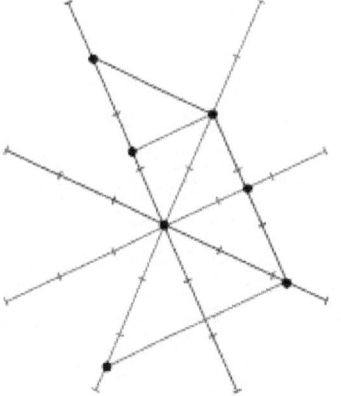

Time UV of a clock in S is shorter compared to Ux' in S', and time UW of a clock in S' is shorter compared to Ux in S.

Clock C in relative motion between two synchronized clocks A and B. C meets A at $d$, and B at $f$.

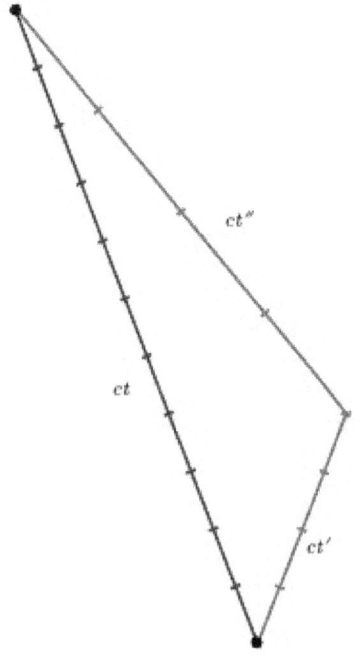

Twin paradox. One twin has to change frames, leading to different proper times in the twin's world lines.

Common sense would dictate that if time passage has slowed for a moving object, the moving object would observe the external world to be correspondingly "sped up". Counterintuitively, special relativity predicts the opposite.

A similar oddity occurs in everyday life. If Sam sees Abigail at a distance she appears small to him and at the same time Sam appears small to Abigail. Being very familiar with the effects of perspective, we see no mystery or a hint of a paradox in this situation.[14]

One is accustomed to the notion of relativity with respect to distance: the distance from Los Angeles to New York is by convention the same as the distance from New York to Los Angeles. On the other hand, when speeds are considered, one thinks of an object as "actually" moving, overlooking that its motion is always relative to something else – to the stars, the ground or to oneself. If one object is moving with respect to another, the latter is moving with respect to the former and with equal relative speed.

In the special theory of relativity, a moving clock is found to be ticking slowly with respect to the observer's clock. If Sam and Abigail are on different trains in near-lightspeed relative motion, Sam measures (by all methods of measurement) clocks on Abigail's train to be running slowly and similarly, Abigail measures clocks on Sam's train to be running slowly.

Note that in all such attempts to establish "synchronization" within the reference system, the question of whether something happening at one location is in fact happening simultaneously with something happening elsewhere, is of key importance. Calculations are ultimately based on determining which events are simultaneous. Furthermore, establishing simultaneity of events separated in space necessarily requires transmission of information between locations, which by itself is an indication that the speed of light will enter the determination of simultaneity.

It is a natural and legitimate question to ask how, in detail, special relativity can be self-consistent if clock C is time-dilated with respect to clock B and clock B is also time-dilated with respect to clock C. It is by challenging the assumptions built into the common notion of simultaneity that logical consistency can be restored. Simultaneity is a relationship between an observer in a particular frame of reference and a set of events. By analogy, left and right are accepted to vary with the position of the observer, because they apply to a relationship. In a similar vein, Plato explained that up and down describe a relationship to the earth and one would not fall off at the antipodes.

In relativity, temporal coordinate systems are set up using a procedure for synchronizing clocks. It is now usually called the *Poincaré-Einstein synchronization procedure*. An observer with a clock sends a light signal out at time $t_1$ according to his clock. At a distant event, that light signal is reflected back, and arrives back at the observer at time $t_2$ according to his clock. Since the light travels the same path at the same rate going both out and back for the observer in this scenario, the coordinate time of the event of the light signal being reflected for the observer $t$E is $t$E = $(t_1 + t_2)$ / 2. In this way, a single observer's clock can be used to define temporal coordinates which are good anywhere in the universe.

However, since those clocks are in motion in all other inertial frames, these clock indications are thus not synchronous in those frames, which is the basis of relativity of simultaneity. Because the pairs of putatively simultaneous moments are identified differently by different observers, each can treat the other clock as being the slow one without relativity being self-contradictory. Symmetric time dilation occurs with respect to coordinate systems set up in this manner. It is an effect where another clock is measured to run more slowly than one's own clock. Observers do not consider their own clock time to be affected, but may find that it is observed to be affected in another coordinate system.

### 29.3.1   Proper time and Minkowski diagram

This symmetry can be demonstrated in a Minkowski diagram (second image on the right). Clock C resting in inertial frame S′ meets clock A at $d$ and clock B at $f$ (both resting in S). All three clocks simultaneously start to tick in S. The worldline of A is the ct-axis, the worldline of B intersecting $f$ is parallel to the ct-axis, and the worldline of C is the ct'-axis. All events simultaneous with $d$ in S are on the x-axis, in S' on the x'-axis.

The proper time between two events is indicated by a clock present at both events.[15] It is invariant, i.e., in all inertial frames it is agreed that this time is indicated by that clock. Interval $df$ is therefore the proper time of clock C, and is shorter with respect to the coordinate times $ef=dg$ of clocks B and A in S. Conversely, also proper time $ef$ of B is shorter with respect to time $if$ in S', because event $e$ was measured in S' already at time $i$ due to relativity of simultaneity, long before C started to tick.

From that it can be seen, that the proper time between two events indicated by an unaccelerated clock present at both events, compared with the synchronized coordinate time measured in all other inertial frames, is always the *minimal* time interval between those events. However, the interval between two events can also correspond to the proper time of accelerated clocks present at both events. Under all possible proper times between two events, the proper time of the unaccelerated clock is *maximal*, which is the solution to the twin paradox.[15]

## 29.4   Overview of formulae

### 29.4.1   Time dilation due to relative velocity

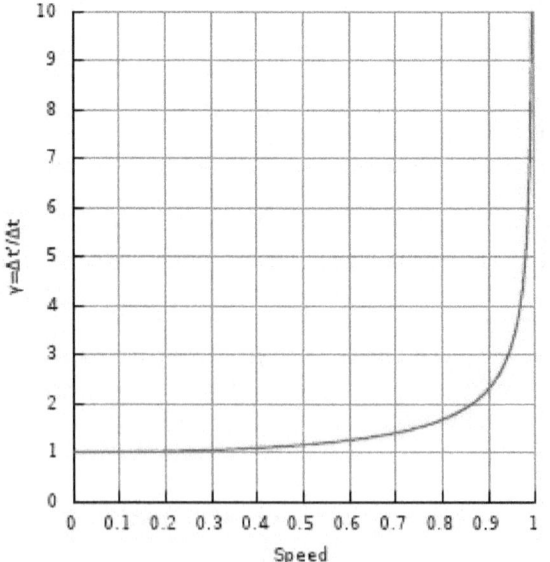

*Lorentz factor as a function of speed (in natural units where $c = 1$). Notice that for small speeds (less than 0.1), $\gamma$ is approximately 1*

The formula for determining time dilation in special relativity is:

$$\Delta t' = \gamma \, \Delta t = \frac{\Delta t}{\sqrt{1 - \frac{v^2}{c^2}}}$$

where $\Delta t$ is the time interval between *two co-local events* (i.e. happening at the same place) for an observer in some inertial frame (e.g. ticks on his clock), known as the *proper time*, $\Delta t'$ is the time interval between those same events, as measured by another observer, inertially moving with velocity $v$ with respect to the former observer, $v$ is the relative velocity between the observer and the moving clock, $c$ is the speed of light, and the Lorentz factor (conventionally denoted by the Greek letter gamma or $\gamma$) is

$$\gamma = \frac{1}{\sqrt{1 - \frac{v^2}{c^2}}} \, .$$

Thus the duration of the clock cycle of a moving clock is found to be increased: it is measured to be "running slow". The range of such variances in ordinary life, where $v \ll c$, even considering space travel, are not great enough to produce easily detectable time dilation effects and such vanishingly small effects can be safely ignored for most purposes. It is only when an object approaches speeds on the order of 30,000 km/s (1/10 the speed of light) that time dilation becomes important.

Time dilation by the Lorentz factor was predicted by Joseph Larmor (1897), at least for electrons orbiting a nucleus. Thus "... individual electrons describe corresponding parts of their orbits in times shorter for the [rest] system in the ratio : $\sqrt{1 - \frac{v^2}{c^2}}$ " (Larmor 1897). Time dilation of magnitude corresponding to this (Lorentz) factor has been experimentally confirmed, as described below.

### 29.4.2   Time dilation due to gravitation and motion together

High accuracy time keeping, low earth orbit satellite tracking, and pulsar timing are applications that require the consideration of the combined effects of mass and motion in producing time dilation. Practical examples include the International Atomic Time standard and its relationship with the Barycentric Coordinate Time standard used for interplanetary objects.

Relativistic time dilation effects for the solar system and the earth can be modeled very precisely by the Schwarzschild solution to the Einstein field equations. In the Schwarzschild metric, the interval $dt$E is given by[17][18]

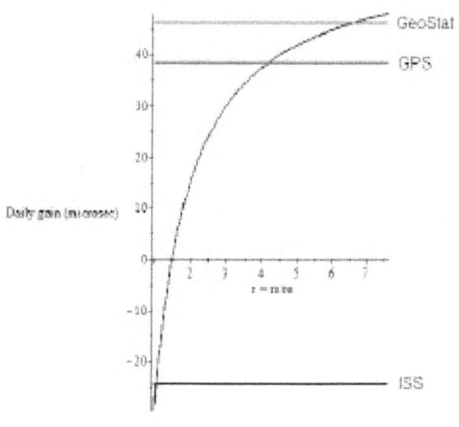

*Daily time dilation (gain or loss if negative) in microseconds as a function of (circular) orbit radius $\mathfrak{r} = \mathfrak{r}s/\mathfrak{r}e$, where $\mathfrak{r}s$ is satellite orbit radius and $\mathfrak{r}e$ is the equatorial Earth radius. At $\mathfrak{r} \approx 1.497$ [Note 1] there is no time dilation. Here the effects of motion and reduced gravity cancel. ISS astronauts fly below, whereas GPS and Geostationary satellites fly above.[11]*

$$dt_{\mathrm{E}}^2 = \left(1 - \frac{2GM_i}{r_i c^2}\right) dt_c^2 - \left(1 - \frac{2GM_i}{r_i c^2}\right)^{-1}$$

$$\frac{dx^2 + dy^2 + dz^2}{c^2}$$

where:

$dt$E is a small increment of proper time $t$E (an interval that could be recorded on an atomic clock);

$dt_c$ is a small increment in the coordinate $t_c$ (coordinate time);

$dx$, $dy$ and $dz$ are small increments in the three coordinates $x$, $y$, $z$ of the clock's position; and

$GM_i/r_i$ represents the sum of the Newtonian gravitational potentials due to the masses in the neighborhood, based on their distances $r_i$ from the clock. This sum $GM_i/r_i$ includes any tidal potentials, and is represented as $U$ (using the positive astronomical sign convention for gravitational potentials). The coordinate velocity of the clock is given by

$$v^2 = \frac{dx^2 + dy^2 + dz^2}{dt_c^2}.$$

The coordinate time $t_c$ is the time that would be read on a hypothetical "coordinate clock" situated infinitely far from all gravitational masses ($U = 0$), and stationary in the system of coordinates ($v = 0$). The exact relation between the rate of proper time and the rate of coordinate time for a clock with a radial component of velocity is

$$\frac{dt_{\mathrm{E}}}{dt_c} = \sqrt{1 - \frac{2U}{c^2} - \frac{v^2}{c^2} - \left(\frac{c^2}{2U} - 1\right)^{-1} \frac{v_{\shortparallel}^2}{c^2}}$$

where:

$v_\shortparallel$ is the radial velocity, and

$U = GM_i/r_i$ is the Newtonian potential, equivalent to half of the escape velocity squared.

The above equation is exact under the assumptions of the Schwarzschild solution.

## 29.5 Experimental confirmation

Time dilation has been tested a number of times. The routine work carried on in particle accelerators since the 1950s, such as those at CERN, is a continuously running test of the time dilation of special relativity. The specific experiments include:

### 29.5.1 Velocity time dilation tests

- Ives and Stilwell (1938, 1941). The stated purpose of these experiments was to verify the time dilation effect, predicted by Larmor–Lorentz ether theory, due to motion through the ether using Einstein's suggestion that Doppler effect in canal rays would provide a suitable experiment. These experiments measured the Doppler shift of the radiation emitted from cathode rays, when viewed from directly in front and from directly behind. The high and low frequencies detected were not the classically predicted values

$$\frac{f_0}{1 - v/c} \qquad \text{and} \qquad \frac{f_0}{1 + v/c}.$$

The high and low frequencies of the radiation from the moving sources were measured as[19]

$$\sqrt{\frac{1 + v/c}{1 - v/c}} f_0 = \gamma (1 + v/c) f_0$$

and

$$\sqrt{\frac{1 - v/c}{1 + v/c}} f_0 = \gamma (1 - v/c) f_0,$$

as deduced by Einstein (1905) from the Lorentz transformation, when the source is running slow by the Lorentz factor.

- Rossi and Hall (1941) compared the population of cosmic-ray-produced muons at the top of a mountain to that observed at sea level. Although the travel time for the muons from the top of the mountain to the base is several muon half-lives, the muon sample at the base was only moderately reduced. This is explained by the time dilation attributed to their high speed relative to the experimenters. That is to say, the muons were decaying about 10 times slower than if they were at rest with respect to the experimenters.

- Hasselkamp, Mondry, and Scharmann[20] (1979) measured the Doppler shift from a source moving at right angles to the line of sight. The most general relationship between frequencies of the radiation from the moving sources is given by:

$$f_{\text{detected}} = f_{\text{rest}} \left(1 - \frac{v}{c}\cos\phi\right) / \sqrt{1 - v^2/c^2}$$

as deduced by Einstein (1905).[21] For $\phi = 90°$ ($\cos\phi = 0$) this reduces to $f_{\text{detected}} = f_{\text{rest}}\gamma$. This lower frequency from the moving source can be attributed to the time dilation effect and is often called the transverse Doppler effect and was predicted by relativity.

- In 2010 time dilation was observed at speeds of less than 10 meters per second using optical atomic clocks connected by 75 meters of optical fiber.[22]

## 29.5.2   Gravitational time dilation tests

- In 1959 Robert Pound and Glen A. Rebka measured the very slight gravitational red shift in the frequency of light emitted at a lower height, where Earth's gravitational field is relatively more intense. The results were within 10% of the predictions of general relativity. In 1964, Pound and J. L. Snider measured a result within 1% of the value predicted by gravitational time dilation.[23] (See Pound–Rebka experiment)

- In 2010 gravitational time dilation was measured at the earth's surface with a height difference of only one meter, using optical atomic clocks.[22]

## 29.5.3   Velocity and gravitational time dilation combined-effect tests

- Hafele and Keating, in 1971, flew caesium atomic clocks east and west around the earth in commercial airliners, to compare the elapsed time against that of a clock that remained at the U.S. Naval Observatory. Two opposite effects came into play. The clocks were expected to age more quickly (show a larger elapsed time) than the reference clock, since they were in a higher (weaker) gravitational potential for most of the trip (c.f. Pound–Rebka experiment). But also, contrastingly, the moving clocks were expected to age more slowly because of the speed of their travel. From the actual flight paths of each trip, the theory predicted that the flying clocks, compared with reference clocks at the U.S. Naval Observatory, should have lost 40±23 nanoseconds during the eastward trip and should have gained 275±21 nanoseconds during the westward trip. Relative to the atomic time scale of the U.S. Naval Observatory, the flying clocks lost 59±10 nanoseconds during the eastward trip and gained 273±7 nanoseconds during the westward trip (where the error bars represent standard deviation).[24] In 2005, the National Physical Laboratory in the United Kingdom reported their limited replication of this experiment.[25] The NPL experiment differed from the original in that the caesium clocks were sent on a shorter trip (London–Washington, D.C. return), but the clocks were more accurate. The reported results are within 4% of the predictions of relativity, within the uncertainty of the measurements.

- The Global Positioning System can be considered a continuously operating experiment in both special and general relativity. The in-orbit clocks are corrected for both special and general relativistic time dilation effects as described above, so that (as observed from the earth's surface) they run at the same rate as clocks on the surface of the Earth.[26]

## 29.5.4   Muon lifetime

A comparison of muon lifetimes at different speeds is possible. In the laboratory, slow muons are produced, and in the atmosphere very fast moving muons are introduced by cosmic rays. Taking the muon lifetime at rest as the laboratory value of 2.197 μs, the lifetime of a cosmic ray produced muon traveling at 98% of the speed of light is about five times longer, in agreement with observations.[27] In the muon storage ring at CERN the lifetime of muons circulating with $\gamma = 29.327$ was found to be dilated to 64.378 μs, confirming time dilation to an accuracy of $0.9 \pm 0.4$ parts per thousand.[28] In this experiment the "clock" is the time taken by processes leading to muon decay, and these processes take place in the moving muon at its own "clock rate", which is much slower than the laboratory clock.

## 29.6 Space flight

Time dilation would make it possible for passengers in a fast-moving vehicle to travel further into the future while aging very little, in that their great speed slows down the passage of on-board time relative to that of an observer. That is, the ship's clock (and according to relativity, any human traveling with it) shows less elapsed time than the clocks of observers on earth. For sufficiently high speeds the effect is dramatic.[2] For example, one year of travel might correspond to ten years at home. Indeed, a constant 1 g acceleration would permit humans to travel through the entire known Universe in one human lifetime.[29] The space travelers could return to Earth billions of years in the future. A scenario based on this idea was presented in the novel *Planet of the Apes* by Pierre Boulle.

A more likely use of this effect would be to enable humans to travel to nearby stars without spending their entire lives aboard a ship. However, any such application of time dilation during interstellar travel would require the use of some new, advanced method of propulsion. The Orion Project has been the only major attempt toward this idea.

Current space flight technology has fundamental theoretical limits based on the practical problem that an increasing amount of energy is required for propulsion as a craft approaches the speed of light. The likelihood of collision with small space debris and other particulate material is another practical limitation. At the velocities presently attained, however, time dilation occurs but is too small to be a factor in space travel. Travel to regions of spacetime where gravitational time dilation is taking place, such as within the gravitational field of a black hole but outside the event horizon (perhaps on a hyperbolic trajectory exiting the field), could also yield results consistent with present theory.

### 29.6.1 Time dilation at constant force

In special relativity, time dilation is most simply described in circumstances where relative velocity is unchanging. Nevertheless, the Lorentz equations allow one to calculate proper time and movement in space for the simple case of a spaceship which is applied with a force per unit mass, relative to some reference object in uniform (i.e. constant velocity) motion, equal to $g$ throughout the period of measurement.

Let $t$ be the time in an inertial frame subsequently called the rest frame. Let $x$ be a spatial coordinate, and let the direction of the constant acceleration as well as the spaceship's velocity (relative to the rest frame) be parallel to the $x$-axis. Assuming the spaceship's position at time $t = 0$ being $x = 0$ and the velocity being $v_0$ and defining the following abbreviation

$$\gamma_0 = \frac{1}{\sqrt{1 - v_0^2/c^2}},$$

the following formulas hold:[30]

Position:

$$x(t) = \frac{c^2}{g}\left(\sqrt{1 + \frac{(gt + v_0\gamma_0)^2}{c^2}} - \gamma_0\right).$$

Velocity:

$$v(t) = \frac{gt + v_0\gamma_0}{\sqrt{1 + \frac{(gt + v_0\gamma_0)^2}{c^2}}}.$$

Proper time:

$$\tau(t) = \tau_0 + \int_0^t \sqrt{1 - \left(\frac{v(t')}{c}\right)^2}\, dt'.$$

In the case where $v(0) = v_0 = 0$ and $\tau(0) = \tau_0 = 0$ the integral can be expressed as a logarithmic function or, equivalently, as an inverse hyperbolic function:

$$\tau(t) = \frac{c}{g}\ln\left(\frac{gt}{c} + \sqrt{1 + \left(\frac{gt}{c}\right)^2}\right) = \frac{c}{g}\operatorname{arsinh}\left(\frac{gt}{c}\right).$$

### 29.6.2 Spacetime geometry of velocity time dilation

The green dots and red dots in the animation represent spaceships. The ships of the green fleet have no velocity relative to each other, so for the clocks on board the individual ships the same amount of time elapses relative to each other, and they can set up a procedure to maintain a synchronized standard fleet time. The ships of the "red fleet" are moving with a velocity of $0.866c$ with respect to the green fleet.

The blue dots represent pulses of light. One cycle of light-pulses between two green ships takes two seconds of "green time", one second for each leg.

As seen from the perspective of the reds, the transit time of the light pulses they exchange among each other is one second of "red time" for each leg. As seen from the perspective of the greens, the red ships' cycle of exchanging light pulses travels a diagonal path that is two light-seconds

*Time dilation in transverse motion*

## 29.7 See also

- Mass dilation
- Length contraction
- Albert Einstein

## 29.8 Footnotes

[1] Average time dilation has a weak dependence on the orbital inclination angle (Ashby 2003, p.32). The $r \approx 1.497$ result corresponds to [16] the orbital inclination of modern GPS satellites, which is 55 degrees.

## 29.9 References

[1] Ashby, Neil (2003). "Relativity in the Global Positioning System" (PDF). *Living Reviews in Relativity* **6**: 16. Bibcode:2003LRR.....6....1A. doi:10.12942/lrr-2003-1.

[2] Toothman, Jessika. "How Do Humans age in space?". *HowStuffWorks*. Retrieved 2012-04-24.

[3] Lu, Ed. "Expedition 7 – Relativity". *Ed's Musing from Space*. NASA. Retrieved 2012-04-24.

[4] Lu, Ed. "Expedition 7 – Relativity". *Ed's Musing from Space*. NASA. Retrieved 2015-01-20. In fact it mentions as result 0.007 seconds, but it is easily seen that this is due to crude intermediate rounding.

[5] For sources on special relativistic time dilation, see Albert Einstein's own popular exposition, published in English translation (1920) as Einstein, Albert (1920). "On the Idea of Time in Physics". *Relativity: The Special and General Theory*. Henri Holt. ISBN 1-58734-092-5. and also in sections 9–12. See also the articles Special relativity, Lorentz transformation and Relativity of simultaneity.

[6] "spaceplace.nasa.gov".

[7] "Hafele and Keating Experiment". *NA*. Retrieved 2015-02-04.

[8] Overbye, Dennis (2005-06-28). "A Trip Forward in Time. Your Travel Agent: Einstein.". *The New York Times*. Retrieved 2015-12-08.

[9] Gott, J., Richard (2002). *Time Travel in Einstein's Universe*. p. 75.

[10] Cassidy, David C.; Holton, Gerald James; Rutherford, Floyd James (2002). *Understanding Physics*. Springer-Verlag. p. 422. ISBN 0-387-98756-8.

[11] Cutner, Mark Leslie (2003). *Astronomy, A Physical Perspective*. Cambridge University Press. p. 128. ISBN 0-521-82196-7.

long. (As seen from the green perspective the reds travel 1.73 ( $\sqrt{3}$ ) light-seconds of distance for every two seconds of green time.)

One of the red ships emits a light pulse towards the greens every second of red time. These pulses are received by ships of the green fleet with two-second intervals as measured in green time. Not shown in the animation is that all aspects of physics are proportionally involved. The light pulses that are emitted by the reds at a particular frequency as measured in red time are received at a lower frequency as measured by the detectors of the green fleet that measure against green time, and vice versa.

The animation cycles between the green perspective and the red perspective, to emphasize the symmetry. As there is no such thing as absolute motion in relativity (as is also the case for Newtonian mechanics), both the green and the red fleet are entitled to consider themselves motionless *in their own frame of reference*.

Again, it is vital to understand that the results of these interactions and calculations reflect the real state of the ships as it emerges from their situation of relative motion. It is not a mere quirk of the method of measurement or communication.

[12] Lerner, Lawrence S. (1996). *Physics for Scientists and Engineers, Volume 2*. Jones and Bartlett. pp. 1051–1052. ISBN 0-7637-0460-1.

[13] Ellis, George F. R.; Williams, Ruth M. (2000). *Flat and Curved Space-times* (2n ed.). Oxford University Press. pp. 28–29. ISBN 0-19-850657-0.

[14] Adams, Steve (1997). *Relativity: An introduction to space-time physics*. CRC Press. p. 54. ISBN 0-7484-0621-2.

[15] Edwin F. Taylor, John Archibald Wheeler (1992). *Spacetime Physics: Introduction to Special Relativity*. New York: W. H. Freeman. ISBN 0-7167-2327-1.

[16] Ashby, Neil (2002). "Relativity in the Global Positioning System". *Physics Today* **55** (5): 45. Bibcode:2002PhT....55e..41A. doi:10.1063/1.1485583.

[17] See equations 2 & 3 (combined here and divided throughout by $c^2$) at pp. 35–36 in Moyer, T. D. (1981). "Transformation from proper time on Earth to coordinate time in solar system barycentric space-time frame of reference". *Celestial Mechanics* **23**: 33–56. Bibcode:1981CeMec..23...33M. doi:10.1007/BF01228543.

[18] A version of the same relationship can also be seen at equation 2 in Ashbey, Neil (2002). "Relativity and the Global Positioning System" (PDF). *Physics Today* **55** (5): 45. Bibcode:2002PhT....55e..41A. doi:10.1063/1.1485583.

[19] Blaszczak, Z. (2007). *Laser 2006*. Springer. p. 59. ISBN 3540711139.

[20] Hasselkamp, D.; Mondry, E.; Scharmann, A. (1979). "Direct observation of the transversal Doppler-shift". *Zeitschrift für Physik A* **289** (2): 151–155. Bibcode:1979ZPhyA.289..151H. doi:10.1007/BF01435932.

[21] Einstein, A. (1905). "On the electrodynamics of moving bodies". Fourmilab.

[22] Chou, C. W.; Hume, D. B.; Rosenband, T.; Wineland, D. J. (2010). "Optical Clocks and Relativity". *Science* **329** (5999): 1630–1633. Bibcode:2010Sci...329.1630C. doi:10.1126/science.1192720. PMID 20929843.

[23] Pound, R. V.; Snider J. L. (November 2, 1964). "Effect of Gravity on Nuclear Resonance". *Physical Review Letters* **13** (18): 539–540. Bibcode:1964PhRvL..13..539P. doi:10.1103/PhysRevLett.13.539.

[24] Nave, C. R. (22 August 2005). "Hafele and Keating Experiment". *HyperPhysics*. Retrieved 2013-08-05.

[25] "Einstein" (PDF). *Metromnia*. National Physical Laboratory. 2005. pp. 1–4.

[26] Kaplan, Elliott; Hegarty, Christopher (2005). *Understanding GPS: Principles and Applications*. Artech House. p. 306. ISBN 1-58053-895-9. Extract of page 306

[27] Stewart, J. V. (2001). *Intermediate electromagnetic theory*. World Scientific. p. 705. ISBN 981-02-4470-3.

[28] Bailey, J. et al. Nature 268, 301 (1977)

[29] Calder, Nigel (2006). *Magic Universe: A grand tour of modern science*. Oxford University Press. p. 378. ISBN 0-19-280669-6.

[30] See equations 3, 4, 6 and 9 of Iorio, Lorenzo (2004). "An analytical treatment of the Clock Paradox in the framework of the Special and General Theories of Relativity". *Foundations of Physics Letters* **18**: 1–19. arXiv:physics/0405038. Bibcode:2005FoPhL..18....1I. doi:10.1007/s10702-005-2466-8.

## 29.10   Further reading

- Callender, C.; Edney, R. (2001). *Introducing Time*. Icon Books. ISBN 1-84046-592-1.

- Einstein, A. (1905). "Zur Elektrodynamik bewegter Körper". *Annalen der Physik* **322** (10): 891. Bibcode:1905AnP...322..891E. doi:10.1002/andp.19053221004.

- Einstein, A. (1907). "Über die Möglichkeit einer neuen Prüfung des Relativitätsprinzips". *Annalen der Physik* **328** (6): 197–198. Bibcode:1907AnP...328..197E. doi:10.1002/andp.19073280613.

- Hasselkamp, D.; Mondry, E.; Scharmann, A. (1979). "Direct Observation of the Transversal Doppler-Shift". *Zeitschrift für Physik A* **289** (2): 151–155. Bibcode:1979ZPhyA.289..151H. doi:10.1007/BF01435932.

- Ives, H. E.; Stilwell, G. R. (1938). "An experimental study of the rate of a moving clock". *Journal of the Optical Society of America* **28** (7): 215–226. doi:10.1364/JOSA.28.000215.

- Ives, H. E.; Stilwell, G. R. (1941). "An experimental study of the rate of a moving clock. II". *Journal of the Optical Society of America* **31** (5): 369–374. doi:10.1364/JOSA.31.000369.

- Joos, G. (1959). "Lehrbuch der Theoretischen Physik, Zweites Buch" (11th ed.). |chapter= ignored (help)

- Larmor, J. (1897). "On a dynamical theory of the electric and luminiferous medium". *Philosophical Transactions of the Royal Society* **190**: 205–300. Bibcode:1897RSPTA.190..205L. doi:10.1098/rsta.1897.0020. (third and last in a series of papers with the same name).

- Poincaré, H. (1900). "La théorie de Lorentz et le principe de Réaction". *Archives Néerlandaises* **5**: 253–78.

- Puri, A. (2015). "Einstein versus the simple pendulum formula: does gravity slow all clocks?". *Physics Education* **50** (4): 431. Bibcode:2015PhyEd..50..431P. doi:10.1088/0031-9120/50/4/431.

- Reinhardt, S.; et al. (2007). "Test of relativistic time dilation with fast optical atomic clocks at different velocities" (PDF). *Nature Physics* **3** (12): 861–864. Bibcode:2007NatPh...3..861R. doi:10.1038/nphys778.

- Rossi, B.; Hall, D. B. (1941). "Variation of the Rate of Decay of Mesotrons with Momentum". *Physical Review* **59** (3): 223. Bibcode:1941PhRv...59..223R. doi:10.1103/PhysRev.59.223.

- Weiss, M. "Two way time transfer for satellites". National Institute of Standards and Technology.

- Voigt, W. (1887). "Über das Doppler'sche princip". *Nachrichten von der Königlicher Gesellschaft der Wissenschaften zu Göttingen* **2**: 41–51.

## 29.11   External links

- Online Time Dilation Calculator

- Proper Time

- Merrifield, Michael. "Lorentz Factor (and time dilation)". *Sixty Symbols*. Brady Haran for the University of Nottingham.

# Chapter 30

# Super Minkowski space

In mathematics and physics, **super Minkowski space** or **Minkowski superspace** is a supersymmetric extension of Minkowski space, sometimes used as the base manifold for superfields. It is acted on by the super Poincaré algebra.

## 30.1   Definition

The underlying supermanifold of super Minkowski space is isomorphic to a super vector space given by the direct sum of ordinary Minkowski spacetime in $d$ dimensions (often taken to be 4) and a number $N$ of real spinor representations of the Lorentz algebra. (When $d$ is 2 mod 4 this is slightly ambiguous because there are 2 different real spin representations, so one needs to replace $N$ by a pair of integers $N=N_1+N_2$, though some authors use a different convention and take $N$ copies of both spin representations.)

However this construction is misleading for two reasons: first, super Minkowski space is really an affine space over a group rather than a group, or in other words it has no distinguished "origin", and second, the underlying supergroup of translations is not a super vector space but a nilpotent supergroup of nilpotent length 2. This supergroup has the following Lie algebra. Suppose that $M$ is Minkowski space, and $S$ is a finite sum of irreducible real spinor representations. Then there is an invariant symmetric bilinear map [,] from $S \times S$ to $M$ that is positive definite in the sense that the image of $s \times s$ is in the closed positive cone of $M$, and is nonzero if $s$ is nonzero. This bilinear map is unique up to isomorphism. The Lie superalgebra has $M$ as its even part, $S$ as its odd or fermionic part, and the Lie bracket is given by [,] (and the Lie bracket of anything in $M$ with anything is zero).

The dimensions of the irreducible real spinor representations for various dimensions $d$ of spacetime are given by the following table:

The table repeats whenever the dimension increases by 8, except that the dimensions of the spin representations are multiplied by 16.

## 30.2   Notation

In the physics literature, Minkowski spacetime is often specified by giving the dimension $d$ of the even bosonic part, and the number of times $N$ that each irreducible spinor representation occurs in the odd fermionic part. In mathematics, Minkowski spacetime is sometimes specified in the form $M^{m|n}$ where $m$ is the dimension of the even part and $n$ the dimension of the odd part. The relation is as follows: the integer $d$ in the physics notation is the integer $m$ in the mathematics notation, while the integer $n$ in the mathematics notation is a power of 2 times the integer $N$ in the physics notation, where the power of 2 is the dimension of the irreducible real spinor representation (or twice this if there are two irreducible real spinor representations). For example, the $d=4$, $N=1$ Minkowski spacetime is $M^{4|4}$ while the $N=2$ Minkowski spacetime is $M^{4|8}$. When the dimension $d$ or $m$ is 2 mod 4 there are two different irreducible real spinor representations, and authors use various different conventions.

In physics the letter $P$ is used for a basis of the even bosonic part of the Lie superalgebra, and the letter $Q$ is often used for a basis of the complexification of the odd fermionic part, so in particular the structure constants of the Lie superalgebra may be complex rather than real. Often the basis elements $Q$ come in complex conjugate pairs, so the real subspace can be recovered as the fixed points of complex conjugation.

## 30.3   References

Deligne, Pierre; Morgan, John W. (1999), "Notes on supersymmetry (following Joseph Bernstein)", in Deligne, Pierre; Etingof, Pavel; Freed, Daniel S.; Jeffrey, Lisa C.; Kazhdan, David; Morgan, John W.; Morrison, David R.; Witten., Edward, *Quantum fields and strings: a course for mathematicians, Vol. 1*, Providence, R.I.: American Mathematical Society, pp. 41–97, ISBN 978-0-8218-1198-6,

MR 1701597

# 30.4 Text and image sources, contributors, and licenses

## 30.4.1 Text

- **Minkowski space** *Source:* https://en.wikipedia.org/wiki/Minkowski_space?oldid=709121334 *Contributors:* XJaM, William Avery, Stevertigo, Patrick, PhilipMW, Michael Hardy, Tim Starling, Karada, Stupidmoron, Hawthorn, Charles Matthews, Kbk, Zoicon5, LMB, Phys, Fvw, Josh Cherry, Jheise, Marc Venot, Decumanus, Giftlite, Gene Ward Smith, BenFrantzDale, Lethe, MathKnight, Fropuff, Dratman, Anythingyouwant, Frau Hitt~enwiki, Hidaspal, Bender235, Syp, Rgdboer, I9Q79oL78KiL0QTFHgyc, Phils, Anthony Appleyard, NukWik, Egg, Ringbang, Japanese Searobin, Stemonitis, Linas, StradivariusTV, Mpatel, Tlroche, Rjwilmsi, Zbxgscqf, R.e.b., Lionelbrits, Mathbot, Chobot, Dylan Thurston, DVdm, Hmonroe, YurikBot, Hairy Dude, Rhythm, Nick, Arthur Rubin, KasugaHuang, That Guy, From That Show!, Sardanaphalus, KnightRider~enwiki, SmackBot, Turbos10~enwiki, Vald, ZerodEgo, Xie Xiaolei, Bluebot, Alexwagner, Complexica, Waprap, Jbergquist, Andrei Stroe, Gregapan, Lambiam, Eliyak, Jim.belk, NongBot~enwiki, Bosons, Xenure, Loadmaster, Dan Gluck, JRSpriggs, CRGreathouse, Cydebot, Michael C Price, Thijs!bot, Martin Hogbin, Headbomb, D.H, Noclevername, Gökhan, JAnDbot, Jyotirmoyb, Sullivan.t.j, First Harmonic, Maurice Carbonaro, TomyDuby, Goutui, WaiteDavid137, Fylwind, Equazcion, XCelam, JohnBlackburne, Red Act, Nxavar, DWP17, PaulTanenbaum, Geometry guy, Zain Ebrahim111, StevenJohnston, YohanN7, Juanmantoya, Paradoctor, Happysailor, DaveBeal, Henry Delforn (old), Udirock, Mr. Stradivarius, Renata500, Martarius, ClueBot, RODERICKMOLASAR, Bastien Sens-Méyé~enwiki, Sun Creator, Carriearchdale, Forbes72, Whizmd, Addbot, Mortense, Kwvan, Gregz08, 84user, Tide rolls, Cesiumfrog, Luckas-bot, Yobot, Amirobot, AnomieBOT, Sfaefaol, Illegal604, Corwin323, ArthurBot, Nanog, NOrbeck, RibotBOT, Ashi009, MeDrewNotYou, Paine Ellsworth, Sławomir Biały, Guitarstud101, Tsester, Serols, Stephen Henry Davies, Tcnuk, Trappist the monk, Retired user 0001, Dinamik-bot, John of Reading, WikitanvirBot, AlexUT, Netheril96, Hhhippo, Quondum, Thine Antique Pen, Maschen, JFB80, ClueBot NG, Jack Greenmaven, CaroleHenson, Helpful Pixie Bot, BG19bot, F=q(E+v^B), Aisteco, ChrisGualtieri, Deltahedron, Makecat-bot, Twhitguy14, TwoTwoHello, CsDix, Frinthruit, JaconaFrere, Monkbot, Biblioworm, WillemienH, KasparBot, Arnaud Dorthe, Baking Soda and Anonymous: 98

- **Spacetime** *Source:* https://en.wikipedia.org/wiki/Spacetime?oldid=720073768 *Contributors:* Paul Drye, The Cunctator, Dreamyshade, Bryan Derksen, Malcolm Farmer, Josh Grosse, XJaM, Karl Palmen, Stevertigo, Patrick, Infrogmation, Smelialichu, Michael Hardy, Wshun, Pit~enwiki, Dcljr, Karada, Mcarling, Looxix~enwiki, William M. Connolley, Snoyes, Kingturtle, Glenn, Loren Rosen, Hollgor, Adam Bishop, Dcoetzee, Reddi, Jay, E23~enwiki, Omegatron, Fvw, Robbot, Kristof vt, Goethean, Ashley Y, Sverdrup, Blainster, DHN, Papadopc, Tea2min, Finlander, Matt Gies, Giftlite, ByteCoder, Wolfkeeper, Herbee, Tom Radulovich, Everyking, Snowdog, Michael Devore, Niteowlneils, Yekrats, Eequor, Utcursch, Beland, Karol Langner, Wikimol, JimWae, Karl-Henner, Adashiel, ELApro, Chris Howard, Juan Ponderas, Discospinster, Rich Farmbrough, Cacycle, Ascánder, Dolda2000, Bender235, Ben Standeven, El C, Rgdboer, Lankiveil, Shoujun, Teorth, Che090572, Rbj, Tobacman, I9Q79oL78KiL0QTFHgyc, Como, Obradovic Goran, Free Bear, Keenan Pepper, Sourcer66~enwiki, Riana, Geoff-codes, ReyBrujo, Arag0rn, DonQuixote, Eddie Dealtry, DominicC13, H2g2bob, Loxley~enwiki, Camw, StradivariusTV, TheNightFly, Pkeck, ^demon, Doran, Jeff3000, Mpatel, GregorB, Palica, Graham87, Deltabeignet, Li-sung, Mkn1234, MekaD, Rjwilmsi, KYPark, Kinu, Vary, MarSch, FayssalF, Lebha, Mathbot, Alexjohnc3, Jrtayloriv, Exelban, Pete.Hurd, Tardis, Chobot, Tene, DVdm, VolatileChemical, YurikBot, Wavelength, Splintercellguy, Wolfmankurd, CanadianCaesar, Yamara, NawlinWiki, Mipadi, Trovatore, Schlafly, JocK, Crasshopper, Tony1, T, Zythe, Gadget850, Sahands, Light current, Zzuuzz, StuRat, KGasso, JoanneB, Heathhunnicutt, Anclation~enwiki, RG2, Teply, Mejor Los Indios, Qero, Eigenlambda, Sardanaphalus, SmackBot, RDBury, Formativ, Maksim-e~enwiki, Forteller~enwiki, RaulMiller, Ashill, Kurochka, Lestrade, InverseHypercube, KnowledgeOfSelf, C.Fred, AndreasJS, Jaytan, Alex earlier account, JeffieAlex, Yamaguchi先生, Gilliam, NickGarvey, JMiall, Oli Filth, TheScurvyEye, Silly rabbit, Complexica, Dabigkid, Jerome Charles Potts, Nbarth, Sbharris, Bryan Truitt, Can't sleep, clown will eat me, Tamfang, Chlewbot, Rrburke, Celarnor, Tsop, CanDo, Dylanrush, RaCha'ar, Mtmelendez, Looris, Richard001, Hammer1980, Romanski, Sayden, Kuru, MagnaMopus, Hernoor, Homan2006, 16@r, Loadmaster, Lampman, Hypnosifl, Ace Frahm, Inquisitus, FVP, Shoeofdeath, Newone, Yourstruly, Andrew Hampe, Lxl, Aeons, Xammer, Paolodm, CalebNoble, Robinhw, JForget, Twipie, Blve23, Jsd, Jnoa, WeggeBot, Myasuda, Azakreski, Joshua BishopRoby, Cydebot, AniMate, Kanags, Fl, MC10, Llort, Eu.stefan, Palindromica, Manfroze, DarkLink, Ameliorate!, DBaba, TarquiniusWikipedius, Kylewriter, Raoul NK, Letranova, Thijs!bot, Wikid77, Jedibob5, HappyInGeneral, Gamer007, Headbomb, Vertium, RolanGaros, Pigalle, Washingtonlerias, Ubuthustra, D.H, Nick Number, Klausness, Sam42, DarthNemesis, Northumbrian, Escarbot, WikiSlasher, AntiVandalBot, Seaphoto, Maxibons, Tim Shuba, Braindrain0000, Tempest115, Jrw@pobox.com, Narssarssuaq, Husond, MER-C, Andrewdolby, RogierBrussee, Bongwarrior, VoABot II, Bakken, Appraiser, Faizhaider, Cuardin, Stijn Vermeeren, Trebor1, Catgut, Cardamon, NMarkRoberts, IkonicDeath, MetsBot, Mwasim1, JaGa, GuelphGryphon98, NatureA16, FisherQueen, Flowanda, MartinBot, TechnoFaye, Wikeepeedier, Player 03, Tgeairn, HEL, J.delanoy, Bobvinson, Maurice Carbonaro, Foober, 3halfinchfloppy, Lantonov, NewEnglandYankee, LeighvsOptimvsMaximvs, KylieTastic, Ja 62, Vinsfan368, Izno, Idioma-bot, Makewater, 28bytes, VolkovBot, XCelam, JohnBlackburne, AlnoktaBOT, Philip Trueman, Oshwah, Zidonuke, Red Act, Anonymous Dissident, Yilloslime, Fizzackerly, PaulTanenbaum, PhilyG, Wingedsubmariner, Hotmoklet, Eubulides, Zhongsan, SmileToday, Falcon8765, Cubed mass, LachlanSosa, StevenJohnston, Hunter826242, PSSnyder, Hobojaks, YohanN7, SieBot, ShiftFn, Paradoctor, Dawn Bard, Vanished user 82345ijgeke4tg, Flyer22 Reborn, Csblack, Henry Delforn (old), Nuttycoconut, MrWikiMiki, Hjelmerus, Dposte46, Jeanlovecomputers, Mátyás, OKBot, Fedosin, Coldcreation, Fuddle, Mike2vil, Anchor Link Bot, MarkMLl, VanishedUser sdu9aya9fs787sads, ImageRemovalBot, Martarius, De728631, ClueBot, The Thing That Should Not Be, EoGuy, Exploto, Razimantv, IshanAlmazi, Shinpah1, JFlav, Noca2plus, Eeekster, Tam 66 7, DPCU, Cenarium, Mozart21, Mentor364, Themantyke, McXX, Tin Whistle Man, Galor612, MalWilley, NERIC-Security, Rror, Avoided, Richard-of-Earth, Whizmd, Addbot, Derivator, Gravitophoton, DOI bot, Gul e, Startstop123, Gustavo José Meano Brito, Vishnava, CanadianLinuxUser, CarsracBot, RTG, Monypooh12, DFS454, Glane23, Tod.davidson, Mcsploogerson, AnnaFrance, Favonian, Doniago, West.andrew.g, Numbo3-bot, Lightbot, Zorrobot, Legobot, Yobot, Ht686rg90, THEN WHO WAS PHONE?, Allowgolf~enwiki, Synchronism, AnomieBOT, Jim1138, Materialscientist, Citation bot, Maxis ftw, ChristianH, Expooz, Xqbot, Tjcheckley, Δ?, Anna Frodesiak, Shindamaru, False vacuum, Omnipaedista, Frankie0607, RibotBOT, Gsard, MerlLinkBot, Bearnfæder, 7575474087ALBERT, CES1596, FrescoBot, H.W. Clihor, Paine Ellsworth, Dogposter, Tj2691, Majopius, Mouselarry, Haeinous, Vbrcat, Citation bot 1, Deadtotruth, Pinethicket, Hypernovic, Lesath, 10metreh, Tom.Reding, Smuckola, Rushbugled13, A8UDI, RedBot, Sjb13, Elvis633, December21st2012Freak, Fredkinfollower, IVAN3MAN, Double sharp, Euriditi, TobeBot, 0x30114, LogAntiLog, Jonkerz, Dinamik-bot, Capt. James T. Kirk, Aribashka, Briann MacAmhlaidh, Seahorseruler, Bj norge, The Utahraptor, Scipioafficanus, Reagster, Thslackliner19, Theslackliner19, EmausBot, WikitanvirBot, AlexUT, Pekka.virta, Stewiefool, RA0808, Slightsmile, Wikipelli, Scalable Vector Raccoon, Thecheesykid, Hhhippo, JSquish, Fæ, StringTheory11, Int21hexster, Quondum, Ellie Rickett, Donner60, Nelium12, Maheshbahadur, Ihardlythinkso, AndyTheGrump, RockMagnetist, Winston7, Rmashhadi, ClueBot NG, MelbourneStar, Irisrune, Lanthanum-138, Doorsrocklikerocks, Frietjes, Helpful Pixie Bot, HMSSolent, Gob Lofa, Bibcode Bot, Transscientific, Bobc3, BG19bot, Hz.tiang, Fw0116, Davidiad, Piguy101, Giarcea, Naveedyaykhan, Cadiomals, Kaaalbert, MrBill3, Wiki2103, Penguinstorm300, RiseUpAgain, The1337gamer, Nirmal kumar 9, SteveBM, Stigmatella aurantiaca, Cyberbot II, Zachhansonhart, U5ard, Nkzf, Khazar2, Ekren, Larskk101, Vanished user 23i4hjwrjfiij4t, Kevinfrank17, Twhitguy14, CuriousMind01, Mike.leivers, Kingcircle, DmVdx, Telfordbuck, Cadillac000, Asad shayan, Epicgenius, Bantennyson, Eachandall, AnthonyJ Lock, B14709, Ericgermate, Wedgeline, RogrMexico, Jwratner1, NottNott, Frinthruit, Iliketrains1234567890, Skyshad4w, AnonymousAuthority, TuxLibNit, Da.pro1, Monkbot, Yikkayaya, BethNaught, Zacwill, 97dc, Neeraj Bhakta, KH-1, Vedic Earthian, AHusain3141, Aaronfranke, 39Debangshu, Tetra quark, Isambard Kingdom, Epictacotree, Anand2202, Kbap2002, Djniew, Hriton, KasparBot, Edgar-lausanne, Alout, Baking Soda, Tramrattan, Chemistry1111, Sundaram108, Boomer Vial, Spaceviewer, Tolibonboni06, Virtumanity and Anonymous: 576

• **Minkowski plane** *Source:* https://en.wikipedia.org/wiki/Minkowski_plane?oldid=694268597 *Contributors:* Apyule, MFH, Chris the speller, YK Times, Sullivan.t.j, CommonsDelinker, Ramses68, Luckas-bot, Yobot, AnomieBOT, FrescoBot, MondalorBot, Quondum, Wcherowi, Ag2gaeh, K9re11 and WillemienH

• **Four-dimensional space** *Source:* https://en.wikipedia.org/wiki/Four-dimensional_space?oldid=720271742 *Contributors:* Zundark, The Anome, SJK, William Avery, Patrick, Michael Hardy, Dcljr, Delirium, Ahoerstemeier, PeterBrooks, Charles Matthews, Reddi, Dysprosia, Robbot, Fredrik, Gandalf61, AceMyth, Dina, Tea2min, Adam78, Giftlite, Mporter, Cobaltbluetony, Lethe, Fropuff, Bovlb, Daniel Brockman, Jossi, Bumm13, Tomruen, Bodnotbod, Icairns, Sam Hocevar, Mare-Silverus, Discospinster, Pak21, FT2, Bender235, JoeSmack, Ben Standeven, Aecis, Rgdboer, Lankiveil, AJP, Bobo192, Army1987, Shlomital, Physicistjedi, Ultra megatron, Alansohn, SemperBlotto, Cjthellama, Hu, Wtmitchell, Jheald, Mikeo, Axeman89, Gamiar, Oleg Alexandrov, Feezo, OwenX, Linas, LOL, Yansa, Kmg90, Maartenvdbent, Jon Harald Søby, Marvelvsdc, KyuuA4, Jclemens, Zoz, Phoenix-forgotten, Chipuni, Attitude2000, MarSch, Salix alba, NeonMerlin, DoubleBlue, Algebra, FlaBot, TiagoTiago, SiriusB, Nihiltres, RexNL, Gurch, 8q67n4tqr5, TeaDrinker, LeCire~enwiki, Cpcheung, TheSun, King of Hearts, DaGizza, Jared Preston, DVdm, VolatileChemical, Amaurea, Algebraist, YurikBot, Wavelength, Maelin, Splintercellguy, Sceptre, Retodon8, Michael Slone, KamuiShirou, SpuriousQ, Hellbus, Shawn81, Yamara, MightyGiant, Ihope127, Havok, NawlinWiki, Tailpig, MacGyver07, Deckiller, Mgnbar, Tetracube, Cspalletta, PBurns, Arthur Rubin, Th1rt3en, CWenger, Fram, Geoffrey.landis, AndrewWTaylor, Attilios, Havocrazy, SmackBot, Honza Záruba, GoldenXuniversity, Vkyrt, Unyoyega, Canthusus, Alex earlier account, Gilliam, Kevinalewis, Quinsareth, Oli Filth, Magicindark, Moshe Constantine Hassan Al-Silverburg, Nbarth, Robth, MaxSem, Can't sleep, clown will eat me, Tamfang, Onorem, Koolone0, Jwy, Nakon, Snakeyes (usurped), Jbergquist, Aaker, Dogears, Clicketyclack, Meetarnav, Mike1901, Accurizer, Bjankuloski06en~enwiki, 041744, Xenure, Hotblaster, Werdan7, Kazikame, Cerealkiller13, Animedude360, Nehrams2020, Iridescent, FVP, Use4d, Mimicat, Fsotrain09, Octane, Phoenixrod, Tawkerbot2, Robinhw, CmdrObot, Lavateraguy, JasonHise, Zureks, Nadyes, PeanutCheeseBar, Arnavion, Skrapion, Skybon, Cydebot, Reywas92, Red Director, Pedro Fonini, Robertinventor, Krm500, Ryan Gittins, Brink14, Aazn, JodyB, Nol888, Mailerdaemon, Cheveyo, JamesAM, Thijs!bot, Ulnevets, GentlemanGhost, Protocoi, Pampas Cat, Mojo Hand, AgentPeppermint, D.H, AntiVandalBot, Luna Santin, Ryttu3k, Oatmealcookiemon, The Dan, Figma, Fusionshrimp, JAnDbot, Husond, Pipedreamergrey, Xeno, Hut 8.5, J-stan, Stuffchanger, Bennybp, Bongwurrior, VoABot II, AuburnPilot, Littlewood~enwiki, Catgut, Jbav1278, David Eppstein, JMyrleFuller, Glen, Wdflake, Donrad, Rettetast, Nono64, Akronym, LedgendGamer, Asalt2233~enwiki, Tgeairn, Bogey97, Uncle Dick, Extransit, Thaurisil, Century0, Lantonov, Zoxxi, McSly, Starnestommy, Supuhstar, LittleHow, NewEnglandYankee, Rominandreu, Richard Wolf VI, Juliancolton, Cometstyles, DH85868993, MoForce, Homo logos, Xiahou, RJASE1, Xnuala, Hamzabahaa, Wikieditor06, BowToChris, Seldon1, Chaos5023, JohnBlackburne, Soliloquial, Veddan, Philip Trueman, Zidonuke, Anonymous Dissident, Crohnie, Sean D Martin, Berchin, IronMaidenRocks, Ian Strachan, Thomas s. briggs, AgentCDE, Dmcq, Subh83, SieBot, RageGarden, Pi is 3.14159, Flyer22 Reborn, Paolo.dL, Prestonmag, Faradayplank, Harry-, Yone Fernandes, Nick90210, Lightmouse, Hobartimus, Svick, Pinkadelica, ImageRemovalBot, ClueBot, The Thing That Should Not Be, ArdClose, Njbh9, Mild Bill Hiccup, SuperHamster, Blanchardb, JohnTheCrow, Alexbot, Brews ohare, NuclearWarfare, Arjayay, Razorflame, Frozen4322, Thfo, Error −128, Drippy knees, JDPhD, Questsong, Versus22, SoxBot III, Christianw7, XLinkBot, Hotcrocodile, Zzubiri, WikHead, SilvonenBot, Alexius08, Jedi870, Addbot, Proofreader77, Theothersteve7, Jafeluv, Tcncv, Ronhjones, Jncraton, Adam08, Glane23, Glass Sword, Favonian, Farmercarlos, Tassedethe, Pbryant7, Tide rolls, Teles, Zorrobot, Quantumobserver, Luckas-bot, Roflcopter1000, Yobot, TaBOT-zerem, Sarrus, MassimoAr, Tempodivalse, AnomieBOT, Wickedmangroves, Turul2, AdjustShift, Citation bot, ArthurBot, Xqbot, Intelati, Blackknight7429, JFBGT, RadiX, Bradjuhasz, Papercutbiology, HedonismBot2911, Basket of Puppies, Locobot, Bigger digger, WaysToEscape, FrescoBot, 123456789riitta, 123456789gary, Sky Attacker, Majopius, A little insignificant, Åkebråke, DrilBot, Pinethicket, I dream of horses, JranZu, Atm2177, Daclyff, LordxDeath, Jschnur, Btilm, RandomStringOfCharacters, RockSolidCosmo, Double sharp, Puzl bustr, Vrenator, 4, Juana1990, DARTH SIDIOUS 2, Whisky drinker, Swishaa218, Regancy42, Fredgds, WikitanvirBot, Gfoley4, Pekka.virta, Vish-aero, Dualitynature, GoingBatty, Vanished user zq46pw21, 4dimention, Slightsmile, Tommy2010, Wikipelli, Fafaerer, ZéroBot, Deehack, Azuris, Bamyers99, Brandmeister, L Kensington, Alborzagros, Tyughj2, MPLSboarder, Orange Suede Sofa, Brianumeda, DASHBotAV, ClueBot NG, Wcherowi, Joefromrandb, Vacation9, Bryce Albe-Quenzer, Bakrnl, LaszloSimon, Frietjes, Fujicapesta, Helpful Pixie Bot, Ernest3.141, BG19bot, Waqsajaz, Longbyte1, Nospildoh, Piguy101, Mark Arsten, Zeke, the Mad Horrorist, Puzzle314, Muthafucker, Snow Blizzard, Tesseract4d, Isacdaavid, CeraBot, Alkagl, NGC 2736, JYBot, Rainbowking5, Davidcdunne, Samwick6, Chakravarti1997, Hillbillyholiday, Uniquestman, 4-Dimensional0034, Neve2004, Prokaryotes, Mrkicky, Martixy, Newold123, Technetium-99, Jackscd, Artist.poet, Loraof, Infernus 780, ChaoticDequix, Dalvarezso, Maddiemmm, Ubernachten, Gustavo noise, HeidiShaban, Peshwavignesh, Babul446 and Anonymous: 602

• **Minkowski diagram** *Source:* https://en.wikipedia.org/wiki/Minkowski_diagram?oldid=721317060 *Contributors:* Patrick, SebastianHelm, Charles Matthews, Robbot, Aetheling, Wolfgangbeyer, Lupin, Ablewisuk, Rgdboer, Teorth, Apyule, Cpcjr, Burn, Bart133, DonQuixote, Dmitry Brant, PoccilScript, Mo-Al, Mathbot, DVdm, Siddhant, KSchutte, Escuerdo, Krea, JocK, Roger wilco, That Guy, From That Show!, Crystallina, SmackBot, Andyflux, Colonies Chris, OrphanBot, Serenity-Fr, JBel, Rook wave, Josh-Levin@ieee.org, Cronholm144, Dr Greg, CmdrObot, Myasuda, Marek69, D.H, Openlander, Ais523, JAnDbot, Swpb, Beżet, Belovedfreak, Biglovinb, HiraV, TXiKiBoT, Someguy1221, Pennstatephil, Sai2020, Jean-Christophe BENOIST, Juanmantoya, Smsarmad, Myrikhan, PlantTrees, Wwheaton, VQuakr, Namreh ekim, Anticipation of a New Lover's Arrival, The, Addbot, D1d4, 84user, Yobot, Bobw52, NOrbeck, FrescoBot, Paine Ellsworth, Duschi, Dewritech, Fisicamartin, Quondum, Maschen, Ollyoxenfree, Frietjes, Bibcode Bot, Acmedogs, BattyBot, CarrieVS, DmVdx, Frinthruit, 7Sidz, Myrral, Loraof, Tetra quark, Ari Ljd and Anonymous: 44

• **Minkowski space (number field)** *Source:* https://en.wikipedia.org/wiki/Minkowski_space_(number_field)?oldid=604300753 *Contributors:* Ironholds, Deltahedron and Anonymous: 1

- **Poincaré group** *Source:* https://en.wikipedia.org/wiki/Poincar%C3%A9_group?oldid=717756936 *Contributors:* AxelBoldt, Zundark, The Anome, XJaM, Mbecker, Stevertigo, Patrick, Michael Hardy, Marco Krohn, AugPi, Stupidmoron, Charles Matthews, Phys, Anupamsr, Giftlite, Lethe, Fropuff, DefLog~enwiki, Jossi, Rich Farmbrough, Bender235, Rgdboer, Aronbeekman, Danski14, Keenan Pepper, Gene Nygaard, Oleg Alexandrov, JFG, Mpatel, Allen3, Rjwilmsi, DVdm, YurikBot, That Guy, From That Show!, SmackBot, Incnis Mrsi, Nbarth, Tsca.bot, Kcordina, Cybercobra, JRSpriggs, Cydebot, Headbomb, Nearyan, JAnDbot, Fetchcomms, Yill577, SHCarter, Sullivan.t.j, Cuzkatzimhut, XCelam, Drschawrz, YohanN7, VVVBot, Phe-bot, Addbot, Luckas-bot, Ptbotgourou, AnomieBOT, Omnipaedista, Paine Ellsworth, Thinking of England, Meaghan, Jowa fan, Skater00, ZéroBot, Quondum, Git2010, Maschen, JFB80, Dexbot, CsDix, Prokaryotes, Kfitzell29, Ryanexler, Unknown111111 and Anonymous: 34

- **Euclidean space** *Source:* https://en.wikipedia.org/wiki/Euclidean_space?oldid=718598581 *Contributors:* AxelBoldt, Mav, Zundark, Tarquin, XJaM, Youandme, Tomo, Patrick, Michael Hardy, Dcljr, Karada, Looxix~enwiki, Angela, Charles Matthews, Dysprosia, Grendelkhan, David Shay, MathMartin, Tea2min, Tosha, Giftlite, Lethe, Fropuff, Sriehl, DefLog~enwiki, Andycjp, Tomruen, Iantresman, Tzanko Matev, JohnArmagh, Rich Farmbrough, Paul August, Rgdboer, Msh210, Jimmycochrane, PAR, Eddie Dealtry, Dirac1933, Woohookitty, Isnow, Qwertyus, MarSch, MZMcBride, VKokielov, Kolbasz, Fresheneesz, NevilleDNZ, Chobot, Bgwhite, JPD, Wavelength, Hede2000, Epolk, KSmrq, SpuriousQ, ENeville, Mgnbar, Arthur Rubin, Brian Tvedt, RG2, JDspeeder1, SmackBot, Iamhove, Incnis Mrsi, Reedy, Mhss, JoeKearney, Silly rabbit, Hongooi, Tamfang, SashatoBot, Jim.belk, DabMachine, Dan Gluck, Kaarel, Yggdrasil014, Heqs, CmdrObot, GargoyleMT, Rudjek, Philomath3, Aiko, Guy Macon, Orionus, Salgueiro~enwiki, JAnDbot, Bencherlite, CrizCraig, Magioladitis, TheChard, Avicennasis, Nucleophilic, Oderbolz, R'n'B, Reedy Bot, Policron, Trigamma, The enemies of god, Cerberus0, VolkovBot, IWhisky, Philip Trueman, Richardohio, WereSpielChequers, Da Joe, Caltas, Paolo.dL, MiNombreDeGuerra, Lightmouse, Denisarona, Tomas e, Mild Bill Hiccup, Gwguffey, Vsage, DhananSekhar, SilvonenBot, SkyLined, The Rationalist, Addbot, Nilesj, AkhtaBot, Pmod, Tide rolls, Legobot, Yobot, 陳, Collieuk, Materialscientist, Citation bot, Sandip90, Xqbot, St.nerol, Nfr-Maat, Deadclever23, RoyLeban, Ksuzanne, Mineralquelle, FrescoBot, Sławomir Biały, Alxeedo, RandomDSdevel, Gapato, Mikrosam Akademija 2, Yunesj, Wikivictory, EmausBot, John of Reading, Quondum, Gbsrd, ClueBot NG, Wcherowi, Master Uegly, Cntras, Frank.manus, ElectricUvula, ElphiBot, MRG90, FeralOink, Userbot12, Lugia2453, Brirush, Limit-theorem, Eyesnore, Yardimsever, Tentinator, Fentonville, Mgkrupa, BemusedObserver, OrganicAltMetal, Ro4sho, Preethambittu, KasparBot, Dan6233, DatGuy and Anonymous: 107

- **Euclidean group** *Source:* https://en.wikipedia.org/wiki/Euclidean_group?oldid=674549735 *Contributors:* Zundark, Edward, Patrick, Michael Hardy, Charles Matthews, Henrygb, Giftlite, BenFrantzDale, Fropuff, Tomruen, Rgdboer, Oleg Alexandrov, Woohookitty, Salix alba, MikeJ9919, Mathbot, Dmharvey, SmackBot, TimBentley, Joefaust, Nbarth, E946, Daqu, Hetar, Chetvorno, YK Times, Albmont, Stdazi, C quest000, JohnBlackburne, Kyle the bot, Red Act, Arcfrk, JerroldPease-Atlanta, Anchor Link Bot, Mr. Stradivarius, PixelBot, SchreiberBike, Addbot, Snaily, ArthurBot, Xqbot, Anne Bauval, Erik9bot, ZéroBot, Quondum, Solomon7968, Brirush, Misra120sourav and Anonymous: 8

- **Pseudo-Euclidean space** *Source:* https://en.wikipedia.org/wiki/Pseudo-Euclidean_space?oldid=704977187 *Contributors:* Michael Hardy, Tea2min, Tosha, Rgdboer, Oleg Alexandrov, Larsobrien, Incnis Mrsi, Magioladitis, MystBot, Addbot, LilHelpa, GrouchoBot, FrescoBot, ErikvanB, EmausBot, Dewritech, Quondum, JFB80, Helpful Pixie Bot, Monkbot, Yikkayaya and Anonymous: 3

- **Time** *Source:* https://en.wikipedia.org/wiki/Time?oldid=720822475 *Contributors:* AxelBoldt, The Cunctator, Lee Daniel Crocker, Archibald Fitzchesterfield, Mav, MarXidad, The Anome, Rjstott, Andre Engels, Eclecticology, XJaM, Matusz, William Avery, SimonP, DavidLevinson, Ben-Zin~enwiki, Heron, Karl Palmen, Stevertigo, Spiff~enwiki, Quintessent, Lir, Patrick, Drahflow, Michael Hardy, Tim Starling, Kwertii, Aholstenson, Gabbe, DaVinci, Ixfd64, Kalki, Cyde, Qaz, Fwappler, Karada, Delirium, Kosebamse, Tregoweth, CesarB, Looxix~enwiki, Ihcoyc, Ahoerstemeier, DavidWBrooks, Mac, Ronz, William M. Connolley, Bluelion, Suisui, Den fjättrade ankan~enwiki, Brettz9, Kevin Baas, Glenn, Susurrus, Jiang, Evercat, Harry Potter, Lee M, Smack, Hashar, Jengod, Disdero, RodC, Charles Matthews, Timwi, Dysprosia, Andrewman327, Wik, Selket, Patrick0Moran, Maximus Rex, Hyacinth, Saltine, Paul-L~enwiki, Joy, Fvw, Optim, Jerzy, Johnleemk, BenRG, Banno, Finlay McWalter, Pollinator, RickBeton, Branddobbe, Robbot, Sander123, Pigsonthewing, Fredrik, Chris 73, Aliter, Goethean, Lowellian, Arashi, Postdlf, Lsy098~enwiki, Rfc1394, Cornellier, Texture, Meelar, DHN, Rasmus Faber, SvavarL, Hadal, Mark Krueger, Borislav, Lupo, Wile E. Heresiarch, Alan Liefting, Enochlau, Dave6, Marc Venot, Ancheta Wis, Decumanus, Giftlite, Andries, ShaunMacPherson, Wikilibrarian, BenFrantzDale, Lee J Haywood, Hagedis, Fastfission, Peoplesyak, Dissident, Monedula, Bradeos Graphon, Peruvianllama, COMPATT, Mishac, Curps, Michael Devore, Markus Kuhn, Joe Kress, SteffenB~enwiki, Wikisux, Eequor, Matt Crypto, YapaTi~enwiki, Jrdioko, Fishal, Chowbok, Gadfium, Andycjp, Alexf, J~enwiki, Antandrus, Timlane, Beland, OverlordQ, Kaldari, Jossi, JimWae, Maximaximax, Gauss, Latitude0116, Mysidia, Icairns, Gscshoyru, Jarjar, Lindberg G Williams Jr, Neutrality, Izzycat, Paniq, Usrnme h8er, Joyous!, Claude girardin, M.C. ArZeCh, Intrigue, The stuart, Lacrimosus, Jimaginator, Mike Rosoft, Shiftchange, Kune, CALR, Jiy, EugeneZelenko, Discospinster, Rich Farmbrough, FT2, Cacycle, Pjacobi, Vsmith, Dbachmann, Bender235, ESkog, Kaisershatner, GabrielAPetrie, Zamboni~enwiki, Laurascudder, Art LaPella, RoyBoy, Underdog~enwiki, Bobo192, Smalljim, Michael614, BrokenSegue, Elipongo, AncientHaemovore, Maureen, I9Q79oL78KiL0QTFHgyc, Jojit fb, Nk, Tritium6, Famousdog, MARQUIS111, MPerel, Dmanning, Nsaa, Stephen Bain, Elerium~enwiki, Ogress, Jumbuck, Storm Rider, Danski14, Alpharic, Alansohn, FrankP, Guy Harris, Arthena, Andrewpmk, Leonhart~enwiki, Riana, Sade, AzaToth, Batmanand, Wdfarmer, Hu, Radical Mallard, RPellessier, Wtmitchell, Velella, Almafeta, Paul Martin, Stephan Leeds, RainbowOfLight, DV8 2XL, TheCoffee, HenryLi, Bookandcoffee, Kazvorpal, Ceyockey, Ott, Wharrel, Feezo, Novacatz, Thryduulf, Hoziron, Mel Etitis, OwenX, Woohookitty, Georgia guy, Jason Palpatine, David Haslam, Daniel Case, StradivariusTV, BillC, Ruud Koot, WadeSimMiser, K Lepo, -Ril-, Trapolator, GregorB, Plrk, Wayward, 百家姓之四, Christopher Thomas, Vossanova, Pfalstad, LimoWreck, Ashmoo, Graham87, EthanVox, Philipp Miller, Cutepuff, DePiep, NebY, Jclemens, Dpv, Coneslayer, Sjakkalle, Rjwilmsi, Nightscream, Strait, MarSch, VF, PinchasC, Jiohdi, Dennis Estenson II, Trlovejoy, Brucelee, Tawker, Mike Peel, Haya shiloh, Crazynas, Ems57fcva, The wub, DoubleBlue, Olessi, Fred Bradstadt, GregAsche, Baddox, FlaBot, Daderot, RobertG, Jak123, Alhutch, Hiding, RexNL, Ewlyahoocom, TeaDrinker, Carrionluggage, BlkStarr, Alphachimp, Greenozzy, Kri, BradBeattie, Physchim62, Gareth E. Kegg, Phoenix2~enwiki, Spencerk, Sobriquet~enwiki, King of Hearts, Chobot, Madden, Krishnavedala, Moocha, DVdm, Digitalme, Gwernol, UkPaolo, The Rambling Man, Wavelength, TexasAndroid, RobotE, Deeptrivia, Retodon8, MattWright, Adam1213, JarrahTree, Geologician, Captaindan, Eyeon, Epolk, JabberWok, Yamara, MightyGiant, Stephenb, Gaius Cornelius, Chaos, Alex Bakharev, Rsrikanth05, Neilbeach, Wimt, RadioKirk, Big Brother 1984, Shanel, NawlinWiki, Rick Norwood, Bloodofox, AlMac, ONEder Boy, Shaun F, Journalist, JocK, SCZenz, Haoie, Misza13, Doubleg, Bancroftian, Deucalionite, Foofy, Gadget850, Wangi, DeadEyeArrow, Elkman, Acetic Acid, Maunus, Alpha 4615, Wknight94, TransUtopian, FF2010, Paul Magnussen, Enormousdude, Zzuuzz, Theda, Closedmouth, Arthur Rubin, E Wing, Esprit15d, BorgQueen, GraemeL, JoanneB, Digfarenough, LeonardoRob0t, HereToHelp, Willtron, Scrapdog, ArielGold, Nixer, Archer7, Ybbor, Katieh5584, Bala2252, Junglecat, Sheepdude, RG2, Wwzeitler, Maxamegalon2000, Iago Dali, Mejor Los Indios, DVD R W, That Guy, From That Show!, Eog1916, Yvwv, Sardanaphalus, Crystallina, SmackBot, RDBury, Looper5920, Aim Here,

Cubs Fan, Bobet, Lestrade, Herostratus, Prodego, KnowledgeOfSelf, JohnSankey, Ominae, DCDuring, CyclePat, Rokfaith, Jagged 85, Fidocancan~enwiki, ZerodEgo, Fnfd, David.c.h, Timotheus Canens, Amoore5000, CrypticBacon, Xaosflux, Yamaguchi謎謎, Peter Isotalo, Gilliam, Ohnoitsjamie, Skizzik, Hraefen, Christophernandez, JRSP, Master Jay, Busterbros, Bluebot, EncephalonSeven, Rrscott, Trebor, Jerry Ritcey, Jprg1966, Oli Filth, MalafayaBot, Domthedude001, Complexica, Jammycakes, Sadads, Ctbolt, DHN-bot~enwiki, Dual Freq, Darth Panda, A. B., Zachorious, Stedder, Scwlong, Can't sleep, clown will eat me, RyanEberhart, Alphabravotango~enwiki, Matt2h, Ioscius, Saberlotus, Vanished User 0001, MeekSaffron, Snowmanradio, Atomist, Yidisheryid, Homestarmy, Addshore, Chcknwnm, Mr.Z-man, Amazins490, SundarBot, Klimov, Eraseti, COMPFUNK2, Aremond, Model Citizen, Theonlyedge, Lhf, Cybercobra, Nakon, TedE, Richard001, Invincible Ninja, Ollsun, Weregerbil, Sljaxon, Lessthanthree, Crd721, Jose and Ricardo, Ultraexactzz, Metamagician3000, Acdx, Sigma 7, Josephhaley, Kukini, Ged UK, Es547, SirIsaacBrock, Byelf2007, CIS, The undertow, SashatoBot, Skiasaurus, Lambiam, Erimus, Kuru, Khazar, Plugimi, Scientizzle, Wtwilson3, Ascend, Gobonobo, Glooper, Accurizer, Goodnightmush, Hernoor, Genedial, IronGargoyle, DSC~enwiki, Physis, Ckatz, Fernando S. Aldado~enwiki, A. Parrot, Defyn, Loadmaster, Senseiwa, Slakr, Zelaron, Bumatic, Noah Salzman, Hypnosifl, AxG, Metao, Larrymcp, InedibleHulk, Yaddar, Doczilla, TastyPoutine, RichardF, Yodaat, Vsf3000, Billy500, Lee Carre, Hu12, Burto88, SimonD, Levineps, Hetar, Jijithnr, ChazYork, Vanished user, Paul Koning, Xinyu, Cybrarian88, Joseph Solis in Australia, JoeBot, FVP, Wjejskenewr, OMGITSANOOB797, Chiaman280, Igoldste, Courcelles, QuantumOne, Pathosbot, Tawkerbot2, MarylandArtLover, Johnny Lomax, JRSpriggs, Kybenal, Flubeca, Mostly Zen, Conn, Kit, Porterjoh, Van helsing, Scohoust, JohnCD, Randalllin, NickW557, Shakeelmahate, MarsRover, Ken Gallager, Gdbiederman, Gregbard, Funnyfarmofdoom, Aovechkin, Cydebot, Hamid Hassani, Lesqual, Reywas92, Peterdjones, Monachushibernus, Gogo Dodo, Bonojohn, Zginder, Mattergy, Llort, ST47, Goldencako, Bibliope, Michael C Price, DumbBOT, MuratK~enwiki, Kozuch, Daven200520, Ebyabe, JodyB, Daniel Olsen, Maziotis, Satori Son, Krylonblue83, PKT, SummonerMarc, Letranova, Oldpantsnewjersey, Epbr123, Barticus88, Rpba, Ph.D.Nikki, Coelacan, Anshuk, Martin Hogbin, Kahriman~enwiki, Mojo Hand, Berria, Headbomb, Pjvpjv, Marek69, Missvain, Crzycheetah, Bobblehead, SGGH, RFerreira, Muaddeeb, Nick Number, Blathnaid, ThePeg, Rriegs, Bryce byerley, Escarbot, Hmrox, AntiVandalBot, Meoka2368, Luna Santin, Seaphoto, Cacahuate, JHFTC, Beck13, Tangerines, Edokter, Tyco.skinner, Yanickborg, Tim Shuba, RK 808, Spencer, Istartfires, Spartaz, LegitimateAndEvenCompelling, Altamel, Jimleonard, Ghmyrtle, DarylWaite, Asemi, Res2216firestar, Sluzzelin, .alyn.post., JAnDbot, Barek, Skomorokh, CosineKitty, Mitul123, Dsp13, Mcorazao, Liamd330, Rearete, Andonic, Ikanreed, Sitethief, J-stan, 100110100, PhilKnight, Danglegin, Ô, Disto~enwiki, Advany, Moni3, Rareram, Dward Fardhard, Magioladitis, Connormah, Unused0029, Horology, Bongwarrior, VoABot II, Alvatros~enwiki, AuburnPilot, JNW, Mbc362, BobTheMad, JCNSmith, Stijn Vermeeren, Tonyfaull, SwiftBot, Rockinthefarm, Animum, Cgingold, Ldecola, Shamblesguru, Xatso, 28421u2232nfenfcenc, Brian Fenton, Cpl Syx, Vssun, Jacobko, TehBrandon, Glen, Talon Artaine, JaGa, Otritos, Nevit, Hbent, Chefette1223, Steevven1, Ediseiple, Drcaldev, Rickard Vogelberg, NatureA16, MartinBot, Anarchia, Mtevfrog, Rettetast, DHIRENDRA SINGH, R'n'B, CommonsDelinker, Pomte, GstrOSx, Mausy5043, Tgeairn, Jira123, Master of Tofu, Slash, Manticore, J.delanoy, Pharaoh of the Wizards, Nev1, Gotyear, Trusilver, Elizabethrhodes, EscapingLife, Bogey97, Uncle Dick, Maurice Carbonaro, Hoath, Jreferee, Extransit, Drkhan77, Hodja Nasreddin, Sandwichbob, Cpiral, Acalamari, IdLoveOne, Tmtoulouse, Russjb, FrummerThanThou, PhiloNysh, Markgraeme, Aboutmovies, Xyzaxis, RickardV, Fishwristwatch, ElectricValkyrie, Berserkerz Crit, Floaterfluss, NewEnglandYankee, MKoltnow, Aeonimitz~enwiki, Potatoswatter, KylieTastic, Juliancolton, Tangmo~enwiki, Cometstyles, Hunterbertoson, Kinky Kingy, Greatestrowerever, Jamesontai, Oleksiy.n, Wbrito, Vanished user 39948282, Gtg204y, Inter16, Bonadea, TWCarlson, Rockinthe, BernardZ, Beezhive, Idioma-bot, Funandtrvl, Spellcast, Wolkond, Ottershrew, Goosedude, Signalhead, Jmcdon10, Makewater, Lights, VolkovBot, TreasuryTag, CWii, ABF, AlexBG, Jeff G., Jim4444qw, Katydidit, AndperseAndy, Mapsurfer, LeilaniLad, Philip Trueman, Bitnine, PGSONIC, JuneGloom07, TXiKiBoT, Oshwah, The monkey one, A4bot, Evansm2, Anonymous Dissident, DrTimer, Mark Maloney, IPSOS, Qxz, Someguy1221, Vanished user ikijeirw34iuaeolaseriffic, Whalin, JhsBot, IronMaidenRocks, BwDraco, Abdullais4u, Fbs. 13, PDFbot, MrASingh, Bearian, Wiae, Modocc, Aliasd, Rumiton, Damërung, Ohsochevy7, Randallwetzig, Nazar, G2star, Eubulides, Meters, Lamro, Tom Atwood, Cjc0013, Dwetherow~enwiki, Richwil, Synthebot, Falcon8765, Enviroboy, Pooppaste123, 謎謎, Insanity Incarnate, Teetaweepo, Heroandgloom, HiDrNick, Empesey, AlleborgoBot, Logan, Carlodn6, MrChupon, Vitalikk, Finnrind, FlyingLeopard2014, EmxBot, Thor666, Floydclaptonblues, MatthewTStone, SieBot, Chimin 07, PlanetStar, Tiddly Tom, Scarian, BotMultichill, ToePeu.bot, Winchelsea, Gerakibot, Jsc83, YourEyesOnly, Caltas, Mathnerd1212, Spongeguy122, Hyper oxane, Lunoma, SteveThePhysicist, Yintan, Vinaymangal, Americansushi, Weswammy, Carrliadiere, Andersmusician, GlassCobra, Maximo90, Keilana, Flyer22 Reborn, Exert, Oda Mari, Tirosmellsbad, Aombk, Dhatfield, Telcourbanio, Mattmeskill, Oxymoron83, Antonio Lopez, Faradayplank, Beast of traal, Hello71, Yassertariq, Steven Crossin, Lightmouse, RW Marloe, Mayalld, Iain99, Crisis, Fbarw, Nadavatik, Alex.muller, Deejaye6, Fratrep, Redmon, Chain funds, Torchwoodwho, SpartanJames, StaticGull, Cbulletproof, Tasha777, Gdfgdgd, Spartacus3, Mygerardromance, Worthedges1234567890, Huku-chan, Arathviel, JacksonBoyd, Fishnet37222, Denisarona, Jimmy326830263, Escape Orbit, Lovewagon, Explicit, Kookro, Leranedo, Atif.t2, Loren.wilton, Ratemonth, ClueBot, Jlammens, Amaamaddq, Narom, Hamid2007, The Thing That Should Not Be, Sammmttt, LAgurl, All Hallow's Wraith, Rodhullandemu, Rjd0060, DionysosProteus, BobMill, Hult041956, Packhorse, Drmies, Little saturn, TheOldJacobite, UrHawt, Timberframe, CounterVandalismBot, Namnoops, Niceguyedc, Lbeben, Blanchardb, Richerman, Jerainseltran, Mematootoo, Thom.pane, Bennirubber, Neverquick, Griffin Jones, DragonBot, RedvsBlue11, Niknavin, Awickert, Excirial, Nymf, Quercus basaseachicensis, Alexbot, Midgetdan, Watchduck, Deathsi911, Eeekster, Azxsdcvf, Gwguffey, Ybenharim, Zaharous, Emirc, AZatBot~enwiki, Brews ohare, NuclearWarfare, Cenarium, Dmyersturnbull, Obnoxin, Jr beals, Kencurran, Tnxman307, Fredtothe8, Analbumface, Elizium23, Dekisugi, Stegner3, SchreiberBike, Audaciter, HELOON, Dutchartlover, Thehelpfulone, Jfioeawfjdls453, USA-art-check, Thingg, Redrocketboy, Lukiboy10, Aitias, Versus22, Alex10alex10, Josh mcgrath, Arrrrrrrrrrdvark, Berean Hunter, Apparition11, Vanished User 1004, GKantaris, Lockeandkey, XLinkBot, AgnosticPreachersKid, Tw3435, Zrgt, Ghetto fresh thug, Pfhorrest, Lumenos, Rror, Robvsamerica, WillAddy, Feinoha, Chabit, AndreNatas, Little Mountain 5, Avoided, Rreagan007, WikiHead, Ougner, SilvonenBot, Borock, Frood, Cmr08, Kevinmann675, WikiDao, Eleven even, Whizmd, Time keeps on slippin, Kingoftigers, HexaChord, Addbot, TheNightRyder, TheDestitutionOfOrganizedReligion, Stevo21, Ryururu, Some jerk on the Internet, Flare210, DOI bot, Horrorsbiggestfan, Hellboy2hell, Jgedsudki, OLOF KAHDINSKI, JOSEPH HERRERA, Youre dreaming eh?, Fgnievinski, Pazdelamente, Zahd, SpellingBot, Marx01, Njaelkies Lea, Vishnava, CanadianLinuxUser, Fluffernutter, Kapaleev, BaronVonWingmaster, Cst17, Freemasonx, Download, PranksterTurtle, AArdVarK899, Mikeydude91, Angellugo, Chzz, Debresser, Favonian, Timgainpower, 5 albert square, Firedrop, Andrus Kallastu, Wikimassiker, Justpassin, Dayfield, Jasonfitz, Numbo3-bot, Fdorville, F Notebook, Ely ely82, Tide rolls, Lightbot, OlEnglish, Krano, Ammar gerrard117, Superboy112233, Gail, Atulocal, Arbitrarily0, Tortillategra, Chasfagan, Espers, Libreroj, Legobot, Ajpj999, Roflcopter1000, Yobot, Fraggle81, Lars1965, II MusLiM HyBRiD II, EnochBethany, ArchonMagnus, Ajh16, KamikazeBot, Knownot, Flankk2, Jkrsdhhhh, Tempodivalse, AnomieBOT, Valueyou, DemocraticLuntz, Sertion, Papa Johns78, Jim1138, IRP, Galoubet, Tucoxn, Piano non troppo, 90, LiieaWill, Ipatrol, Kingpin13, M00npirate, Crecy99, Nojh333, Bluerasberry, Materialscientist, The High Fin Sperm Whale, Citation bot, E2eamon, Micogobodo, HiP Cavallo, Frankenpuppy, Neurolysis, Sturm009, LilHelpa, W W Wiki, Xqbot, Arieldreyer, Xzxdooglasxzx, Gromble~enwiki, Timir2, Sionus, Transity, Tensioncore, Capricorn42, Jhcruz96, Robbofosso,

Behan, R'n'B, E.Shubee, Thegreenj, Smite-Meister, Dispenser, Jeepday, Ross Fraser, Squids and Chips, Cuzkatzimhut, TreasuryTag, XCe-lam, LokiClock, Optokinetics, Thurth, Philip Trueman, Yakeyglee, Red Act, Viridiflavus~enwiki, 0nlyth3truth, Sargon3, TobiasS, Neparis, YohanN7, SieBot, Tiddly Tom, ToePeu.bot, RatnimSnave, Jdaloner, Udirock, Crazyjimbo, Randallbsmith, Francvs, Bschaeffer~enwiki, Arti-choker, Wwheaton, VQuakr, Multipole~enwiki, Muhandes, Tassos Kan., Arjayay, Rathemis, SchreiberBike, Mestizo777, Boethius65, Timothy-Rias, Forbes72, Davidslindsey, Ost316, Fiskbil, Addbot, DOI bot, Delaszk, TStein, 84user, Stanmorgan, Grandfatherclok, Lightbot, Legobot, Tedtoal, Yobot, Ptbotgourou, Niout, Estudiarme, Kilom691, Johanvdv, AnomieBOT, Götz, Jo3sampl, Messier35, Aaagmnr, Whit537, Arthur-Bot, LilHelpa, Xqbot, Ahandrich, NOrbeck, CES1596, FrescoBot, Grav-universe, Paine Ellsworth, Martlet1215, Citation bot 1, I dream of horses, Alan.poindexter, Michael Lenz, Mercy11, Aoidh, Earthandmoon, 🔲🔲, Yoctobarryc, EmausBot, John of Reading, WikitanvirBot, Fly by Night, Netheril96, ZéroBot, Dario Gnani, Quondum, SporkBot, AManWithNoPlan, Erianna, Iiar, Maschen, Mentibot, Maxdlink, Isocliff, Clue-Bot NG, Starshipenterprise, El Roih, Jj1236, Bopomofo, Stefan Neumeier, CasualVisitor, Helpful Pixie Bot, Bibcode Bot, BG19bot, Davidiad, Ugncreative Usergname, DaleSpam, F=q(E+v^B), Tishchen, Bobo123456, Halfblt, Tm14, TheAmbsAce, Totallyweb, JYBot, Nunibad, Mo-gism, $Z = z^2 + c$, Mark viking, ArtCorpse, Tentinator, Blackbombchu, Friedlicherkoenig, Sharik007, Kix.swamp, Frinthruit, Yut23, Yikkayaya, InfoTheorist, Loraof, Chestax, Doulph88, Breakjohn, TJC663 and Anonymous: 211

- **Derivations of the Lorentz transformations** *Source:* https://en.wikipedia.org/wiki/Derivations_of_the_Lorentz_transformations?oldid=712020880 *Contributors:* Patrick, Anders Feder, Mpatel, Vegaswikian, DVdm, Roman Spinner, JRSpriggs, Headbomb, Marek69, D.H, Magiola-ditis, Vukkarak, YohanN7, Ironholds, Yobot, Pvkeller, Paine Ellsworth, Achandrasekaran99, Normvcr, Maschen, Frietjes, BG19bot, Bakkedal, AmericanLemming, Tentinator, Orioncons, Nigellwh, ZvikaFriedman, Rhodeshart, Nicophil, Robodile and Anonymous: 16

- **Causal structure** *Source:* https://en.wikipedia.org/wiki/Causal_structure?oldid=712570187 *Contributors:* The Anome, Michael Hardy, Reddi, Rgdboer, Mpatel, R.e.b., Vonkje, Gadget850, SmackBot, Lambiam, P199, Richaraj, Cydebot, Raoul NK, Noclevername, Lantonov, Pleas-antville, StevenJohnston, Wmpearl, CristianCantoro, Addbot, Luckas-bot, Yobot, ETST, False vacuum, Phn229, Sławomir Biały, Throwmeaway, SaksithJaksri, BG19bot, Xin.xiong0, Mark viking, Chethan.krishnan and Anonymous: 15

- **Light cone** *Source:* https://en.wikipedia.org/wiki/Light_cone?oldid=718409484 *Contributors:* Toby Bartels, Hephaestos, Stevertigo, Edward, Boud, Snoyes, AugPi, Rednblu, Phys, BenRG, Tea2min, Fropuff, Sigfpe, ConradPino, Karol Langner, Rich Farmbrough, Rgdboer, Jason One, Quaestor~enwiki, Rhialto, Mpatel, Heptapod, Ems57fcva, Alphachimp, DVdm, Hillman, RussBot, Iamdalto, JocK, Obey, SmackBot, Inc-nis Mrsi, Thumperward, Silly rabbit, Mgiganteus1, Illythr, Hypnosifl, RekishiEJ, Fdssdf, Mct mht, Equendil, Cydebot, Thijs!bot, Jq4bagiq, JAnDbot, Mujokan, PhilKnight, Powerinthelines, Oren0, Brendan gibat, Maurice Carbonaro, 3halfinchfloppy, Gzkn, Lantonov, VolkovBot, K. Aainsqatsi, Jauerback, Dhatfield, OsamaBinLogin, Martarius, Mild Bill Hiccup, 51kwad, Addbot, Fgnievinski, Leszek Jańczuk, Bob K31416, Barak Sh, AnomieBOT, Mlpearc, Omnipaedista, Sławomir Biały, Casimir9999, Bounce1337, Simon96DH, ZéroBot, WeijiBaikeBianji, Chuis-pastonBot, Helpful Pixie Bot, Makecat-bot, Rezarmac, Frinthruit, Levi12349, Loraof, Sir Cumference and Anonymous: 41

- **Length contraction** *Source:* https://en.wikipedia.org/wiki/Length_contraction?oldid=721168992 *Contributors:* The Anome, Ken Arromdee, SebastianHelm, Wik, Selket, Fvw, Rasmus Faber, Giftlite, Jyril, Mikez, Dmmaus, JimWae, Bender235, RJHall, El C, Rgdboer, John Vanden-berg, Alex.g, Weyes, Pauwel~enwiki, GregorB, Rjwilmsi, Ems57fcva, Dougluce, Fresheneesz, DVdm, Algebraist, YurikBot, Hillman, Pb is an eslut, Schlafly, Krea, Profero, Kingboyk, Pournami, Sardanaphalus, SmackBot, YellowMonkey, Unyoyega, Harald88, Ati3414, Avana~enwiki, Allard Mosk, E4mmacro, Radagast83, Andrei Stroe, Clicketyclack, JzG, Rijkbenik, Beetstra, Brienanni, JoeBot, GyBlop, Matthew Kornya, JRSpriggs, Paulmlieberman, Gregory9, Dub8lad1, Vyznev Xnebara, Abebasal, Thijs!bot, Martin Hogbin, Headbomb, Fallen Seraph, D.H, Tim Shuba, Alphachimpbot, RebelRobot, WolfmanSF, R'n'B, Leyo, Hodja Nasreddin, Jorfer, Optokinetics, Pamputt, Paradoctor, WWStone, Cold-creation, MenoBot, Binksternet, Ideal gas equation, Ndenison, VQuakr, Bob108, Brews ohare, 7&6=thirteen, Tnxman307, DS1000, Boethius65, Crowsnest, Steve D. Gage, Whizmd, Addbot, Delaszk, Mpfiz, Lightbot, Legobot, Luckas-bot, WikiDan61, AnomieBOT, Citation bot, NOrbeck, Charvest, BenzolBot, Citation bot 1, Sergiacid, Vazelinis, Duschi, Hughston, Dhwacky, Oakycoppice, RjwilmsiBot, Franciscouzo, EmausBot, Bennyhava, WikitanvirBot, Solomonfromfinland, Squadsquad22, ZéroBot, Cogiati, Quondum, Drognan231, Chrisman62, ClueBot NG, Duik-bootman, Jshutzman, Helpful Pixie Bot, Bibcode Bot, Rjbuenker, Ghwellsjr, Engineer112, BattyBot, Ejmarcus, Donn300, Thepasta, Reatlas, Desiderata9, I am One of Many, Sauagt rai, Wamiq, Lesser Cartographies, Frinthruit, FitzJellett4, LCcritic, Knownornot, CAPTAIN RAJU and Anonymous: 99

- **Time dilation** *Source:* https://en.wikipedia.org/wiki/Time_dilation?oldid=721244672 *Contributors:* Vicki Rosenzweig, Bryan Derksen, XJaM, Miguel~enwiki, Mjb, Stevertigo, Patrick, Alfio, Ahoerstemeier, Ronz, William M. Connolley, Kbk, Tpbradbury, Francs2000, KeithH, Tlogmer, Goethean, Gandalf61, StefanPernar, Cornellier, Rasmus Faber, Hadal, Xanzzibar, Enochlau, Matt Gies, Giftlite, Barbara Shack, Mikez, Inter, Geeoharee, Meursault2004, MathKnight, Marcika, Wwoods, Rchandra, Decoy, Phe, Piotrus, Khaosworks, Jkliff, Nickptar, Mschlindwein, John-Armagh, Laguna72, Qef, Mike Rosoft, Chris Howard, D6, Poccil, JimJast, Bender235, Kjoonlee, El C, Mdf, Acanon, IDC, Rbj, Cayte, Shehal, Alansohn, JYolkowski, Nik42, MatthewEHarbowy, Hackwrench, Keenan Pepper, ABCD, PAR, MrBudgens, Velella, Knowledge Seeker, Allen McC.~enwiki, LOL, Cleonis, Before My Ken, Jok2000, Pauwel~enwiki, Christopher Thomas, Xiong, Gerbrant, Slgrandson, Rjwilmsi, Strait, Ems57fcva, Bubba73, JYOuyang, Fresheneesz, Zayani, DVdm, YurikBot, ErNa, Reverendgraham, Huw Powell, Hillman, DarkAvenger, Micah-brwn, Yamara, Gaius Cornelius, Salsb, NawlinWiki, ErkDemon, Jeff Carr, Bota47, JustAddPeter, FF2010, Moogsi, Shawnc, Ilmari Karonen, Al-lens, Profero, Roke, That Guy, From That Show!, IslandHopper973, MacsBug, SmackBot, Promatrax161~enwiki, Mdd4696, WildElf, Harald88, Elronxenu, Gilliam, JMiall, Kmarinas86, Ati3414, Chris the speller, Henrisalles, Colonies Chris, Squibman, TheRaven7, Mirshafie, Fairychild, Frap, Jguy101, E4mmacro, Emre D., Kntrabssi, Hgilbert, Evlekis, Yevgeny Kats, Andrei Stroe, Petr Kopač, Harryboyles, BurnDownBabylon, Loadmaster, Andrés D., Andypandy.UK, Sailor for life, Stizz, Inquisitus, Quaeler, DouglasCalvert, Clarityfiend, Jwalte04, Tophtucker, Twas Now, Nerfer, G-W, CalebNoble, Robinhw, Gregory9, Thermochap, CmdrObot, TunaSushi, Vyznev Xnebara, MFlet1, ShelfSkewed, Penbat, Thaabomb, Icek~enwiki, Abebasal, Cakeman, Kozuch, Omicronpersei8, Zalgo, Calvero JP, EvocativeIntrigue, Epbr123, King Bee, Gamer007, Nonagonal Spider, Headbomb, Lotte Monz, Davidhorman, Jonny-mt, D.H, Opelio, Xenophon (bot), Tyco.skinner, Scepia, Penjr, Tim Shuba, Qwerty Binary, Kiwichipster, RJFerret, Dirkshelley, Andonic, MegX, Pedro, Bongwarrior, VoABot II, JNW, Dkracht, Greg park avenue, Car-damon, EagleFan, LorenzoB, JaGa, Oroso, R'n'B, Maurice Carbonaro, Lantonov, Zedmelon, Anonywiki, NatureBoyMD, CanonRAP, Warut, DorganBot, Moroder~enwiki, TopGun, Valence Band, Mike.albrecht, Redtigerxyz, TreasuryTag, Bendav, Jlaramee, MagicBanana, DoorsAjar, Red Act, Jarvism, Selain03, JhsBot, UnitedStatesian, Modocc, Eubulides, VanishedUserABC, NinjaRobotPirate, Praphull8888, FlyingLeop-ard2014, DionysiusThrax, Paradoctor, Caltas, Flyer22 Reborn, RSStockdale, The-G-Unit-Boss, Coldcreation, Anchor Link Bot, Winterheat, Deciwill, Trick McKaha, Velvetron, ClueBot, Helenabella, KRhodesian, Dominorose, TwPx, Awickert, Jusdafax, Sun Creator, Brews ohare, MrTsunami, DS1000, 1ForTheMoney, Erodium, XLinkBot, Forbes72, Terry0051, Little Mountain 5, Facts707, Arseny1992, Brdrpunk, Ad-dbot, Tman1997al, Deadgood86, Howard Landman, M.nelson, ShanePOneilYI6054331415, Vega2, Delaszk, SamatBot, Tide rolls, Lightbot,

Krano, Amadeus50, Soyyo10, Frehley, Luckas-bot, Yobot, NotARusski, Wikipedian2, Babbar.ankit, Jim1138, Materialscientist, Citation bot, Elfsborgarn, Bwik2051, Gap9551, Srich32977, NOrbeck, Amaury, Mnmngb, Aaron Kauppi, SD5, FrescoBot, Grav-universe, LucienBOT, Poliwog7, Citation bot 1, Swordsmankirby, Tom.Reding, Pmokeefe, Boobarkee, Reconsider the static, Banan14kab, Nashpur, Aribashka, RjwilmsiBot, Tesseract2, Skamecrazy123, EmausBot, WikitanvirBot, Gupta himanhsu, Nuujinn, Fly by Night, Gebinsk, Slightsmile, Bollyjeff, Cyberia3, Physicslovercperiod, Quondum, Rexprimoris, Thevictor99, L Kensington, Donner60, ChuispastonBot, Raguna Ash, ClueBot NG, Duikbootman, KirtZJ, MelbourneStar, Registrator photon, Jj1236, Widr, Helpful Pixie Bot, Sandmantreefitty, Gob Lofa, Bibcode Bot, DBigXray, BG19bot, Czyx, Augustinosc, Mark Arsten, Aristidesfl, Beneyameen, Anchit singla, Viraj nadkarni, BattyBot, Charlessdarwinn, Pete.clat, Khazar2, EuroCarGT, AbhishekChakravartty, Steven Schierman, Carpediemspain, Trevor.karstens, Manstein1942, Graphium, Jamesx12345, TreasureTrove, Sui docuit, VIPUL MUKUND NEWASKAR, Rbdash21, Tango303, Wamiq, Sethur2, Glaisher, YiFeiBot, Gigantmozg, Ginsuloft, Lily Soso, Varkman, Lyncstaa, Ibensis, Dred37, Frinthruit, FrB.TG, TheBurgerShot, Zyxyea, KentuckyJohnson, Monkbot, Fjmfarley, Knyttl, Loraof, EvilMossman, Tetra quark, Arghya Chakraborty (Mathematician), Vivek Selvaraju, Matmatpenguin, Unit388, Chickenpox333, Inalak, Furious Mythical Beast, Kurousagi, Sooryakiran Pallikulathil, Billycoskun and Anonymous: 432

- **Super Minkowski space** *Source:* https://en.wikipedia.org/wiki/Super_Minkowski_space?oldid=707946054 *Contributors:* Rgdboer, R.e.b., Headbomb, AnomieBOT, Jim1138, Baking Soda and Anonymous: 2

## 30.4.2 Images

- **File:120-cell_graph_H4.svg** *Source:* https://upload.wikimedia.org/wikipedia/commons/c/c1/120-cell_graph_H4.svg *License:* Public domain *Contributors:* Own work *Original artist:* Tomruen
- **File:24-cell_graph.svg** *Source:* https://upload.wikimedia.org/wikipedia/commons/8/8a/24-cell_graph.svg *License:* Public domain *Contributors:* I (Tom Ruen (talk)) created this work entirely by myself. *Original artist:* Tom Ruen (talk)
- **File:4-cube_t0.svg** *Source:* https://upload.wikimedia.org/wikipedia/commons/6/67/4-cube_t0.svg *License:* Public domain *Contributors:* Own work *Original artist:* self
- **File:4-cube_t3.svg** *Source:* https://upload.wikimedia.org/wikipedia/commons/4/4c/4-cube_t3.svg *License:* Public domain *Contributors:* Own work *Original artist:* self
- **File:4-simplex_t0.svg** *Source:* https://upload.wikimedia.org/wikipedia/commons/b/b9/4-simplex_t0.svg *License:* Public domain *Contributors:* Own work *Original artist:* Tomruen
- **File:45,_-315,_and_405_co-terminal_angles.svg** *Source:* https://upload.wikimedia.org/wikipedia/commons/5/56/45%2C_-315%2C_and_405_co-terminal_angles.svg *License:* Public domain *Contributors:* Own work *Original artist:* Adrignola
- **File:600-cell_graph_H4.svg** *Source:* https://upload.wikimedia.org/wikipedia/commons/f/f4/600-cell_graph_H4.svg *License:* Public domain *Contributors:* Own work *Original artist:* Tomruen
- **File:8-cell-simple.gif** *Source:* https://upload.wikimedia.org/wikipedia/commons/5/55/8-cell-simple.gif *License:* Public domain *Contributors:* Transferred from en.wikipedia to Commons. *Original artist:* JasonHise at English Wikipedia
- **File:Affine_subspace.svg** *Source:* https://upload.wikimedia.org/wikipedia/commons/8/8c/Affine_subspace.svg *License:* CC BY-SA 3.0 *Contributors:* Own work *Original artist:* Jakob.scholbach
- **File:Ambox_important.svg** *Source:* https://upload.wikimedia.org/wikipedia/commons/b/b4/Ambox_important.svg *License:* Public domain *Contributors:* Own work, based off of Image:Ambox scales.svg *Original artist:* Dsmurat (talk · contribs)
- **File:BoysSurfaceTopView.PNG** *Source:* https://upload.wikimedia.org/wikipedia/commons/2/2d/BoysSurfaceTopView.PNG *License:* CC-BY-SA-3.0 *Contributors:* ? *Original artist:* ?
- **File:CDel_3.png** *Source:* https://upload.wikimedia.org/wikipedia/commons/c/c3/CDel_3.png *License:* Public domain *Contributors:* Own work *Original artist:* User:Tomruen
- **File:CDel_4.png** *Source:* https://upload.wikimedia.org/wikipedia/commons/8/8c/CDel_4.png *License:* Public domain *Contributors:* Own work *Original artist:* User:Tomruen
- **File:CDel_5.png** *Source:* https://upload.wikimedia.org/wikipedia/commons/1/16/CDel_5.png *License:* Public domain *Contributors:* Own work *Original artist:* User:Tomruen
- **File:CDel_node.png** *Source:* https://upload.wikimedia.org/wikipedia/commons/5/5e/CDel_node.png *License:* Public domain *Contributors:* Own work *Original artist:* User:Tomruen
- **File:CDel_node_1.png** *Source:* https://upload.wikimedia.org/wikipedia/commons/b/bd/CDel_node_1.png *License:* Public domain *Contributors:* Own work *Original artist:* User:Tomruen
- **File:CMB_Timeline300_no_WMAP.jpg** *Source:* https://upload.wikimedia.org/wikipedia/commons/6/6f/CMB_Timeline300_no_WMAP.jpg *License:* Public domain *Contributors:* Original version: NASA; modified by Ryan Kaldari *Original artist:* NASA/WMAP Science Team
- **File:ChipScaleClock2_HR.jpg** *Source:* https://upload.wikimedia.org/wikipedia/commons/e/e0/ChipScaleClock2_HR.jpg *License:* Public domain *Contributors:* ? *Original artist:* ?
- **File:Circle_manifold_chart_from_slope.svg** *Source:* https://upload.wikimedia.org/wikipedia/commons/2/2f/Circle_manifold_chart_from_slope.svg *License:* CC BY-SA 3.0 *Contributors:*
- Circle_manifold_chart_from_slope.png *Original artist:*
- derivative work: Pbroks13 (talk)
- **File:Circle_with_overlapping_manifold_charts.svg** *Source:* https://upload.wikimedia.org/wikipedia/commons/6/64/Circle_with_overlapping_manifold_charts.svg *License:* CC BY-SA 3.0 *Contributors:*

### 30.4.3 Content license